Systems and Software Process

Systems and Software Process
324 pages

Brijendra Singh
Department of Computer Science
University of Lucknow
Lucknow

Shikha Gautam
Department of Information Technology
Indian Institute of Information Technology
Lucknow

NAROSA PUBLISHING HOUSE PVT. LTD.

22 Delhi Medical Association Road, Daryaganj, New Delhi 110 002
35-36 Greams Road, Thousand Lights, Chennai 600 006
306 Shiv Centre, Sector 17, Vashi, Navi Mumbai 400 703
2F-2G Shivam Chambers, 53 Syed Amir Ali Avenue, Kolkata 700 019

www.narosa.com

Printed from the camera-ready copy provided by the Authors.

ISBN 978-81-8487-661-1

Published by N.K. Mehra for Narosa Publishing House Pvt. Ltd.,
22 Delhi Medical Association Road, Daryaganj, New Delhi 110 002

Printed in India

Dedicated to

My family members, who taught me how to love this world

&

My gurus, who taught me how to face this world

Preface

During the past decades, increasing attention has been focused on "Systems and Software Process". Systems and software process succeeds, when it meets the needs of the people who use it, when it performs flawlessly over a long period of time, when it is easy to modify and even easier to use, it can and does change things for the better. The business of many companies and organisations is essentially based on software. Software intensive systems are automotive, telecommunication systems and services. We all want to build system that makes things better, avoiding the bad things that lurk in shadow of the failed efforts. To succeed, we need discipline when system is designed and built, we need an engineering approach. During case study research in systems and software process that the research is thorough is not easy. Although there is a long history of research in software development field, it has been difficult to translate their research guidelines into the software process domain. Systems and software process is now such a huge area that it is impossible to cover the whole subject in one book. Our focus, therefore, is on key research topics based on literature survey that are fundamental to all software development processes and topics concerned with the development of systems. This book is very useful for researchers wanting to undertake any research project and help them to avoid some of the problems already experienced by the authors and other researchers. The specific objectives of the text book are:

- To provide comprehensive coverage for researchers in systems and software process.
- To be easy for the readers to understand and to serve as a reference for the researches working in software industry and university.

This book contains twelve chapters.

Chapter 1 provides an introduction to the subject with a brief history and a general outline of systems and software process. Software process used to create and maintain software product. It is used to achieve quality in software product within given time and cost. This chapter analysed history of software failure based on output of the researches in literature. Finally, discuss the software process.

There are various software processes but all must include basic set of activities to develop software product such as: Feasibility study, Requirements elicitation and specification, Software architecture and design, Coding and testing, and Delivery and maintenance. Each and every activity has its own role and importance to develop a quality product. Chapter 2 covers the activities of software development process.

Quality is concerned with developing software products with required quality and assessing the level of software quality. Software processes are significant assets in achieving and assessing the software quality, therefore Chapter 3 is based on software quality.

Chapter 4 is based on product quality vs. process quality. Software process used to create software product and achieve quality in software product within given time and cost. In this chapter, software process quality and software product quality are discussed from several perspectives to see the relationship of software process quality and software product quality. This chapter also analysed software process and software product based on output of the researches in literature. Finally, discuss the changing nature of software process and software product.

Chapter 5 discusses the software process models as a means of instructing software organisation on how to achieve software development, business objective, and improvement goals. Situational factors are important to consider during development process to develop a quality product. Chapter 6 provides the overview of situational factors that affect software process and analyse how situational factors affect software process.

Objective of the Chapter 7 is to explain the impact of software development process on software quality. Process of software development is used to create and achieve quality in software products. In this chapter various quality attributes have been considered during analysis of development process to see the impact on software quality. Chapter 8 discusses the effects of software process maturity on software development.

In a fast and dynamic competitive environment it is not easy to survive and maintain a credit in market. It can be possible when user trust on product quality and its performance. This can be possible when software process is well defined because quality of software product directly related to the quality of software process, Therefore Chapter 9 introduced software process model and knowledge management.

Chapter 10 describe design pattern, software coding, the impact of programming languages on software quality, overview of programming languages, the code quality analysis, the comparison of programming languages and discuss the impact analysis of software coding on software quality.

Chapter 11 covers the integrating usability through software process. This chapter also discussed the usability framework to achieve usability and software usability requirement analysis as well as usability evaluation. Chapter 12 discussed software testing fundamental, test planning, test process, test tools, and quality improvement.

We thank Prof. S. P. Singh, Vice chancellor, University of Lucknow, Lucknow and Prof. Arun Mohan Sherry, Director, Indian Institute of Information Technology, Lucknow (IIIT Lucknow) for fostering the excellent academic atmosphere at the campus. We express our deepest sense of regards to Prof. R. R. Singh, former Dean, Faculty of Science, University of Lucknow. We have read several books, and many research articles on the web, and in various journals and magazines. They all have contributed to this work. We have acknowledged most of them as references and further readings. The omissions, if any, are inadvertent and not deliberate. Many people have directly or indirectly contributed to the development of this book. We greater to all of them. We also wish to thank the numerous MCA, M.Tech., and Ph.D. students who enthusiastic participation in discussions helped us to present many ideas and concepts, as discussed in this book, with greater clarity.

Finally, we would like to thank the publishers, particularly Mr. Narendra K. Mehra managing director and their production and publication department for all understanding, cooperation in bringing out timely publication of the book.

Brijendra Singh
Shikha Gautam

Contents

Chapter - 1

SOFTWARE PROCESS

Software process used to create and maintain software product. It is used to achieve quality in software product within given time and cost. In this chapter, user interface and interactive systems are discussed from several perspectives to see the relationship between us. This chapter analyzed history of software failure based on output of the researches in literature. Next, discuss the software process. It is essential to choose right software for particular objective. Section I describes the introduction of this chapter. Section II describes the interactive systems and user interface. Section III describes the history of software failure. Section IV describes the concept of software process. Section V describes the right software and systems quality. Section VI describes the summary of this chapter.

1.1 INTRODUCTION

Currently, the business of many companies and organizations is basically based on software. Software intensive systems and services such as: telecommunication or automotive systems, banking system, hospital, hotel, and financial services, increasingly depend on software. Software adds considerable value to many services and products and allows for competitive discrimination in the market. The escalating importance of software as well as novel software development methodology such as model driven, lean software development, and future software based applications force many challenges and difficulty on software development, operation, and maintenance. Normally, software and software intensive systems are developed with hundreds or thousands of people in teams. Team based software development has numerous characteristics that are demanding to deal with when conducting software projects. They perform a huge amount of different activities called processes. Systematic management and collaboration mechanisms are needed in order to fruitfully create user value and fulfill project objective and goals under specified project constraints such as cost limitations and time deadlines. Descriptions of processes called software process models are a compulsory means for coordinating such accomplishments. Software process models can be used to describe work procedures, recommend the interfaces between tasks, maintain the organization of work products, and sustain the management of necessary resources.

1.2 INTERACTIVE SYSTEMS AND USER INTERFACE

In interactive systems the user and system exchange information regularly and dynamically. Norman's evaluation/execution model is a useful approach of understanding the nature of interaction:

1. User has a goal (something to achieve)
2. User looks at system and attempts to exercises how he would perform a series of tasks to accomplish the goal.
3. User carries out several events (providing input to the system by touching a screen, pressing buttons, speaking words etc)
4. System responds to the events, actions and presents results to the user. System can use text, graphics, sounds, and speech etc.
5. User looks at the results of his action and attempts to assess whether or not the goals have been achieved.

A good interactive system is one where:

- User can effortlessly work out how to control the system in an attempt to accomplish his goals.
- User can simply assess the results of his action on the system.

What interactive systems do we use in our day-to-day life? However, the term interactive system can be applied to a greatly broader collection of devices, such as:

- The World Wide Web
- Mobile phones
- Cash dispensing machines
- Windows operating systems
- Car navigation systems
- Data entry systems
- Video recorders
- Machines driven call centre (e.g. for telephone banking)
- Workflow system to co-ordinate a team's work-efforts.

The original interactive systems were command line systems, which strongly controlled the interaction between the human and the computer. The user was compulsory to know the commands that might be issued and how the arguments were to be controlled. UNIX operating system and DOS (Disk Operating System) are classic examples of this category. Users were mandatory to enter data in a particular sequence [1]. The options for the output of data were also strongly controlled and usually limited. Such systems normally put a high demand on the user to memorize commands and the syntax for issuing these commands.

Command line systems progressively gave way to a second generation of menu-based systems, form-based systems, and dialog-based systems that eased some of the load on memory. An automatic teller machine (ATM) is a good example of a form-based system where users are given a tightly controlled set of promising actions. Data entry systems are commonly form or dialog oriented systems offer the user a limited set of choices but deeply relieving the memory demands of the earlier command line systems [2].

Next, third generation of interactive computing was introduced by Xerox Corporation in 1980. The Xerox Star was the outcome of a half dozen years of research and development through which the desktop metaphor, mouse, windows, icons, and bit-mapped displays were all brought together and completed to function. The Xerox Star was simulated in the Lisa and Macintosh first presented by Apple Computer Inc. in the mid 1980s. The windows, icon, menu, and pointer (WIMP) approach was made worldwide by Microsoft in the Windows family of operating systems introduced in the 1990s [3]. With the maturation of WIMP interfaces, also known as graphical user interfaces (GUIs), interaction moved from command-based to direct manipulation.

In command based systems, the user specifies an action and then an object on which that action is to be performed. In a direct manipulation system, an object is selected, and then the user specifies the action to be performed on that object. The most recent developments in interactive systems have focused on visualization, virtualization, and agents. During the 1980s and 1990s, there were many efforts to take benefit of the human capability to process information visually. At the simplest level, consider that a human looking at a picture on a television screen has no problem in sharp a pattern that consists of millions of individual pixels per second, changing in both time and space [4]. Visualization systems manipulate information at high levels of aggregation, making the information additional reachable to users.

In the 1990s, researchers began to experiment with extending interactive systems from symbolic interaction such as: mice, icons, and pointers to virtual systems. In these systems, every effort was made to allow the user to explore a virtual world with small or no translation to symbolic form. Thus, using visualization techniques and novel forms of input devices, such as data gloves, hand movements could be used to manipulate virtual objects represented graphically in a virtual world. This virtual environment was presented to the user using two display screens, each of which provided a somewhat different perspective, giving the user a stereoscopic view of a virtual space that appeared to have strength [5]. Work on virtual and artificial realism continues on a number of particular fronts, including a field known as telemedicine.

The next generation of interactive systems, represented by agents in embedded systems, will yet again change how humans and computers interact. Direct manipulation environments will still be around for many years to come. At the same time, we have begun to see both agents and embedded systems make their manifestation. Embedded systems can be as easy as the analog sensor systems that open a department store door, or turn on lights when someone enters a room. At a more complex level, most cars being built today include air bag deployment systems and antilock brakes that operate invisibly by gathering data from the environment and inserting computer control between our actions and the environment. As air bag deployment systems become more difficult, they react based not simply on acceleration data, but also based on the weight of the individuals occupying the seat and their relative position (leaning forward or back) on the seat [6].

The basic programming paradigm had to change from the process driven approach to an event driven perspective. In earlier systems, the program's main process would control what the user could do. Now, it was possible for the user to initiate a broad series of actions by selecting an object, a window, an icon, a text box. This required some method for collecting events and handling them. The X Window System on UNIX was one of the early famous systems for doing this. Each graphical component of the interface was able of producing one or more events. For example: a window might be opened or closed generating an event. Similarly, a button might be pressed, or the text in a text box might be changed. The programmer's task is to show a coordinated set of components that can generate events. The programmer is also required to write code that will start some action when an event occurs. These code fragments are called event handling functions. In object based and object oriented programming (OOP) environments, this task of handling events is made easier through object classes which associate default event handling methods with specific classes of objects [7]. For example: The code for how the look of a button is changed when it is pressed may be provided as a default method of the button objects.

The user interface (UI), in the software industrial design field of human computer interaction, is the area where interactions between humans and systems occur. The aim of this interaction is to agree to effective operation and control of the system from the human end, whilst the system simultaneously feeds back information that aids the operator's decision making process. Examples of this wide concept of user interfaces include the interactive aspects of computer operating systems, heavy machinery operator controls, hand tools, and process controls. The design considerations applicable when creating user interfaces are related to or occupy such disciplines as ergonomics and psychology. Usually, the goal of user interface design is to produce a user interface which makes it easy (self explanatory), enjoyable (user friendly) and efficient to operate a system in the way which produces the preferred result [8]. This generally means that the operator needs to give minimal input to achieve the desired output, and also that the system minimizes undesired outputs to the human.

With the increased use of personal computers and the relative decline in common awareness of heavy machinery the term user interface is generally assumed to mean the graphical user interface, while industrial control panel and machinery control design discussions more commonly refer to human machine interfaces. Next terms for user interface are man machine interface (MMI) and when the machine in question is a computer human computer interface. All effective interfaces share eight quality or characteristics:

1. Clarity: The interface avoids uncertainty by making everything clear through language, hierarchy, flow, and metaphors for visual elements.
2. Concision: It's easy to make the interface obvious by over clarifying and labeling everything, but this leads to interface bloat, where there is just too much stuff on the screen at the same moment. If too many things are on the screen, finding what you're looking for is tricky, and so the interface becomes dull to use. The real challenge in making a good interface is to make it concise and clear at the same time.
3. Consistency: Keeping your interface consistent across your application is vital because it allows users to recognize usage patterns.
4. Efficiency: Time is money and a good interface should make the user more productive through shortcuts and good design.
5. Familiarity: Even if someone uses an interface for the first time, certain elements can still be familiar. Real life similes can be used to communicate meaning.

6. Responsiveness: A good interface should not experience slow. This means that the interface should provide good feedback to the user about what's happening and whether the user's input is being successfully processed.
7. Aesthetics: While we don't need to make an interface attractive for it to do its work, making something look good will make the time users spend using application more enjoyable; and happier users can only be a good thing.
8. Forgiveness: A good interface should not punish users for their mistakes but should instead provide the resource to cure them [9].

User Interface (UI) design focuses on anticipating what users might need to do and ensuring that the interface has elements that are easy to access, understand, and use to facilitate those events. UI brings together concepts from interaction design, visual design, and information architecture. Everything stems from knowing users, including understanding their goals, preferences, skills, and tendencies. Once we know about user, make sure to consider the following when designing interface:

- Keep the interface simple and easy. The best interfaces are nearly invisible to the user and avoid unnecessary elements
- Create consistency and use general UI elements. By using common elements in UI, users feel more comfortable and are able to get things done more quickly. It is also imperative to create patterns in language, layout and design throughout the site to help ease efficiency. Once a user learns how to do something, they should be capable to transfer that skill to other parts of the site.
- Be focused in page layout. Consider the spatial relationships between items on the page and structure the page based on importance. Careful placement of items can help illustrate attention to the mainly significant pieces of information and can assist scanning and readability.
- Purposefully use color and texture. Use direct concentration toward or redirect attention away from items using color, contrast, light, and texture to advantage.
- Use typography to develop hierarchy and clarity. Carefully consider how to use typeface. Different fonts, sizes, and arrangement of the text to help increase legibility, scanability, and readability.
- Make sure that the system communicates what's happening. Always inform users of actions, location, changes in state, or errors. The use of various UI elements to communicate status and, if necessary next steps can decrease irritation for user.
- Think about the defaults. By suspiciously thinking about and anticipating the goals people bring to site; we can create defaults that decrease the load on the user [10].

1.3 HISTORY OF SOFTWARE FAILURE

The history of software development is an incredible success. Just look around us for evidence of that. But that success has a long, dark shadow that we don't talk about very much: it's littered with huge failures. What's particularly disturbing is that the vast failures keep recurring year after year. The names and dollar amounts may vary, but the story is otherwise the similar. Most software projects fail fully or partially because they don't meet all their requirements. These requirements can be the cost, time, quality, and software functional or non-functional requirements. According to many studies, failure rate of software projects ranges between 50% - 80%. To cope these problem software development process is used. Software development process easily utilizes the resource and minimizes the risk.

A latest report, notes that success in 68 percent of technology projects is unbelievable. Low requirements analysis causes many of these failures, meaning projects are damned right from the start. According to IBM study, only 40% of projects meet schedule, budget, and quality goals. Further, they found that the biggest barriers to success are people factors. Geneca, a software development company, noted from its studies that fuzzy business objectives, out of sync stakeholders and excessive rework mean that 75% of project participants need confidence that their projects will succeed [11].

On Mars mission in 1998 the Climate Orbiter spacecraft was eventually lost in space. Although the failure confused engineers for some time it was exposed that a sub contractor on the engineering team failed to build a simple conversion from English units to metric. An uncomfortable lapse that sent the $125 million craft deadly close to Mars surface after attempting to stabilize its orbit too low. Flight controllers think the spacecraft ploughed into Mars atmosphere where the connected stresses crippled its interactions, leaving it hurtling on during space in an orbit around the sun.

In 1996, Arian-5 space rocket, developed at the cost of $7000 million over a time of 10 years was destroyed within less than a minute after its launch. The crash occurred because there was a software bug in the rocket guidance system. In 1996, one of the largest banks of US credited accounts of nearly 800 customers with approximately $924 lacs. Later, it was detected that the problem occurred due to a programming bug in the banking software [12].

Year 2000 (Y2K) problem refers to the extensive snags in processing dates after the year 2000. The roots of Y2K problem can be traced back to 1960-80 when developers shortened the 4-digit date format like 1972 to a 2-digit format like 72 because of limited memory. At that time we did not realize that year 2000 will be shortened to 00 which is less than 72. In the 1990s, experts began to understand this main shortcoming in the computer application and then millions were spent to grip this problem [13].

The Northeast blackout in 2003 has been one of the main power system failures in the history of North America. This blackout involved failure of 100 power plants due to which almost 50 million customers faced power loss that resulted in financial loss of approximately $6 billion. Afterward, it was determined that the main reason behind the failure was a software bug in the power monitoring and management system [14].

In 2004, EDS introduce a highly complex IT system to the U.K.'s child support agency (CSA). At the same time, the department for work and pensions (DWP) decided to reorganize the entire agency. The two pieces of software were totally incompatible and permanent errors were introduced as a result. The system someway managed to overpay 1.9 million people, underpay another 700,000, had $7 billion in uncollected child support payments, 36,000 latest cases stuck in the system, a backlog of 239,000 cases, and has cost the UK taxpayers over $1 billion to date.

Bitcoin Hack, Mt. Gox Launched in 2010, Japanese bitcoin exchange, Mt. Gox, was the main in the world. Once being hacked in June, 2011, Mt. Gox stated that they'd lost over 850,000 bitcoins (value around half a billion US dollars at the time of writing). Although around 200,000 of the bitcoins were recovered, Mark Karpeles admits "We had weaknesses in our system and our bitcoins vanished." Leaving thousands of Mt. Gox clients out of pocket, 32 year old, France born, Mark Karpeles is presently on trial for embezzlement and a number of other charges. Karpeles has pled not guilty but if found guilty, Karpeles could be sent to jail for up to 5 years and fined up to ($4000) [15].

Millions of TSB customers were locked out of their accounts after an IT upgrading led to an online banking outage. A planned system upgrade was estimated to shut internet and mobile banking services down for one weekend in April 2018 but ended up causing months of disturbance. The problems arose from TSB's shift to a novel banking platform following its divide from Lloyds Banking Group. Instantly after the latest system was switched on numerous users experienced problems logging in while others were revealed details from other people's accounts or erroneous credits and debits on their own. Customers remained locked out of their accounts two weeks once the primary outage. In July 2018, TSB was still working its way through the backlog of complaints when a different outage struck, locking customers away of their online accounts once again. TSB claimed that the crisis was resolved later that day but the debacle will crack the bank's relationship with parent company Sabadell.

In 2018, hospital staff and doctors of the Wales NHS qualified a widespread computer failure that led to them being incapable to access patient files. According to the national cyber security centre, the failure was owed to technological issues as opposed to a cyber attack however it still caused wide disturbance as unable to access blood, X-Ray results, and etc. It also caused a backlog as patients could not be contacted to cancel appointments and notes could not be typed up and saved on NHS systems [16].

The reasons for software failure refer to the lack of presence of success factor for the software project. Some important factors for software failure:

1. Inadequate Project Planning: Project planning is a central part of project management and it is the responsibility of the project manager to set an appropriate plan for the project.
2. Scope Creep: Scope creep refers to change in scope of the project and also known as requirement creep or feature creep. The scope is the work required for a project. Scope creep refers to how the requirements of a project keep on varying over a project lifecycle.
3. Use of Unpracticed Tools and Techniques: Good tools and techniques are required for the success of a software project. A universal illusion is made by

the project manager and team leader to utilize unpracticed tools and techniques at the preliminary of project.

4. Shortage of Resources/Requirements: Every project requires some resources according to requirement and need. The quantity of resources depends on the size and scope of the project. Sometimes, the project is unreachable due to the shortage of resources and necessary requirements.
5. No or Poor Risk Management: At present, we have to deal with some real facts in project management. Poor or no risk management has the capability to influence the project management. Project failure is the worst case of poor risk management. Some of the most vital influences on the poor risk management are given as:
   - Project failure
   - Slow- running projects
   - Risk of reputation damage
   - Superfluity budget
   - Unresponsive customers and less user adoption
   - Inactive benefits

   Poor risk management is one of the major contributors to project failure and had a negative impact on project success.
6. Lack of User Engagement: A project which is aimed at developing various products is going to have especial users i.e. a group of people who does business at the organization.
7. Poor Controlling and Monitoring: Controlling and monitoring are the necessary parts of project management. A project can succeed only when there is suitable governance for the project management. Without proper planning and monitoring, the project may fail. The absence of controlling and monitoring impacts the project in many ways:
   - Difference in cost, scope, and schedule baselines
   - Project may not be finished on time as expected
   - Quality of the output can be degraded
   - Organization figure will damage
   - Opposition between project team can be raised
   - Poor project performance
   - Unsatisfied user
8. Inexperienced Project Managers: Project failure is a widespread term that every project manager wants to split from. No one craves to take the responsibility of project failure as it may blot his career record. But if a project fails, then it simply means that project manager did a unfortunate job. Ensure that the project manager has sufficient knowledge of what the best techniques are because hiring of well skilled project manager can't be mislaid.
9. Ineffective Communication: Project team knows their manager only through his communications. Whether it a project, an operation or personal life, communication plays an important role. Without communication, we are executing tasks in the dark area. Project managers should develop a communication plan. Even time to time meet-ups should be planned to discuss the project performance.

10. Poor Project Management: When there is no proper management for project then the project may fail due to poor management. If a project is decently staffed, proper planning, have a good WBS, availability of resources, proper scheduling, and the support of sponsors but even gets fail then the simply reason behind it is that project management was poor [17].

Fortunately, all is not gone. Here are five steps to improving the people based factors affecting IT delivery:

1. Freeze the technology/business relationship via governance
2. Integrate technology intro strategic planning
3. Set and share a simple, multi-year roadmap for overall business strategy
4. Establish an open planning process
5. Teach and promote communication and relationship skills

We believe if we view IT projects as not just a technology problem and consider the people factors then software organization will boost its implementation success, create better relationships, and maximize its return on investment (ROI). So, it becomes important for the project manager to know which factors can result in project failure [18]. It will help to focus on those factors while managing the software project.

1.4 CONCEPT OF SOFTWARE PROCESS

In the early days of software computing for the difficulty of writing helpful and efficient computer programs in the required cost and time are known as software crisis. The software crisis was due to the quick increases in computer power and the complexity of the real problems that could not be solved. Through the increase in the complexity of the software, numerous software problems arose because existing methods were unsatisfactory. The term software crisis was used by some attendees at the first NATO software engineering conference in 1968 at Garmisch, Germany. Edsger Dijkstra's in 1972 ACM Turing award lecture makes mention to this problem. "The main cause of the software crisis is that the machines have become numerous orders of magnitude more powerful. To set it quite directly: as long as there were none machines, programming was no problem at all; when we had a little weak computers, programming became a gentle problem, and now we have gigantic computers, programming has turn into an similarly gigantic problem" said by Edsger Dijkstra, The Humble Programmer (EWD340), Communications of the ACM

The problems of the software crisis were correlated to the large complexity of hardware and the software development process. The crisis observable itself in numerous ways:

- Projects running over cost
- Projects running over time
- Software was very useless
- Software often did not meet requirements
- Software was of low quality

- Projects were unmanageable and software code hard to maintain
- Software was never delivered

The main reason is that improvements in software computing power had outpaced the ability of software programmers to effectively employ those capabilities. Various software processes and methodologies have been developed over the previous few decades to develop software quality management using procedural programming and object oriented programming [19]. Though software projects that are large, complex, poorly specified, complicated, and occupy unfamiliar aspects, are still vulnerable to huge, unexpected trouble.

When software organization fail to develop a quality product then software process is used to resolve that issue. The process that deals with the technical and managerial issues of software development is called software process. Any software process must include the following four activities:

1. Software requirements (specification): Describe the main functionalities of the software and constrains around them.
2. Software design and code (implementation): The software is to be designed and programmed (software code).
3. Software testing (verification and validation): The software must conform to its specification and meets the user needs.
4. Software maintenance (software evolution): The software is being modified to meet user and software market requirements changes.

The concept of software process is based on the software development lifecycle (waterfall, spiral model, iterative model, and agile model etc.) defines a framework and philosophy of how the process must be carried out. Adopting a lifecycle is not however satisfactory to provide practical assistance and control of a software project as it does not recommend a course of action, an organization, tools, technology, operating procedures, development policies, and constraints. The concept of software process builds on the notion of the software development lifecycle (SDLC) and provides a broader concept which acts as a structure to organize other factors and issues related to software development activities. Thus, a software process provides a number of contributions and concepts that might be categorized as technological and managerial support used in the process, guidelines on how to use tool and technology, and complete software development activities; the science of organizations and people, marketing, and economy [20].

To analysis the process in this broad sense has helped extensively to identify the imperative dimensions of software development and the issues and problems which need to be addressed. The results presented now orient to the science of organizations and people of the software process i.e. how software development activities are carried out by teams of people to organize and manage their organizational structure.

One of the imperative objectives of the software development project should be to develop software product that is simple to maintain means ensures software maintainability. Software testing uses the most resources during software development. Underestimating the software testing effort regularly causes the planners to assign inadequate resources for software testing which, in turn, results in unreliable software product or time slippage. The aim of the software process should not be to decrease the effort of software design and coding but to cut the cost of software maintenance. Both software testing and maintenance depend greatly on the software design and coding of software, and these costs can be significantly reduced if the software is designed and coded well to build software testing and maintenance easier. Hence, during the early phases of the software development process the main issues should be "can it be effortlessly tested" and "can it be simply modified". Errors can arise at any stage of software development [21]. However software error detection and correction should be a continuous process that is done during software development process. Detecting errors quickly after they have been introduced is obviously a purpose that should be supported by the software process. A software process is also not a static unit.

As the productivity (thus the cost of a software project) and software quality are determined mostly by the software process to assure the software engineering objectives of software quality improvement, cost reduction, and the software process must be improved. Having software process improvement as a fundamental goal of the software process implies that the software process used is such that is supports its improvement.

Software applications are complex products that are complicated to develop and test. Very frequently software exhibits undesired and unexpected behaviours that may even cause rigorous problems and damages. For these reasons scientist, researchers, developers, and practitioners have been paying increasing concentration to understanding and improving the quality of the software product being developed. The underlying statement is that there is a straight correlation among the quality of the software process and the quality of the developed software product. So software process is research areas which deal with these issues and develop a quality product within given time and cost [22].

1.5 RIGHT SOFTWARE AND SYSTEMS QUALITY

Basically right software specifies do the deliverables satisfy the user. It is the part of software validation where "have we built the right software" is evaluated. Validation is the process of evaluating a software or component during or at the end of the software development process to verify whether it satisfies user specified requirements. Software validation is evidence by examination and through condition of objective evidence that the software requirements for a particular intended use or purpose have been fulfilled. System quality is a planned effort to manage how business produce the software products and devices they sell. Its main goal is to ensure companies perform responsibly and have the organizational structure, processes, procedures, and resources for implementing rules that keep the software products they make safe. System quality focuses on the performance characteristics of the software under study by researching resource, reliability of devices or software products, investment utilization, response times of employees, human factors, a device's ease of use, design controls, and software accuracy etc.

Software quality measurement quantifies to what amount a software program or system rates beside each of these five structural characteristics (reliability, security, efficiency, maintainability, and size). An aggregated assess of software quality can be computed using a quantitative or a qualitative scoring scheme or using both and then a weighting system reflecting the priorities. This observation of software quality being situated on a linear continuum is supplemented by the investigation of critical programming errors that under exact circumstances can guide to catastrophic outages or performance degradations that make a specified system inappropriate for use regardless of ranking based on aggregated measurements. Such programming errors originate at the system level signify up to 90% of production issues while at the unit level, even if far additional numerous, programming errors account for less than 10% of development issues [23]. As an outcome, code quality without the framework of the whole system as described by W. Edwards Deming, has restricted value.

Software quality is not a complete notion but a relative one and it must be reasonable. The verification and validation (V&V) rules used for verifying architecture and code as well as verifying the business suitability by software testing. If we perform more and more test cases then we can be sure that the effort will rise infinitely. If we execute more and more test cases then we would like to expect improved quality product but this is a myth as long as we do not recognize the coverage of the test cases. But it is certainly the case that if software quality improves then more and more effort has to be spent. If software quality is good then risk will be little and if software quality is poor then risk will be high. Mostly, risk is the probability of damage leading to an index multiplied by the resulting cost when it happens. Risk is therefore a prospect value and is used to evaluate risk levels. Risk management generally provides means for risk avoidance, risk transfer, risk reduction, risk dissemination, risk insurance, and risk acceptance. We can use such methods, procedures, and tools to find out the resultant risk level. All these activities require an assured quantity of money to be conducted because quality is not free. Thus some cost is needed for logical as well as beneficial assurance of software quality. If proper quality actions are taken in early development phases just in development then we would suppose their impacts not only to be seen in this phase but in operations too [24].

1.6 SUMMARY

There is no model includes all the required components defined as part of a process model under the concern of a well defined SDLC process. Each process model has general and unique elements, and each SDLC has its own benefits and limitations. The formulation of newer SDLCs is all subjective by previous software process models. Main drivers to create new software process models are technological (to handle with the challenges of the new technologies), managerial (to effectively manage and organize), and knowledge gaps (to cope with the user's requirement, demands, organizational settings, and environments for a better project management and control). There is not a single unique SDLC that could be applied in all cases. This implies that the application of a detailed SDLC will rely on the particular characteristics of the application under software development, the organization's and team size, the tool and technology used, and the developer's and tester experience. The advantage of using a software process model for the development of software intensive systems has been confirmed and its elimination could lead to non successful software products.

1.7 REFERENCES

1. Interactive Systems. Computer Sciences, Encyclopedia.com http://www.encyclopedia.com/computing/news-wires-white-papers-and books/interactive-systems
2. Bederson, B. B. (2004). Interfaces for staying in the flow. Ubiquity, pp. 1-8.
3. Raskin, J. (2000). *The Humane Interface: New Directions for Designing Interactive Systems*. Addison-Wesley.
4. Greenberg, S.; Witten, I. H. (1989). Directing the user interface: how people use command-based computer systems. In Analysis, Design and Evaluation of Man Machine Systems, pp. 349-355.
5. Ball, T.; Eick, S. G. (1996). Software visualization in the large. Computer, Vol. 4, pp. 33-43.
6. Sangiovanni-Vincentelli, A.; Martin, G. (2001). Platform-based design and software design methodology for embedded systems. IEEE Design & Test of Computers, Vol. 18, Issue 6, pp. 23-33.
7. Dabek, F.; Zeldovich, N.; Kaashoek, F.; Mazières, D.; Morris, R. (2002). Event-driven programming for robust software. In Proceedings of the ACM 10th workshop on ACM SIGOPS European workshop, pp. 186-189.
8. Mayhew, D. J. (1999). The usability engineering lifecycle. In ACM CHI'99 Extended Abstracts on Human Factors in Computing Systems, pp. 147-148.
9. Terry, M.; Mynatt, E. D. (2002). Recognizing creative needs in user interface design. In Proceedings of the ACM 4th conference on Creativity & cognition, pp. 38-44.
10. Bradley, M.; Kristensson, P. O.; Langdon, P.; Clarkson, P. J. (2018). Interaction Patterns: The Key to Unlocking Digital Exclusion Assessment?. In International Conference on Applied Human Factors and Ergonomics, pp. 564-572, Springer, Cham.
11. Charette, R. N. (2005). Why software fails [software failure]. IEEE Spectrum, Vol. 42, No. 9, pp. 42-49.
12. CS2 Software Engineering: Note 1 (2001). Software Failures & Software Standards. CS2Ah Autumn 2004.
13. Kwak, Y. H. (2005). A brief history of project management. The story of managing projects.
14. http://ecomputernotes.com/software-engineering/software-crisis
15. https://raygun.com/blog/costly-software-errors-history/
16. https://www.computerworlduk.com/galleries/infrastructure/top-software-failures-recent-history-3599618/
17. https://www.whizlabs.com/blog/top-10-reasons-for-project-failure/
18. https://faethcoaching.com/it-project-failure-rates-facts-and-reasons/
19. http://www.cs.utexas.edu/users/EWD/transcriptions/EWD03xx/EWD340.html
20. Rönkkö, K.; Dittrich, Y.; Randall, D. (2005). When plans do not work out: How plans are used in software development projects. Computer Supported Cooperative Work (CSCW), Vol. 14, No. 5, pp. 433-468.
21. Korelsky, T.; McCullough, D.; Rambow, O. (1993). Knowledge requirements for the automatic generation of project management reports. In Proceedings of IEEE Eighth Knowledge-Based Software Engineering Conference, pp. 2-9.
22. Rodriguez, L. C.; Mora, M.; Martin, M. V.; O'Connor, R.; Alvarez, F. (2009). Process models of SDLCs: comparison and evolution. In Handbook of Research on Modern Systems Analysis and Design Technologies and Applications, IGI Global, pp. 76-89.

23. Hassan, M. U.; Umair, M., Ali, H. (2017). Novel Approaches to Improve Software Quality. International Journal of Software Engineering and Its Applications, Vol. 11, No. 6, pp.15-24.
24. Wieczorek, M.; Vos, D.; Bons, H. (2014). *Systems and Software Quality*. Springer, New York.

Chapter - 2

ACTIVITIES OF SOFTWARE DEVELOPMENT PROCESS

Software process is a set of related activities that leads to the development of software product. Software process is inter-leaved sequences of technical, managerial, and collaborative activities with the overall goal of specifying, designing, implementing, and testing software. There are various software processes but all must include basic set of activities to develop software product such as: Feasibility study, Requirements elicitation and specification, Software architecture and design, Coding and testing, and Delivery and maintenance. Each and every activity has its own role and importance to develop a quality product. Section I describes the introduction of software development process activity. Section II describes the feasibility study. Section III describes eliciting and specifying requirements. Section IV describes software architecture and design. Section V describes software coding and module testing. Section VI describes integration and system testing. Section VII describes delivery, deployment, and maintenance. Section VIII provides the summary of this chapter.

2.1 INTRODUCTION

Software development process is a large and complex activity. To develop quality product lots of recourses, time, effort, and cost are required. For effectively and efficiently execution of software development various technical and managerial tasks are involve. Successfully completions of software development, these tasks are divided into set of small but inter related activity to fulfill the business and user objective [1].

There are various software development process but all must include basic set of activities to develop software product such as: feasibility study, requirements elicitation and specification, software architecture and design, coding and testing, and delivery and maintenance. Each and every activity has its own role and importance to develop a quality product.

Feasibility study considers whether the proposed system will be cost effective from a business point of view and if it can be developed within existing resource constraints. Eliciting requirements is a task to determine what their requirements are by communicating with customers and users. It is also called requirements gathering. Requirements specification used for translating the information gathered during the analysis activity into a document that defines user and system requirements [2-3].

Software architecture is the structure or structures of the system which consist of software components, the externally noticeable properties of those components, and the relationships between them. Software design is a process of problem solving and planning for a software solution. After the requirement specification software is determined, software developers will design or employ designers to develop a plan for a solution. It includes low level component and algorithm implementation issues as well as the architectural view [4-5].

During coding, the software design is realized as a set of programs or program units. Module testing involves verifying that each module meets its specification and objective. The individual program units or programs are integrated for integration testing and tested as a whole system to ensure that the software requirements have been met which is known as system testing [6].

After testing, the software is delivered to the customer or user. Software deployment is all of the activities that make software available for use. Software maintenance is the modification of a software product after delivery to correct faults, improve performance or other attributes, and adapt the product to a modified environment [7-8].

2.2 FEASIBILITY STUDY

Feasibility is defined as the practical extent to which a project can be performed successfully. To evaluate feasibility of project feasibility study is performed. A feasibility study should be comparatively economical and fast. Estimation is made of whether the recognized user requirements may be satisfied using existing software and hardware technology. The result should notify the verdict of whether or not to go forward with an added comprehensive analysis [9].

It is important to find out the strengths and weaknesses of proposed project and present guidelines of activities which will improve a project and accomplish desired results. Feasibility study is used to review the project to find out the strengths and weaknesses of project. The nature and components of feasibility study depend mainly on the areas in which analyzed projects are implemented. The acronym TELOS refers to the five areas of feasibility such as: Technical, Economic, Legal, Operational, and Schedule [10]. Technical feasibility is one of the most important and first study that must be conducted after a project has been identified.

For sustainable success of projects economic feasibility is an important aspect to consider. Therefore, project component, investment, reinvestment costs of system, and running costs for its operation were analyzed. The project team has to construct a systematic analysis of the legal issues surrounding the project. Operational feasibility study evaluates the project potential for success and schedule feasibility used to assess of how reasonable the project timetable is [11]. These feasibility studies summarized as:

1. **Technical feasibility** focused on gaining an understanding of the current technical resources of the organization and their applicability to the estimated needs of the proposed system. It is an assessment of the hardware and software and how it meets the requirement of the proposed system. For example: A space organization constructs a prototype to determine if a new type of spacecraft propulsion is technically feasible or not.

2. **Economic feasibility** demonstrates the net benefit of a proposed project for accepting or disbursing electronic funds, consider the benefits, costs to the organization, and the common user as a whole. For example: A software organization performs a study to determine if an innovative new software design would be cost competitive given the current cost of required resources, components, machines, technology, and effort.
3. **Legal feasibility** determines whether the proposed system conflicts with legal requirements. For example: A bank is considering introduce an innovative novel financial product. They begin with a feasibility study focused on all legal aspects such as regulatory risks and compliance.
4. **Operational feasibility** is a measure of how well a proposed system solves the problems, takes advantage of the opportunities acknowledged during scope definition, and how it satisfies the requirements identified in the requirements analysis of software development. For example: A data center investigates the feasibility of power self-adequacy using solar panels and battery systems with the network as a backup.
5. **Schedule feasibility** is a measure of how rational the project schedule is. For example: A software project is in the early planning phases with partial scope and no schedule when an executive announces to the world that the project will be completed by July. The project manager rapidly informs the executive that this is might be unfeasible. The executive requests that the project manager make it occur. The project manager responds by conducting a schedule feasibility assessment based on a list of requirements, out-of-scope items, and July deadline.

Feasibility study consists of three basic steps:

Step 1: Evaluation of a project according to selected criteria

In the first step of a feasibility study, an evaluation of a project according to recognized criteria is conducted. There are generally five main criteria such as: technological and system requirements, legal requirements, operational requirements, economic requirements, and requirements related to the schedule.

Step 2: Summary of evaluation results

This part of analysis can be carried out using the components of strengths, weakness, opportunities, and threats (SWOT) analysis. A summary of project evaluation results by presenting the strengths and weaknesses of suggested solutions is another step of feasibility study.

Step 3: Recommendations

Recommendations are the last component of feasibility study. They include guidelines of actions in the project which allow its better and enhanced implementation [12].

Basically there are four techniques of feasibility study. The techniques are: Interviews, Observation, Record inspection, and Investigation.

Interviews are the most common, useful, and satisfactory method of gathering information about objective, distribution of duties, and problems and failures in the previous and existing system. In this method, both source and seeker are active. The objective of an interview is to acquire and provide information to accomplish the goal. Completeness, accuracy, and relevancy of information are essential for software design which is achieved through interview [13]. People are observed by their actions using observation method. This method is used to study people in group, to study process, what is being done, how it is done, why it is done who does it when, and how long it takes. In this method, source is active and seeker is passive.

Using observation, it is better to identify and define what is to be observed, approximate the time, secure management approval, and clarify objectives to parties being observed. This method may be used with interview method for observing the people [14]. Record inspection is a study of organization charts, statistics, and procedure manuals which reveal the information about the procedure. Close study of the record provides the best guide to current practice which may or may not deal with original requirements [15].

In investigation, information is gathered with carefully design questionnaire. Questionnaire is constructed after brainstorming and after taking into consideration the objective of the survey and study [16]. For better result and quality product, there is a need for shifting the traditional method of feasibility study to a new approach that embraces the principles of sustainable development [17]. These techniques are used to conduct feasibility study of projects.

2.3 ELICITING AND SPECIFYING REQUIREMENTS

Requirements elicitation is the process of deriving the system requirements throughout inspection of existing systems, discussions with potential users, task analysis, and so on. This may involve the development of one or more system models and prototypes. These help to recognize the system to be specified [18].

Basically there are two type of elicitation activity such as: facilitated activity and independent activity. Facilitated activity, directly interact with stakeholders to elicit requirements and in independent activity, work on own way to discover information. Facilitated activities mainly focus on discovering user and business requirements. Working directly with users is essential because user requirements include the tasks that users need to achieve with the software. While independent elicitation techniques complement requirements that users describe and expose needed functionality that end users may not be aware of [19].

Requirements elicitation is non-trivial because this never is sure that all requirements from the user and customer by just asking them what the system should do or not do. Requirements elicitation practices include brainstorming, interviews, user observation, questionnaires, use cases, workshops, prototyping, and role playing [20]. All these requirements elicitation methods are used as per specific situation.

Interview is a conventional foundation of requirements input for both commercial products and information system products, across all software development approaches. User observations can be silent or interactive. Questionnaires are a way to survey huge groups of users to recognize their needs and requirements. Use case describes a sequence of relations between system and an external actor that results in the actor being capable to accomplish several result of assessment. Workshops support stakeholder association in defining requirements. It is a structured meeting of selected group of stakeholders in which experts work simultaneously to create, define, refine, and achieve final deliverables [21-22].

Requirements specification is the activity of translating the information gathered during the analysis activity into a document that defines a set of requirements. Two types of requirements may be included in this document such as: 1) user requirements are conceptual statements of the system requirements for the customer and end user of the system and 2) system requirements are a more comprehensive explanation of the functionality to be provided. Effective requirements specification is a necessary input to success. A high level requirements specification is required. Before software is designed and implemented, the requirements have to be specified in sufficient detail to formulate analysis and design possible [23-24]. So a software requirements specification (SRS) is an explanation of a software system to be developed.

A framework is proposed for expressing the relationships between multiple views in requirement specification describes various ways to express requirement [25]. A practical approach describes various viewpoints for requirements elicitation using flexible model and incremental requirements elicitation method [26]. In mathematical model of the requirements elicitation process shows the significant role of knowledge in its performance [27]. Another approach is proposed to partially automate the requirements elicitation and specification process using prototype tool [28].

A unified model of the requirements elicitation process uses iterative nature of elicitation to transforms the existing state of requirements, situation to an improved understanding of the requirements, and potentially a modified situation. Meta process of requirements elicitation is captured in the model for selection of an appropriate elicitation technique which improved understanding of elicitation, helps analysts to improve their elicitation efforts, and improve the ability to perform elicitation to meet planned users' requirements [29]. This model is helpful for collecting requirement effectively.

An improved process for requirements elicitation proposed for main key improvements: (1) to guide the non-technical stakeholders (primarily the users) in the capability and limits of computer software, hardware, and of software developers, (2) identify keywords while interviewing the stakeholders as visually and text form, (3) use keyword mapping to create software requirements, (4) apply the techniques of Capability Maturity Model (CMM) and Quality Function Deployment (QFD)during the elicitation process [30].

An empirical mapping study performed to identify what aspects of Software Requirement Specifications (SRS) are affected, which is evaluated on the basis of 46 identified and categorized primary studies in which understandability is the most frequently evaluated feature of SRS [31]. They found that understandability is the most frequently evaluated aspect of SRS, experiments are the most usually used research method, and the academic environment is where mainly empirical evaluation takes place.

2.4 SOFTWARE ARCHITECTURE AND DESIGN

Software architecture is a high level structure of software, a discipline of creating such structures, and the documentation of these structures. Software design process allocates the requirements to either software or hardware systems by establishing an overall software architecture. Software design involves identifying and describing the essential software abstractions and their relationships [32].

Software design is an explanation of the structure of the software to be implemented, data models, structures used by the system, interfaces between system components, and algorithms used. Designers do not enter at a finished design instantly but develop the design iteratively. They insert requirement and element as they develop their design with stable backtracking to accurate former designs [33]. Activities of design process are as:

1. **Architectural design** exposes the top level design and permits a designer to reason about satisfaction of software requirements in terms of assignment of functionality to design elements and allows designers to develop recurring patterns of software organization [34]. There is different architectural design style such as:
   - Data-Flow Architecture (Pipes and Filters)
   - Data-Flow Architecture (Batch Sequential)
   - Data-Centered Architecture
   - Layered Architecture (OS Type)
   - Layered Architecture (Protocol Layer Type)
   - Call and Return Architecture
   - Client Server Architecture

2. **Interface design** is part of software and is designed as it is expected to provide the user. Interface design focuses on anticipating what users might need to do and ensuring that the interface has elements that are easy to understand, access, and use to facilitate those actions. User interface brings together concepts from interaction design, visual design, and information architecture [35]. User interface (UI) is divided into two categories: command line interface (CLI) and graphical user interface (GUI).

There are numerous tools available using which the designers can develop complete GUI on a mouse click. Some tools can be embedded into the software environment such as integrated development environment (IDE).GUI implementation tools provide dominant array of GUI controls. For software customization, software designers can change the code consequently. There are different segments of GUI tools according to their different use and platform. For example: Computer GUI, Mobile GUI, Touch-Screen GUI etc. Here is a list of some tools which is used to create GUI:

   - FLUID
   - LucidChart
   - AppInventor (Android)
   - Wavemaker
   - Visual Studio

3. **Component design** is a design specification for adaptable components. Using the generation procedure set of these components is implemented for certain products

and software [36]. For example: A component design is a specification for an adaptable component that can be used to assemble a draft application work product. The adaptation specification defines the adaptation parameters for this adaptable component. The interface specification is parameterized where suitable in terms of these adaptation parameters.

4. **Database design** is a process of developing a detailed data model of database which contains all the needed physical and logical design. Physical storage parameters needed to generate a design in a data definition language which can then be used to create a database [37]. For example: The unified modeling language (UML) is used to expressing complex systems in visual way which is created in an object oriented language. For instance, an entity is known as a class in UML.

Software architecture usually refers to the bigger structures of software and it deals with how many software processes assist to carry out their tasks. Software design refers to the smaller structures and it deals with the internal design of a single software process [38]. Software architecture is a program or computing system which depicts the system that aids in understanding how the system will perform and work.

The architecture design process focuses on the decomposition of a system into different components and their relations to satisfy functional and non-functional requirements. The key inputs to software architecture are:

1. The requirements produced by the analysis tasks.
2. The hardware architecture.

The result or output of the architecture design process is an architectural depiction. The basic architecture design process is composed in three steps: understand the problem, evaluate the architecture design, and transform the architecture design. Software architecture is essential because it affects the maintainability, performance, availability, and robustness of software. The non-functional requirements depend on the software architecture in which these components are planned and communicated. It is a creative process because the activities within the process depend on the type of system being developed, the exact requirements for the system, and the background and experience of the system architect [39].

Software design generally involves problem solving and planning a software solution. This includes both a low level component and algorithm design, and a high level component and architecture design [40]. Software architecture serves as the blueprint for both the project and the system developing it, defining the work assignments that must be carried out by design and implementation teams. The architecture is the major carrier of system qualities such as usability, performance, modifiability, reliability, and security. None of which can be achieved without a unifying architectural rules and vision. Architecture is an object for early analysis to make confident that a design approach will give up an acceptable system. By developing an effective architecture, we can identify design problems and risks which mitigate them early in the software development process.

Software architecture is the worthy study of the large scale structures of software systems. From its qualitative descriptions of useful software organizations, software architecture has matured to include wide explorations of notations, analysis techniques, tools, and creation methods. Initially the research area interpreted software practice and offers concrete direction for complex software design. Software architecture interacts and overlaps with the study of software families, software design, domain specific design, specific classes of components, component based reuse, and program analysis. It is not productive to try strict division among these areas [41].

For software design and architecture, to mature into a robust area capable of handling the issues posed by emerging applications, advances in numerous areas are essential. First is sufficient support for moving from architecture to implementation and gracefully moving between design and coding. If design does not eventually support development of a satisfactory implementation then it is an unsuccessful effort. Various existing design approaches fall silent when it comes to coding. Second is capacity to characterize stakeholder's interests. Various interests and multiple perspectives enforce demands for assessing or guaranteeing consistency of design decisions as well as demands for several presentations of select design data. Third, as software applications become gradually interwoven into an organization there is a need to develop co-design of software. Fourth is the design of applications as seen and practiced by users [42].

2.5 SOFTWARE CODING AND MODULE TESTING

Software coding and implementation stage of software development is the process of converting a system specification into an executable system. It always involves programming to develop software product. During the coding, the project team creates the actual product. Software development can be an exciting phase for the user because their idea for the software becomes impressive product. Project developers and programmers commence building and coding the software. If requirements are gathered correctly and software designed accurately, the coding process is more efficient. Project teams are better capable to meet software coding deadlines when the accurate information is gathered straight from the user [43].

In business programming it is general principle that each system application is so unique that it must be designed and coded from the beginning. So, prewritten reusable modules cannot be designed, coded, and reused [44]. A code editor is also called an IDE. An IDE is a software application for formatting code, checking syntax, as well as running, and testing code. Some IDEs can work with many programming languages while some are extremely specific for only one language. Programmers use an IDE for checking syntax, formatting code, and testing programs. The common steps for writing a program include the following:

- Understand the problem you are trying to solve
- Design a solution
- Draw a flow chart
- Write pseudo code
- Then write code
- Test and debug
- Testing with real-world users
- Release program
- Iterate the steps for the next version

Computer code is fundamentally a list of instructions and commands that can be run by a certain program. Mainly code consists of plain text documents so they can be used for numerous different programs. A unique file extension is given to the document to specify the nature of the code. For example: a file created using Python is saved with a .py extension, like 'program.py.' However, the real content of the file is still just simple text. Table 2.1 shows hello word program in C, C++, Java, Python, and Mathematica.

Table 2.1 Hello word program in C, C++, Java, Python, and Mathematica

#include
<stdio.h>
main ()
{
printf("Hello
world");
}
	#include
	<iostream>
	main()
	{
	std::cout << "Hello
	world\n";
	}
		class Hello {
		public static void
		main(String[] args) {
		System.out.println("Hello
		world");}
		}
			print("Hello
			world")
				Print ["Hello
				world"]

Above table shows how hello world is printed using C, C++, Java, Python, and Mathematica.

Software testing is an investigation conducted to provide stakeholders with information about the quality of the software product or service under test. Test techniques contain but are not restricted to the process of executing a program through the aim of finding software bugs. There are black box testing, grey box testing, and white box testing methods that can be used for software testing. Functional and non-functional levels of testing include diverse methodologies that can be used for conducting software testing [45]. There are numerous approaches in software testing. Reviews, walkthroughs, and inspections are referred as static testing whereas essentially executing programmed code with a given set of test cases is called as dynamic testing. Mainly there are three levels of testing such as: unit testing, integration testing, and system testing.

Documentation for testing help estimating the testing effort, test coverage, requirement tracing. Most commonly used documented artifacts are: test case, test plan, test scenario, and traceability matrix. Test cases involve a set of conditions, steps, and inputs that can be used for performing testing tasks. Test plan outlines the plan that will be used to test software. Test scenarios used to guarantee that all process flows are tested from end to end. Traceability matrix (Requirement Traceability Matrix, RTM) is a table which is used to trace the requirements during the development cycle. It can be used for forward tracing (from requirements to design or coding) or backward (from coding to requirements) [46]. In traditional waterfall model, testing is performed by an independent group of testers. This can be done after the functionality is developed but before it is delivered to the user. However even in waterfall model unit testing is regularly done by the software development team still when further testing is done by a separate testing team. In comparison of some emerging software development techniques such as: extreme programming (XP) and agile software development hold to a 'test-driven software development' model. In this software process, unit tests are written initial, by the software engineers. Unit tests are maintained along with the rest of the software code and generally integrated into the build process. The vital goals of this test process are to sustain continuous integration and reduce defect rates. This method increases the testing effort done by development process before reaching any formal testing team. In other development models most of the test execution occurs after the requirements have been defined and coding process has been completed.

Module testing is the testing of entire code objects as created by the compiler when built from source. Modules are completely formed chunks of coherent source code that can usually be tested by driving a few purpose signatures with different stimulus [47]. Module testing is a process of testing the individual subprograms, classes, subroutines, and procedures in a program. As a replacement for of testing whole software program at once, module testing proposes testing the smaller building blocks of the program. Module testing is principally a white box oriented. The objective of module testing is not to exhibit proper functioning of the module but to reveal the existence of an error in the module. Module testing allows to apply parallelism into the testing process by giving the prospect to test several modules simultaneously [48]. Module testing is suggested because probability of identifying bugs or errors on smaller chunks of program becomes high, multiple modules can be tested concurrently thus supports parallel testing and complexity of testing can be effortlessly managed. Table 2.2 shows the difference between module testing and unit testing

Table 2.2 Module Testing vs. Unit Testing

Module Testing	Unit Testing
Module tests are a set of tests written by a tester after some code has been written by a software developer.	Unit tests are a compilation of tests written by a developer through the software development process.
Module testing may be engage combining the units test.	Unit Testing might be testing units in separate form.

So basically module testing is a non-incremental testing which requires more work for debugging test, misunderstanding test doubles, and need to understand the code.

2.6 INTEGRATION AND SYSTEM TESTING

Integration testing, individual software modules are combined and tested as a group. It occurs after unit testing and before validation testing. Integration testing takes as its input modules that have been unit tested, groups them in bigger aggregates, apply tests defined in an integration test plan to those aggregates, and delivers as its output the incorporated system prepared for system testing [49]. Integration testing is a level of software testing where individual units are combined and tested as a group. The use of this level of testing is to expose faults in the interaction among integrated units. Test drivers and test stubs are used to support in integration testing. Figure 2.1 shows the software testing levels.

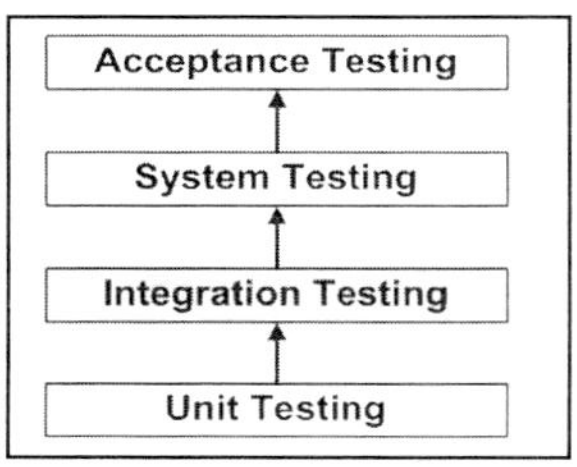

Figure 2.1 Software testing levels

Integration testing that seeks to validate the interfaces among components against a software design. Software components may be integrated in an iterative way or altogether. Integration testing works to find defects in the interfaces and interaction between integrated components (modules). Progressively larger groups of tested software components equivalent to elements of the architectural design are integrated and tested until the software works as a system [50]. Some different types of integration testing are: big bang, risky-hardest, mixed (sandwich), top down, and bottom up. Other integration testing is: layer integration, collaboration integration, backbone integration, client server integration, distributed services integration, and high frequency integration.

Data quality is major problem of testing. For quality evaluation coverage measurement is used. In module testing, algorithms and their data is considering while in integration testing interfaces and relation between modules are consider [51]. Object oriented software development raises vital testing issues. Many of them try to apply theoretical constructs and techniques of traditional software development and testing to object oriented software. There are five levels of object oriented testing, four of these plan nicely into the usually accepted unit, integration, and system levels of traditional software testing [52].

The model (test dependency graph) serves as an origin for ordering classes and methods to be tested for regression and integration purposes. It shows how to construct a model of structural system test dependencies which evolves with the refinement process of the object oriented design [53]. Self-adaptive software is capable to change their own structure and behavior at run time in order to cope with changes. Dynamic creation of plans for conducting the integration testing of self-adaptive software considers architectural reconfiguration to interoperate as specified [54]. Integration testing can depict the problems with the interfaces between programs components by trouble occur in real world program execution. Integration testing is a component of XP, a practical method of software development that takes a particular approach to building a software product by means of repetitive testing and revision.

System testing is testing conducted on whole software by integrated system to evaluate the system's conformity with its specified requirements and need. System testing falls within the area of black box testing. System testing tests a completely integrated system to validate that it meets user requirements. System components are integrated to produce a whole system. This process is apprehensive with finding errors that result from unpredicted interactions between components and module interface problems. It is also concerned with viewing that the system fulfills its functional and non-functional requirements and testing the developing software properties [55]. System testing checks that components are compatible, interact properly, and transfer the correct information at the exact time across their interfaces. For large systems, this may be a multi-phase process where components are included to form subsystems that are independently tested before these sub systems are themselves integrated to form the final system [56].

System testing done by a professional tester on the finished software product before it is introduced to the user and market. Automated system testing using visual GUI testing tools automate the interaction with software systems at the GUI level [57]. Modern test case generation techniques can automatically achieve high code coverage.

By generating test cases at the GUI level, system interface is available and avoid false failures using search based system testing [58]. System testing is done after integration testing. This plays an important role in delivering a high quality software product. It helps to minimize after deployment troubleshooting, error, and support calls.

2.7 DELIVERY, DEPLOYMENT, AND MAINTENANCE

After completing all these activates software is delivered to the user. Software delivery is the process of getting a software product to market. Software delivery includes the main groups of actions such as:

1. Business and user for requirements
2. Technology for software development and deployment
3. Quality assurance for testing and validation [59-60]

Dynamic assets deployment (DAD) is a collection of management policies that promote flexible development of an effective and profitable delivery service [61]. A service process is a strategy used to bring out the necessary activities. Performance concerns how well the activities are performed and encompasses service effectiveness and efficiency. Delivery of the required task means ensuring that the right product (meeting the requirements of quality) is delivered in the right quantity, in the right place, at the right time, from the right source, with the right service (both before and after sale) and finally at the right price [62].

Deployment process consists of numerous interrelated activities with potential transitions between them. These activities can occur at the producer site, at the consumer site, and both side. Every software system is unique, the specific process or procedure within each activity can hardly be defined. Therefore, deployment should be interpreted as a general process that has to be modified according to definite requirements or characteristics. Deployment activities divided into Release, Install and activate, Deactivate, Adapt, Update, Built-In, Version tracking, Uninstall, and Retire [63].

For software vendors the process of release, delivery, and deployment to customers are naturally complex. However, software vendors can significantly improve their product quality and quality of service by applying a model, customer configuration updating (CCU) that can assess the capabilities of software that focus on customer interaction if such a model were available [64]. The deployment process is similar to a method for deploying software by adding steps and variables to software project. Each step contains a detailed action (set of actions) that is executed as part of the deployment process each time software is deployed. Once the initial setup, deployment process shouldn't change between deployments, even though the software being deployed will modify as part of the software development process.

Software maintenance is an extremely wide activity that includes error correction, deletion of obsolete capabilities, enhancements of capabilities, and optimization. Because change is inevitable, mechanisms must be developed for assessment, controlling and assembly modifications. Maintenance categorized into four types such as:

1. **Adaptive Maintenance** - modifies the system to handle changes in the software environment. Adaptive maintenance includes changes to the functionality of the

system developed for particular user requirement and needs. Adaptive maintenance also implies the need for modifications of certain functionalities, although the system works as estimated and there is no fault or error in the system. It generally occurs when there comes to an alteration in legal norms or a shift in the political business users.

2. **Perfective Maintenance** - implement new or changed user requirements which concern functional enhancements to the software. Modification of a software product after delivery to identify and correct hidden faults in the software product before they become effective faults. Enhances the performance of resources by increasing uptime. Enhances user (internal or external) service because maintenance teams have fewer unplanned maintenance and can react quicker to latest problems.
3. **Corrective Maintenance** - diagnose and fix errors, possibly ones found by users. Corrective maintenance most frequently referred to as bugs, is the most usual change associated with maintenance work. Corrective maintenance addresses errors, bugs, and faults in software that could affect various areas of software, design, code, and logic. Most commonly these changes are sprung by bug reports created by users.
4. **Preventive Maintenance** - increase software maintainability or reliability to prevent problems in the future. Preventive maintenance refers to changes made to increase the understanding and maintainability of software in the extended run. Preventive maintenance is focused in decreasing the deterioration of software in the long run. Optimizing code, restructuring, and updating documentation are common preventive changes. Executing preventive maintenance reduces the quantity of unpredictable effects software can have in the extensive term and helps it become stable, scalable, understandable, and maintainable [65-67].

Figure 2.2 show the representation of maintenance types where perfective uses 50%, adaptive 25%, preventive is 5%, and corrective uses 20% area.

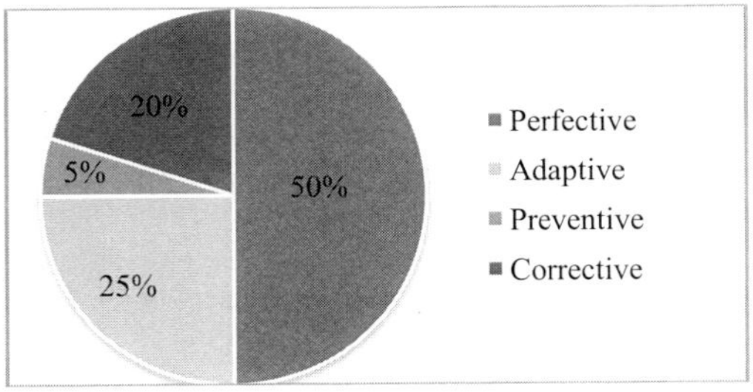

Figure 2.2 Representation of maintenance types

As figure, shows the whole picture of maintenance types and its importance. Successful software requires continuous change that is trigger by evolving requirements, technologies, and stakeholder knowledge. This stable change constitutes software evolution and maintenance [68].

2.8 SUMMARY

Activities of software development process divided into sequential task. Each and every task is related, dependent, and important for software development. Feasibility study is an evaluation of the expediency of a proposed system and software. Requirements elicitation is the process of collecting the requirements of software from users and stakeholders where requirements specification used for translating the information gathered during the analysis into a document that defines user and system requirements. Software architecture is a high level structure of software. Software design is the process of implementing software solutions to one or additional sets of problems. Software coding used for converting software specification into an executable system and using module testing software is tested. Individual software modules are combined and tested as a group using integration testing while system testing is a testing conducted on whole software. Software delivery is the process of getting a software product to market. Deployment activities are used for further processing and software maintenance used for modification and updation in software.

2.9 REFERENCES

1. Fuggetta, A.; Di Nitto, E. (2014). Software process. In Proceedings of the ACM on Future of Software Engineering, pp. 1-12.
2. Boehm, B. W. (1988). A spiral model of software development and enhancement. Computer, Vol. 21, Issue 5, pp. 61-72.
3. Bulusu, S. T.; Laborde, R.; Wazan, A. S.; Barrère, F.; Benzekri, A. (2018). A Requirements Engineering-Based Approach for Evaluating Security Requirements Engineering Methodologies. In Information Technology-New Generations, Springer, Cham, pp. 517-525.
4. Bosch, J. (2004). Software architecture: The next step. In European Workshop on Software Architecture, Springer Berlin Heidelberg, pp. 194-199.
5. Yau, S. S.; Tsai, J. J. P. (1986). A survey of software design techniques. IEEE Transactions on Software Engineering, Vol. 6, pp. 713-721.
6. Reid, S. C. (1997). Module testing techniques - which are the most effective. In Proceedings of Eurostar 97: The Fifth European Conference on Software Testing.
7. Bennett, K. H.; Rajlich, V. T. (2000). Software maintenance and evolution: a roadmap. In Proceedings of the ACM Conference on the Future of Software Engineering, pp. 73-87.
8. Lientz, B. P.; Swanson, E. B.; Tompkins, G. E. (1978). Characteristics of application software maintenance. Communications of the ACM, Vol. 21, No. 6, pp. 466-471.
9. Barreto, H. (2002). Small Business Innovation Research Program Policy Directive. US Small Business Administration.
10. Burke, R. (2013). *Project management: planning and control techniques*. New Jersey, USA.
11. Ackerman, M. S. (2000). The intellectual challenge of CSCW: the gap between social requirements and technical feasibility. Human-computer interaction, Vol. 15, No. 2, pp. 179-203.
12. Feasibility analysis http://www.gis.geo.uj.edu.pl/Teaching_tool_on_knowledge_transfer/eng/wydruki/pdf/Feasibility%20analysis.pdf
13. Van den Berg, H. (2008). Reanalyzing qualitative interviews from different angles: The risk of decontextualization and other problems of sharing qualitative data. Historical Social Research/Historische Sozialforschung, pp. 179-192.
14. Cooper, D. R.; Schindler, P. S.; Sun, J. (2003). Business research methods.

15. Carter, M. P. (1983). Determining information needs. Management Decision, Vol. 21, Issue 4, pp. 45-51.
16. Churchill Jr, G.A.; Surprenant, C. (1982). An investigation into the determinants of customer satisfaction. Journal of marketing research, pp. 491-504.
17. Shen, L. Y.; Tam, V. W.; Tam, L.; Ji, Y. B. (2010). Project feasibility study: the key to successful implementation of sustainable and socially responsible construction management practice. Journal of Cleaner Production, Vol. 18, No. 3, pp. 254-259.
18. Davis, A.; Dieste, O.; Hickey, A.; Juristo, N.; Moreno, A. M. (2006). Effectiveness of requirements elicitation techniques: Empirical results derived from a systematic review. In Requirements Engineering, 14th IEEE International Conference, pp. 179-188.
19. Dieste, O.; Juristo, N. (2011). Systematic review and aggregation of empirical studies on elicitation techniques. IEEE Transactions on Software Engineering, Vol. 37, Issue 2, pp. 283-304.
20. Tsumaki, T.; Tamai, T. (2006). Framework for matching requirements elicitation techniques to project characteristics. Software Process: Improvement and Practice, Vol. 11, No. 5, pp. 505-519.
21. Goguen, J. A.; Linde, C. (1993). Techniques for requirements elicitation. In Proceedings of IEEE International Symposium on Requirements Engineering, pp. 152-164.
22. Zhang, Z. (2007). Effective requirements development - A comparison of requirements elicitation techniques. Software Quality Management XV: Software Quality in the Knowledge Society, E. Berki, J. Nummenmaa, I. Sunley, M. Ross and G. Staples (Ed.) British Computer Society, pp. 225-240.
23. Davis, A.; Overmyer, S.; Jordan, K.; Caruso, J.; Dandashi, F.; Dinh, A.; Kincaid, G.; Ledeboer, G.; Reynolds, P.; Sitaram, P.; Ta, A.(1993). Identifying and measuring quality in a software requirements specification. In Proceedings of IEEE First International Software Metrics Symposium, pp. 141-152.
24. Firesmith, D. (2003). Modern requirements specification. Journal of Object Technology, Vol. 2, No. 2, pp.53-64.
25. Nuseibeh, B.; Kramer, J.; Finkelstein, A. (1994). A framework for expressing the relationships between multiple views in requirements specification. IEEE Transactions on software engineering, Vol. 20, Issue 10, pp.760-773.
26. Sommerville, I.; Sawyer, P.; Viller, S. (1998). Viewpoints for requirements elicitation: a practical approach. In Proceedings of IEEE Third International Conference on Requirements Engineering, pp. 74-81.
27. Hickey, A. M.; Davis, A. M. (2003). Requirements elicitation and elicitation technique selection: model for two knowledge-intensive software development processes. In Proceedings of IEEE 36th Annual Hawaii International Conference on System Sciences, pp. 10.
28. Kassel, N. W.; Malloy, B. A. (2003). An approach to automate requirements elicitation and specification. In Proc. of the 7th Int. Conf. on Software Engineering and Applications, pp. 3-5.
29. Hickey, A. M.; Davis, A. M. (2004). A unified model of requirements elicitation. Journal of Management Information Systems, Vol. 20, No. 4, pp. 65-84.
30. Rajagopal, P.; Lee, R.; Ahlswede, T.; Chiang, C. C.; Karolak, D. (2005). A new approach for software requirements elicitation. In IEEE Sixth International Conference on Software Engineering, Artificial Intelligence, Networking and Parallel/Distributed Computing, 2005 and First ACIS International Workshop on Self-Assembling Wireless Networks. SNPD/SAWN 2005, pp. 32-42.
31. Condori-Fernandez, N.; Daneva, M.; Sikkel, K.; Wieringa, R.; Dieste, O.; Pastor, O. (2009). A systematic mapping study on empirical evaluation of software requirements specifications techniques. In IEEE Proceedings of the 3rd International Symposium on Empirical Software Engineering and Measurement, pp. 502-505.
32. Perry, D. E.; Wolf, A. L. (1992). Foundations for the study of software architecture. ACM SIGSOFT Software Engineering Notes, Vol. 17, No. 4, pp. 40-52.

33. Hofmeister, C.; Kruchten, P.; Nord, R. L.; Obbink, H.; Ran, A.; America, P. (2007). A general model of software architecture design derived from five industrial approaches. Journal of Systems and Software, Vol. 80, Issue 1, pp. 106-126.
34. Garlan, D.; Allen, R.; Ockerbloom, J. (1994). Exploiting style in architectural design environments. ACM SIGSOFT Software Engineering Notes, Vol. 19, No. 5, pp. 175-188.
35. Landay, J. A.; Myers, B. A. (1995). Interactive sketching for the early stages of user interface design. In Proceedings of the ACM SIGCHI conference on Human factors in computing systems, pp. 43-50.
36. Vitharana, P.; Zahedi, F. M.; Jain, H. (2003). Design, retrieval, and assembly in component-based software development. Communications of the ACM, Vol. 46, No. 11, pp. 97-102.
37. Batra, D.; Davis, J. G. (1992). Conceptual data modelling in database design: similarities and differences between expert and novice designers. International Journal of Man-Machine Studies, Vol. 37, Issue 1, pp. 83-101.
38. Sommerville, I. (2011). *Software Engineering*. Ninth Edition ed., Addison-Wesley.
39. Garlan, D. (2000). Software architecture: a roadmap. In Proceedings of the ACM Conference on the Future of Software Engineering, pp. 91-101.
40. Robbins, J. E.; Hilbert, D. M.; Redmiles, D. F. (1996). Extending design environments to software architecture design. In IEEE Proceedings of the 11th Knowledge-Based Software Engineering Conference, pp. 63-72.
41. Clements, P.; Shaw, M. (2006). The golden age of software architecture: A comprehensive survey.
42. Taylor, R. N.; Van der Hoek, A. (2007). Software Design and Architecture: The once and future focus of software engineering. In IEEE Future of Software Engineering, FOSE'07, pp. 226-243.
43. Humphrey, W. S. (1988). Characterizing the software process: a maturity framework. IEEE software, Vol. 5, Issue 2, pp.73-79.
44. Lanergan, R.G.; Grasso, C. A. (1984). Software engineering with reusable designs and code. IEEE Transactions on Software Engineering, Vol. 5, pp.498-501.
45. Beizer, B. (2003). *Software testing techniques*, Dreamtech Press.
46. Software Testing - Documentation, https://www.tutorialspoint.com/software_testing/software_testing_documentation.htm
47. Hayes, I. J. (1986). Specification directed module testing. IEEE Transactions on Software Engineering, Vol. 1, pp.124-133.
48. Hoffman, D. (1989). A CASE study in module testing. In Proceedings of IEEE Software Maintenance Conference, pp. 100-105.
49. Delamaro, M. E.; Maidonado, J. C.; Mathur, A. P. (2001). Interface mutation: An approach for integration testing. IEEE transactions on software engineering, Vol. 27, Issue 3, pp. 228-247.
50. Leung, H. K.; White, L. (1990). A study of integration testing and software regression at the integration level. In Proceedings on IEEE Software Maintenance Conference, pp. 290-301.
51. Linnenkugel, U.; Mullerburg, M. (1990). Test data selection criteria for (software) integration testing. In Proceedings of the IEEE First International Conference on Systems Integration, pp. 709-717.
52. Jorgensen, P. C.; Erickson, C. (1994). Object-oriented integration testing. Communications of the ACM, Vol. 37, No. 9, pp. 30-38.
53. Le Traon, Y.; Jéron, T.; Jézéquel, J. M.;Morel, P.(2000). Efficient object-oriented integration and regression testing. IEEE Transactions on Reliability, Vol. 49, Issue 1, pp. 12-25.
54. da Silva, C.E.; de Lemos, R. (2011). Dynamic plans for integration testing of self-adaptive software systems. In Proceedings of the ACM 6th international symposium on software engineering for adaptive and self-managing systems, pp. 148-157.

55. Murphy, C.; Shen, K.; Kaiser, G. (2009). Automatic system testing of programs without test oracles. In Proceedings of the ACM eighteenth international symposium on Software testing and analysis, pp. 189-200.

56. Gutiérrez, J. J.; Escalona, M. J.; Mejías, M.; Torres, J. (2006). An approach to generate test cases from use cases. In Proceedings of the ACM 6th international conference on Web engineering, pp. 113-114.

57. Borjesson, E.; Feldt, R. (2012). Automated system testing using visual gui testing tools: A comparative study in industry. In IEEE Fifth International Conference on Software Testing, Verification and Validation, ICST 2012, pp. 350-359.

58. Gross, F.; Fraser, G.; Zeller, A. (2012). Search-based system testing: high coverage, no false alarms. In Proceedings of the ACM International Symposium on Software Testing and Analysis, pp. 67-77.

59. Brown, A. W.; Mancisidor, A.L.; Olivia, L.R. (2011). Practical Experiences with Software Factory Approaches in Enterprise Software Delivery, ICSEA2011.

60. Cantor, M.; Royce, W. (2014). Economic governance of software delivery. IEEE Software, Vol. 31, No. 1, pp. 54-61.

61. Cohen, M. A.; Agrawal, N.; Agrawal, V. (2006). Achieving breakthrough service delivery through dynamic asset deployment strategies. Interfaces, Vol. 36, No. 3, pp. 259-271.

62. Kumar, R.; Kumar, U. (2004). A conceptual framework for the development of a service delivery strategy for industrial systems and products. Journal of Business & Industrial Marketing, Vol. 19, No. 5, pp. 310-319.

63. Requirements Analysis www.uacg.bg/filebank/acadstaff/userfiles/publ_bg_397_SDP_activities_and_steps.pdf

64. Jansen, S.; Brinkkemper, S. (2006). Evaluating the release, delivery, and deployment processes of eight large product software vendors applying the customer configuration update model. In Proceedings of the ACM international workshop on Workshop on interdisciplinary software engineering research, pp. 65-68.

65. Lientz, B. P.; Swanson, E. B. (1981). Problems in application software maintenance. Communications of the ACM, Vol. 24, No. 11, pp. 763-769.

66. Singh, B.; Gautam, S. (2017). Impact of Software Evolution on Software Process, National Conference on Recent Advances in Information and Communication Technologies (RAICT-2017), pp. 106-111.

67. Chapin, N.; Hale, J. E.; Khan, K. M.; Ramil, J. F.; Tan, W.G. (2001). Types of software evolution and software maintenance. Journal of software maintenance and evolution: Research and Practice, Vol. 13, No. 1, pp.3-30.

68. Rajlich, V. (2014). Software evolution and maintenance. In Proceedings of the ACM on Future of Software Engineering, pp. 133-144.

Chapter - 3

SOFTWARE QUALITY

Software quality is concerned with developing software products with required quality and assessing the level of software quality. Software processes are significant assets in achieving and assessing the software quality. Additionally, the adherence to defined processes is a main issue to having the software engineering under control and to facilitate process improvement. Software quality assurance (SQA) key process area in the Capability Maturity Model (CMM) is used to improve software process, i.e. activities for monitoring adherence to the processes. Section I describes the introduction of software quality. Section II describes the history of software quality. Section III explains the classification of software quality. Section IV describes the quality attributes. Section V discuss, the quality requirements. Section VI describe, the quality measurement. Section VII explain, the summary of this chapter.

3.1 INTRODUCTION

Software plays an increasingly important role in today's society. 50 years ago, software was used only in limited calculation machines but now most end user electronics as well as safety critical equipment contains software. This evolution has set the spotlight on how to achieve and assure the quality of software product. Software contains a trickle of inevitable trends. For example: website, an app, management system, a database, and an information system. In a very high possibility, its size, and complexity have developed throughout its entire lifetime. Take any software product that is still in energetic use. Likelihoods are the number of its dynamic users and outside dependencies have developed throughout its complete lifetime. Take some segment of software that is still actively continued. Most probable the period between resultant releases is much smaller currently than it was in the previous.

In the past, the software developer and software user was the same person, typically a scientist. The same person did similarly the testing, as a part of his normal flow. Quality assurance was observably, very effective, concentrated, and user-centric, through the fastest potential response cycle [1]. While the usage of software and the number of users raised the new occupation, software developer, was progressively born. Future on emanated, one by one, business analysts, software project managers, software product managers, chief software architects, software engineering leads, software testers, scrum masters, user experience designers, data scientists are required to develop a quality product.

During the 1980's software processes have come into focus as a means for achieving good quality software product. Software quality is the degree of conformance to functional and non-functional requirements, need, and expectations. The processes are needed to conquer the problems connected to the specific nature of software. The Capability Maturity Model (CMM) from the Software Engineering Institute has evolved to be a de facto standard in the area of software processes and assessment of organisations' capability maturity. In CMM, the Software Quality Assurance (SQA) key process area addresses the autonomous software quality assurance activities, in particular for assurance of the faithfulness to defined processes. However, the SQA concept is often misunderstood and intermingled with other definitions and standards for quality assurance, like ISO 9001. For example, testing is a means for quality assurance according to IS0 9001, while the task of CMM SQA is to validate that the testers follow their defined testing procedures. The two standards acquire different perspectives and utilize similar terms for diverse concepts. Thus they are simply misunderstood [2].

3.2 HISTORY OF SOFTWARE QUALITY

The starting point of the software quality history is very much in the mathematical statistics history. The pioneers in quality engineering identified the variations in a production process as a statistical process. By applying statistical methods on the production process, the quality of the products could be set in control. Walter A. Shewart was emphasising the significance of measurements. The vital topic in his publications is how to take care of data and draw conclusions from the min order to supervise and decrease the variation in the production process. Shewart proposed as early as 1924 what now is called "control charts" to calculate the variations of a process in order to control them.

W. Edwards Deming and Joseph M. Juran share Shewhart's statistical point of view. In addition, both of them stress the management role in quality work very powerfully. Only if top management execute themselves to the quality issues, it is possible to achieve continuous quality improvement. Deming's philosophy is presented in his famous "Deming's points".

Deming and Juran were invited to Japan in the 1950s by Kaoru Ishikawa. The Japanese industry adopted the quality philosophy and this is regarded having a significant impact on the Japanese miracle. The quality reputation of the Japanese industry in the 1950s was unfortunate. Ishikawa and others developed techniques, which are known and applied worldwide, like just-in-time production, total quality control (TQC), quality control circles (QC-circles), and the seven management tools (7MT).

- Webster's definition of quality: the totality of functionality and features of a software product that stand on its ability to satisfy fixed or implied needs.
- Crosby's view of quality management's perspective: software quality is conformance of the software solution (software product or IT service) to the requirements.
- Juran's view of quality manufacturer's perspective: software quality is fitness of the software product (IT service) for intended use.
- Deming's view of quality - consumer's perspective: software quality is ability of the software product (IT service) to meet or exceed the end user expectations.

In software engineering, some of these techniques have been considered in improvement initiatives. In the Personal Software Process (PSP) the process control approach is accordingly realised [3]. However, the PSP is rather state of the art than state of the practice in software engineering. Software quality methods have always been insufficient, too slow, and too expensive yet passing through too many defects. It has been widely known for 40 years, at least, that the entire cost of a software defect is cheapest if it is detected and corrected close to the time and place it was made. For contrast, Mr Sakichi Toyoda knew the similar principle already before World War II when he laid the foundations of the world famous Toyota Way.

Conventional knowledge tells that software developers should test their own code previously at the unit level. However, practical knowledge tells that software developers are mainly incompetent and de-motivated when it comes to software testing. Therefore, self-testing rarely happens. Experience has also taught that software developers are constantly busy writing new code and fixing older code so that in practice they never really have the time to do any software testing. The debt of testing accumulates towards the presently stages of the software flow. Still today, late testing is the most general approach to software quality assurance.

Late software testing comes with a group of costly problems. First of all, it requires a large number of software testers. Therefore, testing needs to be planned, organized, and managed professionally which adds a considerable amount of overhead. Second, late testing results in an enormous amount of time spent on waiting and rework, i.e. full waste. Testers wait for a software version to be tested. Developers wait for the test results in order to start debugging and fixing. Testers wait for the fixes to be re-tested. Meanwhile, instead of writing new code, developers spend most of their time fixing defects that could have been found much earlier.

Test automation was invented as a waste reduction solution to the late testing. The general belief was that if test execution were automated the testing times would be shorter and testing costs would be lesser. To a great extent, this is correct. However, in most cases people have learned the hard way that automating tests is very costly and demanding. Therefore, the payback of test automation efforts has, with a little notable exceptions, been negative. As long as software testing as a discipline has been around it has revolved around the same essential questions: how to make developers test early, how to improve efficiency and effectiveness of late testing, and how to automate software testing.

The introduction of agile methods brought along a partial salvation. The whole paradigm of making software was changed. Software was developed by small, self-contained teams in quick cycles, and released frequently to the users. Changing requirements became an asset rather than a vice. It suddenly began to make perfect sense to test early. Because the release cycles were rapid, it also made great sense to automate the tests. It helped, indeed, that the software developers possessed programming skills that were essential for test automation.

The clever agile evangelists and QA gurus came up with believe names and methods for the age-old ideas, such as definition of done, test driven development, and behaviour driven development. Unexpectedly, all the young people were excited about the idea that you can declare a portion of software ready only after you know it works

[4]. The same persons came up with the earth-shaking finding that for each section of software functionality there should be a test case.

Still today, it very humorous when agile aficionados evangelize that it is an actually good idea for people working on the similar thing to speak to each other. This all tells something about the young age of our industry. Parallel with agile, emergence of a new software testing approach called exploratory testing. To be correct, the fundamentals of the approach were introduced in the literature already in the 1970s, under the title error guessing. Exploratory testing is, by far, the well-organized form of software testing. It is also essential in the era of rapid release cycles: automated testing is seldom rapid enough for new functionality while it is superior in regression testing.

The limitation of exploratory testing is that it does not scale. To be a great exploratory tester you need sound software testing skills plus reasonably a bit of business understanding and human understanding. Technical knowledge wouldn't hurt either. Obviously you also need a tester's mindset and an organized, disciplined way of working. Such individuals are uncommon and most of them do not wish for to work as software testers.

The evolutions of agile and exploratory testing reveal something interesting about software industry. They both began as rebellious movements. The message was clear: Let's throw all previous away because it sucks; the only things we need are excellent people, more communication, user focus, and rudimentary method. Gradually, over the years, the rebels have learned that there was a motivation for the existence of the age-old greatest practices. Now they have been brought back under different names and fair to say, often in a greatly improved format.

The salvation brought along by the agile methods was only prejudiced. Still today, less than a half of the agile software population is capable enough to apply the agile methods properly. Moreover, while solutions have become better challenges have mature bigger. Many software assets are so large that they simply cannot be delivered by a particular agile team. Although software teams are nowadays capable to release faster the demand for ever-accelerating release cycles has also grown. The internet, SaaS, and API economy have brought along eco-systems where the software developer cannot truly know, where, by whom, when, in what kind of environments and dependencies, and for what reason his or her software will be used.

Software quality again in front of a novel challenge to tackle. The solution already exists. The idea is easy and it makes sense. Take the people that formerly were in a separate agile development team, use common tools, system QA team, and deployment team together, tell them to talk to each other, and automate all probable. This effort called DevOps.

DevOps has been as big a transform as the introduction of agile methods was 15 years ago. DevOps is definitely the accurate way to go but before the software development mainstream gets there, we will see numerous world-class disasters. A successful DevOps requires the exact tool-chain and its disciplined use, high quality the teamwork, and a lean operational culture. These are all extraordinary virtues.

One of the typical misconceptions about DevOps is that it would change QA and make software testing irrelevant. But DevOps is not a testing method. It is a way of automating, organizing, and measuring the software workflow. Continuous improvement is a necessary part of successful DevOps. DevOps alone does not make good quality product but it puts a high pace requirement on software testing and requires great discipline in mutually exploratory testing and test automation.

The history wouldn't be complete without mentioning software as a service, the business model that broke through big moment with products such as salesforce.com some ten years ago. One of the huge things about software as a service is that there is always accurately one instance of the software publicly available: there are no user installations or end-user installations [5]. This, in turn, means that the software can be updated as often as needed, even numerous times per hour.

Software as a service has enabled to utilize the telemetry approach for software testing. Telemetry in software testing means that the software we run from the cloud is instrumented with a huge number of software sensors that gather data about what we do and how the software behaves. This data is sent back to the cloud and analyzed there. The benefits are marvellous. Testing work is crowd sourced to the end-users without them knowing.

To get information on how the software behaves in several different user environments. We are able to measure not only the functionality and technical performance but also user behaviour. We can also perform A/B testing and decide which users get test versions and which ones don't this also happens without the user knowing.

It is already today promising to use artificial intelligence to analyze the test results collected via the telemetric approach. One can effortlessly imagine a machine-learning system where an artificial intelligence learns from telemetric test results and generates investigative test cases that are readily automated. This is not a sci-fi vision but previously a reality in some few places. If software testing has come this far one may logically ask why programmers haven't been replaced by AI yet [6].

3.3 CLASSIFICATION OF SOFTWARE QUALITY

Software quality classification models are generally based on how accurately they can classify (predict) observations. It has been practical that when comparing two groups of software quality classification models, there are Type I and Type II misclassification error rates. As an effort to utilize an outstanding comparative measure, some researchers have used the overall misclassification error rate of models. Though the Type I, Type II, and overall misclassification error rates have been normally used as accuracy measures, they may not essentially be the most suitable and effective measures [7].

Classification based modelling for software quality evaluation is a proven technique in achieving better software quality control. Some classification techniques used for software quality assessment include optimal set reduction, neural networks, logistic regression, decision trees, and case based reasoning. Software metrics based quality classification models categorize software modules into groups. A two group classification model is the most frequently used, in which program modules are grouped into classes such as: fault prone (FP) and not fault-prone (NFP). Software analysts concerned in predicting whether a software component will be FP or NFP, can utilize classification models to achieve the needed prediction. When associating with such models, two types of misclassification errors are occurred: Type I error, when a NFP component is predicted as FP and Type II error, when a FP component is misclassified as NFP. Practically Type II errors are more rigorous in terms of software development costs and the organization's credit and reputation. They may involve assessment and correction to components once they have been deployed for operations [8].

In various empirical studies which is related to software quality classification modelling, investigated numerous classification techniques, including classification and regression trees (CART), the Treedisc algorithm, tree based classification with S-PLUS, the C4.5 algorithm, logistic regression, the Sprint-Sliq algorithm, and case based reasoning [9].

1. CART

Classification and regression trees extensively used decision tree system with several applications to predictive modelling, data mining, and data pre-processing. CART is a statistical tool of Salford Systems Inc. that automatically sifts huge complex databases, discovering, searching for, and isolating significant patterns and relationships in the data. The algorithms of CART search for questions that split nodes into relatively homogeneous child nodes, such as a group consisting largely of responders, or highrisk components. As the tree evolves, the nodes become increasingly more homogeneous, identifying important segments. The set of predictor variables used to split the nodes into segments, read directly off the tree and summarized in the variable importance tables, are the key drivers of the response variable. The methodology solves a number of performance, accuracy, and operational problems that still plague many current decision-tree methods. Some of innovations of CART include: solving the tree size problem using strictly two-way splitting, incorporating automatic testing and validation, and providing a completely new method for handling missing values in the data set [10-11].

2. S-PLUS

A solution for advanced data analysis, statistical modelling, and data mining, the S-PLUS (Mathsoft Inc.) tool merge an intuitive graphical user interface with an extensive data analysis environment to suggest ease of use and flexibility. Among other data mining functions S-PLUS includes regression tree based models. At the centre of the S-PLUS system is S, a language which is designed specifically for data visualization and exploration, statistical modelling, and programming with data. S provides a prosperous, object oriented environment for interactive data discovery. With a vast library of functions for all aspects of computing with data, S offers excellent extensibility. The predictors are software metrics treated by S-PLUS as ordinal measures which are used to construct regression trees to predict the response variable [12-13].

3. C4.5 Algorithm

The C4.5 algorithm is an inductive supervised learning system which employs decision trees to characterize a quality model. C4.5 is a descendent of another induction program, ID3 and it consists of four primary programs: decision tree generator, decision tree interpreter, production rule generator, and production rule interpreter. The algorithm uses these programs when developing and evaluating classification tree models. Diverse tree models were built by changing parameters: least node size before splitting and pruning percentage. The C4.5 algorithm commands sure pre-processing of data in order for it to build decision tree models. Some of these contain attribute value description type, predefined discrete classes, and satisfactory number of observations for supervised learning. The classification tree is primarily empty and the algorithm begins adding decision and leaf nodes, preliminary with the root node [14-15].

4. Sprint-Sliq algorithm

Sprint-Sliq is an abbreviated version of Scalable PaRallelizable INduction of decision Trees Supervised Learning In Quest. The algorithm can be used to assemble classification tree models that can analyze both categorical and numeric attributes. It is a customized version of the classification tree algorithm of CART which uses a dissimilar pruning technique based on the least description length principle. The algorithm has outstanding scalability and analysis speed. Classification tree modelling using Sprint-Sliq is accomplished in two phases: a tree building phase and a tree pruning phase. Tree building phase recursively partitions the training data until each partition is either pure or meets the stop-splitting rules set by the user. The IBM Intelligent Data Miner tool which implements the Sprint-Sliq algorithm used to build classification trees. Sprint-Sliq uses the Gini Index to assess the integrity of split of all the possible splits. A class assignment rule is needed to classify modules as FP and NFP [16].

5. Case-Based Reasoning

Case-based reasoning (CBR) is a technique that aims to discover solutions to new problems based on past experiences which are represented by cases in a case library. The case library and the associated retrieval and decision rules compose a CBR model. In the perspective of a classification problem, each case in the case library has known attributes and class membership. The functioning hypothesis of CBR for software quality classification modelling is that a module currently under development is possibly FP if a module with related attributes in an earlier release (similar project) was FP. A CBR system can take benefit of availability of new or revised information by adding novel cases or by removing outdated cases from the case library. Its excellent scalability provides quick retrieval even as the size of the case library scales up. CBR systems can be designed to alert users when a latest case is outside the bounds of existing experience [17].

3.4 QUALITY ATTRIBUTES

The conventional and well recognized way to describe software quality is using a quality classification of quality attributes (characteristics). Consequently, the most basic quality aspect hierarchy is to straightly use the quality attributes or characteristics. McCall et al. 1977, started with an amount of 55 quality characteristics which have a significant influence on quality called as factors. For reason of ease, McCall reduced the number of characteristics into eleven such as: efficiency, integrity, reliability, usability, accuracy, maintainability, testability, flexibility, interface facility, reusability, and transferability. Second set of quality factors was defined by Boehm et al. 1978, total nineteen characteristics such as: usability, clarity, efficiency, reliability, modifiability, reusability, modularity, documentation, resilience, correctness, maintainability, portability, interoperability, understandability, integrity, validity, flexibility, generality, and economy. Next in Ghezzi et al. 1991, state that software users only care about the external quality but internal quality deal mainly with the structure of the software that help developers to achieve external qualities. External quality such as: integrity, reliability, usability, and accuracy. Internal quality such as: efficiency, maintainability, testability, flexibility, interface facility, reusability, and transferability. Some important definitions are as:

- **Traceability:** Traceability is concerned with documenting the life of a requirement and providing bi-directional traceability between various associated requirements.
- **Tailorability:** Tailorability is a feature that allows a change of certain aspects of software to drive user characteristics and requirements.
- **Interoperability:** The ability of two or more systems or components to exchange information and to use the information that has been exchanged.
- **Usability:** The ease with which a user can learn to operate, prepare inputs for, and interpret outputs of a system or component.
- **Scalability:** Scalability is the ability to handle increased workload by repeatedly applying a cost effective strategy for extending a system's capacity.
- **Reliability:** The ability of a system or component to perform its required functions under stated conditions for a specified period of time.
- **Readability:** Readability is the ease with which a reader can understand a written text or code.
- **Maintainability:** The ease with which a software system or component can be modified to correct faults, improve performance or other attributes, or adapt to a changed environment.
- **Functionality:** An attributes that relate to the existence of a set of functions and their specified properties. The functions are those that satisfy stated or implied needs.
- **Performance:** The degree to which a system or component accomplishes its designated functions within given constraints.
- **Security:** An ability to prevent unauthorized access, whether accidental or deliberate, to programs and data.
- **Efficiency:** The degree to which a system or component performs its designated functions with minimum consumption of resources.
- **Portability:** The ease with which a system or component can be transferred from one hardware or software environment to another.
- **Evolvability:** Evolvability is defined as the capacity of a system for adaptive evolution.
- **Reusability:** The degree to which a software module or other work product can be used in more than one computer program or software system.
- **Adaptability:** An ability to change something or oneself to fit to occurring changes.
- **Changeability:** Changeability relate to the effort needed for modification, fault removal or for environmental change.

- **Composability:** Composability is a system design principle that deals with the inter-relationships of components.
- **Completeness:** All required component are merge into final making or doing.
- **Effectiveness:** The accuracy and completeness with which users achieve specified goals.
- **Stability:** The capability of the software product to avoid unexpected effects from modifications of the software.
- **User friendliness:** Pertaining to a computer system, device, program, or document designed with ease of use as a primary objective.
- **Understandability:** The degree to which the purpose of the system or component is clear to the evaluator.
- **Flexibility:** The ease with which a system or component can be modified for use in applications or environments other than those for which it was specifically designed.
- **Simplicity:** The degree to which a system or component has a design and implementation that is straightforward and easy to understand.
- **Expandability:** The ease to add new functionality to it.
- **Accountability:** Accountability is a system for which it is possible to measure the usage of the code.
- **Modularity:** The degree to which a system or computer program is composed of discrete components such that a change to one component has minimal impact on other components.
- **Correctness:** The degree to which a system or component is free from faults in its specification, design, and implementation [18-22].

3.5 QUALITY REQUIREMENTS

Software quality requirements are a part of a contractual agreement between acquirer and developer or may be input for a product quality evaluation. No particular development model or measures are approved or assumed. Quality requirements can be rising increasingly different for a software project and the significance given to each quality requirement varies with individual stakeholders. Quality requirements can cooperate in the sense that attempts to achieve one quality can assist or hinder the achievement of other qualities. Such an interaction creates a widespread network of interdependencies and tradeoffs between quality requirements which is not simple to trace or estimate. Software is generally a part of a larger system. Architectural decisions made at an advanced level of system hierarchical structure describe boundaries and interfaces to the software. Architectural decisions regarding which parts of a system will really be implemented in software may not or just partly be made at the point when quality requirements are described. Therefore, it may be impossible to choose at an early time in the software development process whether or not a quality requirement is disturbed with the software. In the process oriented and goal oriented software development approaches, user requirements are recognized based on an analysis of business goals and application domain. Consequently, architectures of the desired systems are designed and implemented. One of the main goals of the architecture of a software system is to develop a system design whereas fulfilling the desired quality requirements. In fact if we consider systems architecture as a set of architectural decisions, the most important ones concern the satisfaction of quality attributes [23].

At the current time, the international standards and technical reports for system and software product quality are dispersed in numerous series of normative documents as: ISO/IEC 9126. These documents are not entirely reliable and do not contain a tools for correct requirements setups. As quality is defined as a degree to which the set of natural characteristic fulfils requirements, the accurate requirement formulation is the key point for the quality measurement evaluation. Requirements are needs or an expectation that is stated, usually implied, or mandatory. One of the weakest points of the today's arena of software product quality standards is the lack of the guidelines for the correct quality requirements formulation. For the unbiased quality evaluation such requirements have to be clear, consistent, complete, and measurable. A correct measurement function represents an interpretation of a software quality property and a goal value of the quality measure represents a quality requirement. Similarly the real value of the quality measurement represents the observed quality of the software.

Software quality requirements as well as all other requirements cannot be seen in isolation but must be view in a broader perspective. Software quality requirements have a particular close relation to functional requirements. Functional requirements play a significant role for specifying software quality requirements. Functionality is one of the most important requirements. Functionality requirements should not be confused with functional requirements. Functionality is the ability of the software to provide functions which meet its functional requirements. Functionality requirements are developed into requirements for the software product to be suitable, interoperable, accurate, secure, and compliant with the relevant functional standards and regulations.

In some situations, it is meaningful to identify a quality requirement for a software product whereas in other situations a quality requirement simply applies to a portion of the software product. For example: some functions are just relevant for particular users and have definite quality requirements, which are dissimilar from quality requirements for other functions proposed for other purposes and other users. It is consequently important to specify which portion of a software product is appropriate for a software quality requirement. In other words, a quality requirement is related to a portion of the software product (a set of functions). For example: some functions may be planned for general end users and may hence need low error tolerance, whereas a different group of functions may be planned for specialists and accordingly authorize a greater error tolerance. In both cases, the error tolerance mechanism and the degree of error tolerance required should be carefully specified. Quality in use is defined as the amount to which a product meets the needs of specified users to accomplish specified goals with effectiveness, safety, productivity, and satisfaction in a specified perspective of use. Therefore quality in use requirements are directly related to other system requirements such as business requirements, hardware requirements, and requirements to end user [24].

In broad, functional requirements indicate what a system does whereas quality requirements describe how well those functions are accomplished. Unfortunately quality requirements can be much additional challenging to implement than functional ones for various reasons. First of all, unless developers and analysts proactively elicit quality requirements, the stakeholders might well consider them to be absolutely understood. Obviously, this can lead to important problems with stakeholder satisfaction of the delivered product. Second, quality requirements tend to display trade-offs that must be suspiciously negotiated and resolved. For example: stakeholders might desire a system to be both very secure and easily accessible to users or they might wish for a system to have very quick response times and support thousands of users but not cost much to develop. Developers must find an architectural solution that balances these conflicting needs in a way that optimizes the delivered product's value. Lastly, quality requirements are regularly harder to measure and track than their functional parts. Whereas functional requirements are either present or not present in a system, quality requirements tend to be achieved at various levels along a continuum. This introduces challenges in their software specification, tracking, and implementation. The benefits of getting the quality requirements right can be substantive and in many safety critical and real time systems the developers have no choice but to deliver systems that assemble very specific quality constraints. However, in most systems, we face a variety of choices that will influence both the delivery schedule and a product's potential impact in the marketplace [25].

The ISO published ISO/IEC 25030, a novel standard on software quality requirements that complements the two IEEE Computer Society standards for software and system requirements. This standard is important that properly identify and specify requirements are main factors in determining a software project's success or failure. Many companies having realized that it is better emphasizing requirements specification. Unfortunately there's a tendency to focus on functional requirements rather than quality issues such as usability, reliability, maintainability, portability, and efficiency. ISO/IEC 25030 can develop and enhance software quality by helping developers identify and specify quality requirements. ISO/IEC 25030 is the only standard in this division that helps to identify and specify quality requirements. Developers can utilize these quality requirements to extract and define quality requirements for a software product to be developed or as input for an evaluation process [26].

When develop a software product for a critical system, specifying quality requirements are essentially important. Quality requirements should be clear based on different stakeholders' needs. Software quality impacts the information system's behaviour and behaviour impacts the behaviour of the external system that contains the information system. Safety is an important issue of the external system. A software product alone is harmless because it can do nothing without computer hardware. However some software quality characteristic such as: security and reliability impacts the external system's safety. One of the most imperative issues of requirements engineering is that the quality model for a target system and the quality model for the related software cannot be the equal. For example: in order to advance the safety of the target system, reliability, security, and usability of the software should be enhanced. Though generally of well-liked software requirements technologies maintain only functional requirements, it is tremendously essential that software quality requirements be clearly defined, especially for critical systems software. Quality requirements should be extensive for all quality characteristics within a quality model because every quality characteristic of software product may have control on system safety. Quality requirements should be also accurate, objective, quantitative, and should also offer evaluation criteria [27].

A key requirements concern in software development is the system's quality requirements. Quality is generally defined in terms of universal properties for a software system such as: usability, reliability, and maintainability. In the perspective of software maintenance they are mostly relevant: maintenance activities are performed to guarantee software quality. Recently an extended view of requirement engineering has been promising, wherein requirements artifacts play a role throughout a system's lifecycle. How significant are quality requirements as the lifecycle progresses? The consequence of quality requirements lies in their inter system comparability. Because of their worldwide nature, quality requirements are tough to build into a design and are often treated post facto in terms of metrics that are applied to the final software product. If requirements are important throughout the life-cycle, an enhanced understanding of requirements after the preliminary release is important [28].

Software quality requirements (SQR) play a fundamental role in software quality (SQ) success. The significance of mastering SQR can be seen as understandable; however, when it comes to user satisfaction, end users are often dissatisfied with SQ [29].

3.6 QUALITY MEASUREMENT

The exercise of software in high assurance and mission critical systems increases the need to develop and quantify measures of software quality. Consequently, software metrics are valuable in the timely prediction of high risk components throughout the software development process. The approach to software quality follows a measurement approach i.e. it is based on measuring the degree to which software properties fulfils specified software requirements. Natural software properties that can be well known quantitatively or qualitatively are called attributes. Quality attributes are measured as the degree to which the software product is able of providing and maintaining its particular level of service. Quality attributes belong to one or more sub-characteristics. Though no widespread software quality measure exists there are consequential quality measures for particular environments. Quality attributes are measured by applying a measurement method. A measurement method is a reasonable sequence of operations used to calculate an attribute with respect to a specified scale. The consequence of applying a measurement method is called a base measure. The quality characteristics and sub-characteristics can be quantified by applying measurement functions. A measurement function is an algorithm used to merge base measures. The effect of applying a measurement function is called a quality measure. In this way quality measures become quantifications of the quality characteristics and sub-characteristics. More than one quality measure may be used to illustrate a quality characteristic or sub-characteristic. Not every software properties are measurable. This is in exacting the case for functional properties [30].

Dependent on how to identify software quality, software quality may be directly measured, indirectly measured, or predicted. The preconditions for the diverse measurement types differ. Under we have applied the measurement preconditions on the measurement of software quality. The preconditions for direct measurement of software quality are that:

1. The empirical relational system of software quality is recognized. In this means at least we should have a universal understanding of the relations same quality as and improved quality than i.e. we must be capable to decide whether software X has better, similar or worse quality than software Y.
2. A numerical or symbolic system with corresponding formal relations to the empirical quality relations is established. For example: the formal relations "=" and ">" representing the relations similar and better quality in the empirical relational system.
3. A measure (mapping) from the attribute software quality to numbers or symbols is definite.
4. A measurement tool or method implementing the measure is implemented.

The preconditions for indirect measurement of software quality are that:

1. The preconditions for measurement of the directly measured software attributes are met.
2. A comprehensive empirical connection between the directly measured attributes and the indirectly measured software quality is recognized. The connection is in case complete when all aspects of the agreed understanding of software quality can be determined from the directly measured attributes.
3. The connection is correctly translated into the formal relational system (through a formula).

Predictions of software quality are similar to indirect measurement. The divergence is that predictions do not necessitate a complete empirical connection or a correct translation into the proper relational system i.e. all the empirical relations are not necessarily preserved into the formal relational system [31]. What the quality of a product really means. There are (at least) two diverse viewpoints: satisfaction of software requirements (according to specifications) and satisfaction of stated and implied needs (fit for purpose). These two viewpoints simply correspond when the requirements specification reflects the stated and implied needs of all stakeholders for all applications. This is typically not the case. Many stakeholders can't expressive or don't even know their actual needs. In addition stakeholders might have incompatible needs.

When software requirements are from a software acquirer or supplier viewpoint then their widespread interest is satisfaction of requirements. The end users and public authorities (for example: in the case of safety critical applications) are interested in satisfaction of stated and implied needs. The first case focuses on the software product while the second case focuses on the system. Both viewpoints are significant so the measure takes both into account. In the ISO quality model, a software quality sub-characteristic is defined as a category of software quality attributes that control software quality. An attribute is a measurable property, for example: size, which can be measured as the number of lines of code (LOC). We can verify a software product's behaviour however its inherent properties and its quality through its inherent quality attribute. So software measurement becomes the connection between the quality model and the software's quality that is used to quantify software quality using software measures. This implies that we can specify software quality requirements by providing a set of quality measures with linked target values. An example is an online administrative system by report generation features. A promising quality requirement for this system is quality characteristic:

- Efficiency: (time behaviour)
- Attribute: Report generation time
- Measure: Average number of seconds to generate report A during normal system use measured 10 times using a stopwatch.
- Target value: 20 seconds (60 seconds is the worst acceptable time).

When the software is finished, we can measure the quality attributes and evaluate actual measurement values to target values to disclose whether the software fulfils its quality requirements. It's important to originate quality requirements such that we can reveal their fulfilment in a reasonable amount of time and with reasonable effort. What's reasonable depends on the severity of the requirements and on the intended application. For a high risk application we're willing to expend more time and effort when evaluating quality [32].

3.7 SUMMARY

Software quality defined as conformance to specification and meeting user needs. Software quality measures how well software is designed refer quality of design and how well the software conforms to that design refer quality of conformance. Whereas quality of conformance is concerned with implementation, quality of design measures how valid the design and requirements are in creating a software product. One of the challenges of software quality is that everyone feels they understand it. Classification based modelling for software quality evaluation is a proven technique for better software quality control. Some classification techniques used for software quality assessment such as classification and regression trees (CART), tree based classification with S-PLUS, the C4.5 algorithm, the Sprint-Sliq algorithm, and case based reasoning. Quality requirements are often harder to measure and track than their functional main parts whereas functional requirements are either present or not present in software, quality requirements tend to be achieved at various levels along a variety. This introduces challenges in their software specification, tracking, and implementation.

3.8 REFERENCES

1. Schulmeyer, G. G.; McManus, J. I. (1992). *Handbook of software quality assurance*. Van Nostrand Reinhold Co.
2. Herbsleb, J.; Zubrow, D.; Goldenson, D.; Hayes, W.; Paulk, M. (1997). Software quality and the capability maturity model. Communications of the ACM, Vol. 40, No. 6, pp. 30-40.
3. Jones, C. C. (1997). *Software quality: Analysis and guidelines for success*. Thomson Learning.
4. Huo, M.; Verner, J.; Zhu, L.; Babar, M. A. (2004). Software quality and agile methods. In Proceedings of the IEEE 28th Annual International Conference on Computer Software and Applications Conference, COMPSAC, pp. 520-525.
5. Shen, P.; Ding, X.; Ren, W.; Yang, C. (2018). Research on Software Quality Assurance Based on Software Quality Standards and Technology Management. In 2018 19th IEEE/ACIS International Conference on Software Engineering, Artificial Intelligence, Networking and Parallel/Distributed Computing (SNPD), pp. 385-390.

6. Hannula, E. (2017). A short history of software quality. Qentinel. https://info.qentinel.com/blog/a-short-history-of-software-quality
7. Laplante, P. A. (2005). Classification of Software Qualities.
8. Mandrioli, D.; Carlo, G.; Mehdi, J. (2004). Software Qualities and Principles.
9. Poth, A.; Heimann, C. (2018, September). How to Innovate Software Quality Assurance and Testing in Large Enterprises? In Springer European Conference on Software Process Improvement, pp. 437-442, Springer, Cham.
10. Galin, D. (2004). *Software quality assurance: from theory to implementation*. Pearson Education India.
11. Steinberg, D.; Colla, P. (1995). CART: A supplementary module for SYSTAT. Salford Systems, San Diego, California.
12. Prahalad, C. K.; Krishnan, M. S. (1999). The new meaning of quality in the information age. Harvard business review, Vol. 77, No. 5, pp.109-118.
13. Clark, L. A.; Pregibon, D. (1992). Tree-based models. In J. M. Chambers and T. J. Hastie, (eds), Statistical Models in S. Wadsworth International Group, Pacific Grove, California, pp. 377-419.
14. Choudhary, V. (2007). Comparison of software quality under perpetual licensing and software as a service. Journal of Management Information Systems, Vol. 24, No. 2, pp. 141-165.
15. Ponnuswamy, V. (2001). Classification of software quality with tree modeling using C4.5 algorithm. Master's thesis, Florida Atlantic University, Boca Raton, Florida USA.
16. Khoshgoftaar, T. M.; Seliya, N. (2002). Software quality classification modeling using the SPRINT decision tree algorithm. In Proceedings: 14th International Conference on Tools with Artificial Intelligence. Washington, DC, USA: IEEE Computer Society, November, pp. 365-374.
17. Khoshgoftaar, T. M.; Seliya, N. (2004). Comparative assessment of software quality classification techniques: An empirical case study. Empirical Software Engineering, Vol. 9, No. 3, pp. 229-257.
18. Deissenboeck, F.; Juergens, E.; Lochmann, K.; Wagner, S. (2009). Software quality models: Purposes, usage scenarios and requirements. In IEEE ICSE Workshop on Software Quality, WOSQ'09, pp. 9-14.
19. Jones, C. (2008). Software Quality in 2002: A Survey of the State of the Art. White Paper, Software Productivity Research.
20. Zhang, Y.; Liu, X.; Liu, Z.; Li, W. (2018). Development and Reconstitution of Software Quality Measurement and Evaluation Standards. In 2018 19th IEEE/ACIS International Conference on Software Engineering, Artificial Intelligence, Networking and Parallel/Distributed Computing (SNPD), pp. 380-384.
21. Iqbal, N.; Qureshi, M. (2012). Improvement of key problems of software testing in quality assurance. arXiv preprint arXiv:1202.2506.
22. Isazadeh, A.; Elgedawy, I.; Karimpour, J.; Izadkhah, H. (2014). An analytical security model for existing software systems. Applied Mathematics & Information Sciences, Vol. 8, No. 2, p. 691.
23. Kassab, M. (2017). A Contemporary View on Software Quality Requirements in Agile and Software Architecture Practices. In IEEE 25th International Requirements Engineering Conference Workshops (REW), pp. 260-267.
24. Vaníček, J. (2006). Software quality requirements. Agriculture Economics, Vol. 52, No. 4, pp. 29-37.
25. Blaine, J. D.; Cleland-Huang, J. (2008). Software quality requirements: How to balance competing priorities. IEEE Software, Vol. 2, pp. 22-24.
26. Bøegh, J. (2008). A new standard for quality requirements. IEEE Software, Vol. 2, pp. 57-63.

27. Azuma, M. (2004). Applying ISO/IEC 9126-1 quality model to quality requirements engineering on critical software. In Proceedings of the 3rd IEEE Int. Workshop on Requirements for High Assurance Systems (RHAS).
28. Ernst, N. A.; Mylopoulos, J. (2010). On the perception of software quality requirements during the project lifecycle. In International Working Conference on Requirements Engineering: Foundation for Software Quality, Springer, Berlin, Heidelberg, pp. 143-157.
29. Liu, X. F. (1998). A quantitative approach for assessing the priorities of software quality requirements. Journal of Systems and Software, Vol. 42, Issue 2, pp. 105-113.
30. Abran, A.; Al-Qutaish, R. E.; Desharnais, J. M.; Habra, N. (2005). An information model for software quality measurement with ISO standards. In Proceedings of the International Conference on Software Development (SWDC-REK), Reykjavik, Iceland, pp. 104-116.
31. Jørgensen, M. (1999). Software quality measurement. Advances in engineering software, Vol. 30, No. 12, pp. 907-912.
32. Schneidewind, N. E. (2002). Body of knowledge for software quality measurement. IEEE Computer, Vol. 35, Issue 2, pp. 77-83.

Chapter - 4

PRODUCT QUALITY VS. PROCESS QUALITY

Software process used to create software product and achieve quality in software product within given time and cost. In this chapter, software process quality and software product quality are discussed from several perspectives to see the relationship of software process quality and software product quality. This chapter analyzed software process and software product based on output of the researches in literature. Finally, discuss the changing nature of software process and software product. It is essential to improve the software process with respect to changing environment and software requirement. Section I describes the introduction of this chapter. Section II discuss, the product-process relationship. Section III describes the software systems evaluation. Section IV describes the product quality. Section V describes the models for software product quality. Section VI describes the process quality. Section VII describes the measurement of process and product quality. Section VIII describes the summary of this chapter.

4.1 INTRODUCTION

Because software more and more rule our world and becomes universal and embedded in almost everything we do. We have to make definite that software work in an enhanced way as we desire. The quality of software involves product and process quality. Process oriented approaches deal with the organization of rules, standards, manuals, principles, guidelines, process definitions, and improvement of software quality processes. The high quality software product is resulted from high quality software development process. The absence of software process and product quality, increase troubles during software development process. Process and product quality are process driven approaches with detailed steps to achieve software development goals. The process and product quality consider requirement, design, development, and production. Quality assurance is protective approach from occurring by providing methods and rules which prevents errors and defects from occurring. It starts in software development process from the early phase of software development lifecycle. It grants self-assurance to user concerning the software development process and the software product. It is a systematic and planned pattern of all events essential to provide sufficient confidence that a software product conforms to recognized technical requirements.

Process and product quality declaration are very significant aspects in development of software. Process and product quality assurance observe the software engineering processes and methods to guarantee software quality. It is the software process of confirming and verifying that whether services and products meet the user requirement and expectation or not. The purpose of process and product quality assurance (PPQA) is to offer management and staff with purpose insight into processes and related work products. Process and product quality have become the key to success in software development lifecycle. For the instance being we identify the fact that software measurement helps us to better evaluate, understand, and control the products, processes, and software projects from the various perspective of evaluating, controlling, tracking, forecasting, and understanding. A suitable measurement process can offer organizations to make improved and appropriate decisions to achieve success in software projects.

4.2 THE PRODUCT-PROCESS RELATIONSHIP

Software process provides a sequence of step for managing and controlling development activities. To achieve quality in finished product well defined process is required because quality of software product directly related to quality of software process. There is various importance of a well defined software process such as: self improving, easy to implement, globally acceptable, and flexible to any size of project [1].

Software process quality refers to the capability of a software process to produce software with increased quality. Software product quality means, there is an underlying quality model that defines a set of product quality requirement and characteristics. Software process quality and software product quality is based on best practices of development activity [2]. Software organizations usually enhance product quality by using software process improvement (SPI).

A systematic approach to measure and improve software process and product quality will help to ensure organization success as depicted in Figure 4.1.

Firstly measure software process and product quality then analyze the results to identify further improvements in process to develop a quality product as shown in Figure 4.1 [3]. Capability Maturity Model Integration (CMMI), ISO 9126, and ISO 12207 hardly address the particular practices used to improve product quality. There is still a need to develop process approaches with which to specify, achieve, and validate software product quality [4].

Figure 4.1 Impact of Software Process Quality on Software Product Quality and Organization Success

Software process quality contributes to software product quality to conform the software requirement and user expectation using monitoring, reviewing, and evaluation of software process. Software process impacts on the software product and its quality such as: usability, performance, reliability, security, availability, functionality, maintainability, and safety. Software process quality is having the highest efficiency and best utilization of all resources in software development. Software product quality is having least defects and higher consumer satisfaction [5-6].

Software process quality is related to how people develop product. There are mainly effectiveness, efficiency, and predictability aspect of software. Software procuct quality associated with the absence of defects in software. Various important quality attributes for software success are usability, efficiency, maintainability, portability, reliability, reusability, and availability [7]. All these software quality totally depends on the quality of software process.

Relationship of software process quality and product quality depends upon various perspectives such as: quality model, conformance, and improvement as shown in Table 4.1.

Software process uses Capability Maturity Model (CMM) is a methodology used to extend and refine an organization's software development process. This model describes a five level evolutionary path of increasingly organized and systematically more mature process such as: initial, repeatable, defined, managed, and optimizing.

Table 4.1 Relationship between Software Process Quality and Product Quality

Quality	Quality Model	Conformance	Improvement
Software Process	CMM, ISO 9001, SPICE (ISO 15504)	ISO 9001, SQA	CMMI, SPICE
Software Product	McCall's model, Dromey's model, Boehm's model, ISO 9126	ISO 9126, ISO 25010	Best Practices

Level 1 - Initial: It is characteristic of process at this level that they are generally undocumented and in a condition of dynamic change, tending to be driven in an ad hoc, uncontrolled, and reactive manner by events or users. This provides a chaotic or unstable environment for the process.

Level 2 - Repeatable: It is characteristic of this level of maturity that some process is repeatable, probably with consistent results. Process control is unlikely to be rigorous, but where it exists it may help to ensure that existing process is maintained during times of pressure.

Level 3 - Defined: It is characteristic of process at this level that there are sets of defined and documented standard process established and matter to some degree of improvement over time. These standard processes are in position. The process may not have been systematically or repeatedly used satisfactory for the users to become capable or the process to be validated in a variety of situations. This could be measured a developmental stage with use in a wider choice of conditions and user capability development the process can develop to next level of maturity.

Level 4 - Managed (Capable): It is characteristic of process at this level that, using process metrics, efficient achievement of the process objectives can be evidenced across a range of working conditions. The suitability of the process in numerous environments has been experienced and the process refined and adapted. Process user has practised the process in various and varied conditions, and is able to exhibit competence. The process maturity enables adjustment to particular projects without measurable losses of quality or deviation from specification. Process Capability is established from this level.

Level 5 - Optimizing (Efficient): It is a characteristic of process at this level that the focus is on repeatedly improving process performance throughout both incremental and innovative technological changes and improvements.

CMM summarized as: initial (chaotic, ad hoc, and individual heroics) - the preliminary point for use of a novel or undocumented repeat process. Repeatable - the process is at slightest documented sufficiently such that repeating the same steps may be attempted. Defined - the process is defined or confirmed as a standard business process. Capable - the process is quantitatively managed in agreement with agreed-upon metrics. Efficient - process management includes purposeful process optimization and improvement [8-9].

The ISO 9000 family tackle various aspects of quality management and include some of ISO's best known standards. The standards provide management and tool for organizations who want to guarantee that their products and services constantly meet user's requirements, and quality is consistently improved. The ISO 9000 series are divided in seven quality management principles (QMP). The seven quality management principles are:

- QMP 1 - Customer focus
- QMP 2 - Leadership
- QMP 3 - Engagement of people
- QMP 4 - Process approach
- QMP 5 - Improvement
- QMP 6 - Evidence-based decision making
- QMP 7 - Relationship management

Principle 1 - Customer focus

Organizations depend on their customers and therefore should understand existing and future customer needs, should meet customer requirements, and struggle to exceed customer expectations.

Principle 2 - Leadership

Leaders create unity of purpose and direction of the organization. They should create and maintain the internal environment in which people can become fully involved in achieving the organization's objectives.

Principle 3 - Engagement of people

People at all levels are the concentrate of an organization and their full participation enables their abilities to be used for the organization's profit.

Principle 4 - Process approach

A desired result is achieved more economically when activities and related resources are managed as a process.

Principle 5 - Improvement

Improvement of the organization's overall performance should be a permanent objective of the organization.

Principle 6 - Evidence-based decision making

Effective decisions are based on the analysis of data and information.

Principle 7 - Relationship management

An organization and its external providers (suppliers, contractors, and service providers) are interdependent and an equally beneficial relationship enhances the ability of both to create value.

Basically the layout of the standard is related to the previous ISO 9001:2008 standard in that it follows the plan, do, check, act cycle in a process based approach but is now further encouraging this to have risk based thinking. The purpose of the quality objectives is to determine the consistency of the requirements (customers and organizations) assist effective deployment and improve the quality management system. ISO 9001:2015 sets out the criteria for a quality management system and is the only standard in the family that can be certified to (though this is not a requirement) [10-11].

ISO/IEC 15504 Information Technology - process assessment also termed Software Process Improvement and Capability Determination (SPICE) is a set of technical standards documents for software development process and related business management functions. The process dimension defines processes divided into the five process categories:

- Customer-supplier
- Engineering
- Supporting
- Management
- Organization

With novel parts being published, the process categories will extend, particularly for IT service process categories and enterprise process categories. ISO 9001 and Software Quality Assurance (SQA) are used as conformance of software process. For software process improvement CMMI and SPICE is used [12-13]. These framework are used to improve or mature the software process.

Software product uses McCall's model attempts to bridge the gap between users and developers by focusing on a number of software quality factor that reflect users view and the developer priorities. There are three main perspectives for defining and identifying the quality of a software product such as: product revision (ability to go through changes), product transition (adaptability to new environment), and product operations (its operation characteristics). Product revision encompasses the revision perspective identifiers quality factors that changes or enhances the ability to change the software product in the future according to the needs and requirements of the user. The category product revision consists of flexibility, maintainability, and testability quality attributes. Product transition perspective enables the software to adapt itself in new environments, the identification of the quality factor which enables the ability of adaption of the software in the new environment. The product transition category consists of reusability, portability, and interoperability quality attributes. The software can run successfully in the system if it according to the specifications of the user and also it should run efficiently without any defects. The product operation perspective influences the scope to which the software fulfils its specifications [14-15]. The product operation category consists of a set of quality attributes that includes reliability, correctness, integrity, usability, and efficiency.

Dromey model establishes the link among tangible product characteristics and less tangible quality attributes. It can support in conducting a systematic search for quality defects. The model guides us where to look for defects and also indicates the properties that will require to be violated to create defects. Dromey presents four quality categories where each category consists of quality attributes. The four categories are correctness, internal, contextual, and descriptive. The four categories divide similar quality attributes along with every own designated attributes. The correctness category includes functionality and reliability. The internal category includes maintainability and efficiency. The contextual category includes reusability and portability. The descriptive category includes usability [16]. This model formulated by associated a set of quality carrying properties with each of the structural forms are used to define the statements and statement components of a programming language.

Boehm's model is a hierarchical quality model which qualitatively defines software quality by a set of attributes and metrics and present structure around high level characteristics, intermediate level characteristics, and primitive characteristics. The model defines the quality of software on the basis of a set of credentials and measurements. It is also elucidates a model of software quality characteristics. The high level of characteristics is as: as-is utility, maintainability, and portability. The intermediate level of characteristics represented by the model displays seven quality factors that altogether signify expected quality from a software system. These are as: flexibility, portability, reliability, testability, efficiency, understandability, and usability [17].

ISO 9126 [18] known as a quality model to develop a quality product. It is used for software product evaluation, quality characteristics, and guidelines for their use standard, also include functionality as well as identifying both internal and external quality characteristics of software products. Internal metrics are those which do not rely on software execution such as: static measure. External metrics are applicable to running software. Quality in use metrics is only available when the final product is used in real conditions. Ideally internal quality determines the external quality and external quality determines quality in use. Software product is defined in a broad sense: it encompasses source code, executables, architecture descriptions, and so on. As a result the notion of user extends to operators as well as to programmers which are users of components such as software libraries. The standard provides a framework for organization's to define a quality model for a software product. On doing so however it leaves up to each organization the task of specifying specifically its own model. This may be completed, for example - by specifying target values for quality metrics which evaluates the degree of existence of quality attributes. Now this standard has been revised by ISO 25010:2011, systems and software engineering: Systems and software Quality Requirements and Evaluation (SQuaRE) [19]. ISO 9126 and ISO 25010 are used as conformance of software product. For software product improvement best practices is used [20]. SQuaRE is a quality in use model composed in five characteristics (some of further subdivided into sub-characteristics) that share to the outcome of interaction when a product is used in a particular situation of use. This model is applicable to the whole human-computer system including both software products in use and computer systems in use.

4.3 SOFTWARE SYSTEMS EVALUATION

Software can be evaluated with respect to different attributes such as: usability, functionality, reliability, efficiency, portability, and maintainability. Evaluation as a general endeavour can be characterised by the following features:

- Evaluation is a task, which results in one or more reported outcomes.
- Evaluation is an assist for planning; therefore the product is an evaluation of diverse potential activities.
- Evaluation is goal oriented. The main goal is to check results of activities or interventions, in order to enhance the quality of the activities or to choose the best action option.
- Evaluation is dependent on the current knowledge of science, technology, and the methodological standards.

In earlier software evaluation of software took place at the end of the development phase, using experimental designs and statistical analysis. Nowadays evaluation is used as a tool for information gathering within iterative design [21].

In software evaluation, the goal can be characterized by three simple questions:

1. "Which one is better?" Aims to evaluate another software systems, e.g. to choose the best proper software tool for given application, for a verdict between several prototypes, or for comparing numerous versions of a software system.
2. "How good is it?" This goal aims at the determination of the degree of desired quality of a software system.
3. "Why is it bad?" The evaluation aims to determine the weaknesses of software such that the result generates suggestions for further development. A typical instance of this procedure is a system developing approach using prototypes or a re-engineering of an existing system.

The first two goals can be subsumed under the concept of summative evaluation, while the third goal is an instance of the formative evaluation approach [22].

In general, summative evaluation is concerned with the global aspects of software development and does not suggest constructive information for changing the design of the software in a straight way. It is performed when the development of the system is approximately or completely accomplished. Summative evaluation is also applied to prototypes in case. There is a need for controlling the effect of design changes in evaluation to a preceding version of the software system [23]. Basically summative evaluation is an assessment of software product where the focus is on the outcome of software program.

In contrast, the goals of formative evaluation are the enhancement of software and design supporting aspects. It is considered the most important part of software evaluation, and plays an important role in iterative system development. In each development cycle, formative evaluation results in:

- Quantitative data for the description of the progress of the realization of quality goals
- Qualitative data which can be used to detect the problems of the system.

The resulting data can be classified by the following criteria:

Objective: Directly visible data; typically user behavior during the use of the interface or the application system.
Subjective: Opinions, normally expressed by the user with respect to the interface or the application system.
Quantitative: Numerical data and results, e.g. user performance ratings.
Qualitative: Non-numerical data, e.g. lists of problems, suggestions for modifications to improve the interaction design.

Evaluation techniques are activities of evaluators which can be specifically defined in behavioral and organizational terms. Evaluation techniques divided into two categories: descriptive evaluation techniques and predictive evaluation techniques, both of which should be present in every evaluation: Descriptive evaluation techniques are used to describe the status and the actual problems of the software in an objective, reliable, and suitable way. These techniques are user based and can be subdivided into several approaches.

Behavior based evaluation techniques record user behavior while working with a system which produces some kind of data. These actions include observational techniques and thinking-aloud protocols. Opinion based evaluation methods aim to elicit the user's (subjective) opinions. Examples are interviews, surveys and questionnaires [24].

The predictive evaluation techniques have as their major aim to make recommendations for future software development and the prevention of errors. These techniques are expert or at least expertise based such as: Walkthrough or inspection techniques. Even though the expert is the energetic power in these methods, users may also contribute in some instances. Predictive evaluation techniques must rely on data. In several predictive evaluation techniques, such data are produced by experts who imitate real users. The criteria objectivity and reliability, which are at the foundation of descriptive techniques, are tough to apply in this setting. Because validity must be the main aim of evaluation procedures, there are attempts to confirm the validity of predictive evaluation techniques directly, e.g. by comparing hits and fake alarm rates of the problems detected by a predictive technique. They are powerful, important, robust, and can be used to estimate prediction uncertainty. They are applied earlier in the software development life cycle [22, 25].

There are various important software evaluation methodologies such as: theory-based evaluation, user-based evaluations, surveys and questionnaires, verbal reports, controlled experimental studies, task-based evaluations, informal design review, formal design analysis - GOMS, and production system analysis. Table 4.2 summarized the evaluation methods and their use.

Table 4.2 Summary of Evaluation Methods and their Use

Technique	Type of Information	Appropriate Use
Questionnaire	Subjective response to question posed. Best with specific questions.	Any time in cycle when clear questions can be posed.
Verbal Report	Record of cognitive processes in system use.	Early in cycle when general information is desired.
Controlled Experiments	Specific measures in a controlled setting.	For specific tests of major alternatives or hypothesis.
Design Review	Common sense review of design to check for general acceptability.	Early in design cycle to filter early decisions
Formal Analysis (GOMS)	Predictions of expert user behavior.	Use in place of user testing for ease of use predictions.
Production System Analysis	Predictions of learning, use, and transfer behavior.	Use to supplement other analysis data.

Evaluation methods cover a variety of techniques to fit a collection of requirements and needs. There are two wide categories: first based on observations of individuals using a real or prototype system and second based on an analysis of the task to be performed. For user observations (User-Based Evaluations), there is a extensive range of techniques which vary in the amount of experimental control placed on the circumstances in which the evaluation takes place and in the type of data collected. The evaluation methods which are based on task analysis (Task-Based Evaluations) presume that understanding the task elements is satisfactory to allocate design decisions to be completed. There are numerous techniques which can be used to gather information from users [26]. Controlled experimental studies used questionnaire or verbal report data as a source of information.

4.4 PRODUCT QUALITY

Product quality: Degree of conformance to specification as defined by the user. Product quality means to integrate features that have a capacity to meet user requirements, needs, and gives user satisfaction by improving software products and making them free from any errors, deficiencies, and defects. There are various significant aspects which define a product quality such as: storage, quality of conformance, quality of design, security, usability, reliability, and safety. Product quality = Function (process maturity, process size, and product design complexity) [27]. Product quality is determined by the amount to which a product or service successfully serves the need, requirement, and purpose of the user during. Cost and time are both transient features whereas the impact of quality is continued long after the desirability or the pain of cost and time has subsided.

The natural characteristics of a product or service created to satisfy user needs, expectations, and requirements. Physical, functional, and non-functional characteristics such as: usability, functionality, speed, capacity, reliability, portability, and security etc. If we define all these characteristics in quantifiable terms and put limits on them. Then we will have defined the standards with which the software product has to conform.

Software quality in numerous cases tough to evaluate; product quality is the overall quality of the product in question: how well it conforms to the user specifications, requirements, and ultimately user expectations. Product quality can be measured by mean time before failure (MTBF).

Mean Time between Failure (MTBF) = Mean Time to Failure (MTTF) + Mean Time to Repair (MTTR)

Where MTTF is the difference of time between two consecutive failures and MTTR is the time required to fix the failure. For example: if MTTF = 100 hours for a software then the software should work for 100 hours of continuous operations. For the same software if the MTTR = 1 hours then MTBF = 100 + 2 = 102.

MTBF is the predicted elapsed time between inherent failures of a system during normal system operation. MTBF can be calculated as the arithmetic mean (average) time between failures of a system. The term is used for repairable systems whereas mean time to failure (MTTF) denotes the expected time to failure for a non repairable system [28-29].

There is no metric can be measured during the organizational process. Not every quality characteristics are of equivalent significance to a software product. Portability may be irrelevant when the development aims at a committed platform and maintainability may not be a concern when it is a product with a tiny life cycle. Therefore, the most important quality (sub) characteristics should be identified and selected by means of a risk evaluation. This can be completed by interviewing stakeholders within the project (such as: product manager, project manager, and software architect) and outside the project (different types of users are imperative) [30-31].

4.5 MODELS FOR SOFTWARE PRODUCT QUALITY

Several software product quality models have been developed to maintain software quality such as: McCall's quality model, Boehm's quality model, Dromey's quality model, and ISO 9126.

One of the most important quality model is McCall's model try to link the gap between developers and users by focusing on a number of software quality factor that reflect the developer priorities and users view. There are three main perspectives for defining and identifying the quality of a software product such as: product revision (ability to go through changes), product transition (adaptability to new environment), and product operations (its operation characteristics). McCall's quality model is the first of the current software product quality models. The model uses a hierarchy of factors, criteria, and metrics to tackle internal and external product quality. Eleven factors define an external or user perspective of quality. Each of these factors is linked to between two and five of 23 criteria that define an internal or development perspective of quality [32-33]. Additional metrics are related with the factors allowing quality to be calculated and managed.

Second basic quality model is Boehm's model. McCall's quality model was followed by Boehm's quality model. Address the shortcoming of existing models automatically and quantitatively evaluate the quality of software. Like McCall's model, Boehm's model presents product quality in a hierarchy with three high level characteristics linked to seven intermediate factors, which are in turn linked to 15 primitive characteristics. Boehm's model qualitatively defines software quality by a set of attributes and metrics and present structure around high level characteristics, intermediate level characteristics, and primitive characteristics [34]. Boehm's model has a wider span than McCall's, with more prominence on the cost-effectiveness of maintenance.

More recently work has been done to develop an international standard for software product quality measurement ISO 9126. This standard is yet again organized in a hierarchy with six characteristics at the top level and 20 sub characteristics with indicators used to measure the sub characteristics. In addition to aspects of internal and external quality, covered by McCall and Boehm's models, ISO 9126 includes quality characteristics with functionality. Internal, external, and functional qualities are also mixed at all levels of the hierarchy. However, ISO 9126 does not visibly state how quality should be measured. It is used for software product evaluation, quality characteristics, and guidelines for their use standard. Presently this standard has been revised by ISO 25010:2011, systems and software engineering: Systems and software Quality Requirements and Evaluation (SQuaRE). A quality in use model composed of five characteristics (some of which are further subdivided into sub characteristics) that relate to the result of interaction when a product is used in a particular context of use. This system model is applicable to the complete human computer system including both computer systems in use and software products in use. A product quality model composed of eight characteristics (which are additional subdivided into sub characteristics) that relate to static properties of software and dynamic properties of the system [35-38]. This model is applicable to both system and software products.

Dromey presents a dissimilar type of model that attempts to tackle some of the issues exist and support developers to develop product quality. Dromey believes that it is impossible to develop high level quality attributes such as: reliability or maintainability into a product, but developers must instead create properties that obvious in achieving these goals. The difference this model makes is important, as using it will verify that it allows the quality required to be achieved. Before Dromey model can be effectively applied, the different groups involved in the development of a software product must agree on what quality attributes should be achieved and to what level [39-40]. This process can be supported using other models.

SQUID project (Software QUality in Development) aimed to develop and automate methods and models for software quality specification, control, and evaluation based on quantitative measures of software quality. SQUID approach requires a method for linking product quality requirements to the characteristics defined in a quality model such as: ISO 9126 model. Table 4.3 list the factors of a modern product quality model. Usage-oriented factors divided as per application and use, while product-oriented factors divided as per system and environment [41]. Basically a usage-oriented factor captures functionality, usability, efficiency, reliability, security, safety, dependability whereas product-oriented factor captures maintainability, portability, and compliance.

Table 4.3 Factors of a Modern Product Quality Model

Usage-oriented Factors	Product-oriented Factors
Functionality	Maintainability
Usability	Portability
Efficiency	Compliance
Reliability	
Security	
Safety	
Dependability	

None of these models present a justification for the selection of characteristics to be included in the quality model and it is not promising to tell if a model presents a complete or reliable definition of quality [34]. Further the assignment of items appears subjective in ISO 9126, with no justification as to why interoperability is not related to portability.

4.6 PROCESS QUALITY

Process quality: Degree of variability in process execution and the minimization of the cost of quality. Process quality is defined as all the steps used in the development of final product. Its focus is on all activities and steps used to accomplish a highest acceptance regardless of its final product. A related relation between software product and function exists in the process and activity [42]. Process quality focuses on how well some part of the process of development that product and getting it into the users hands is working. The process being analyzed in a particular case may have a very huge scope or it may centre in on little details of a single step.

Process quality can be measured by the rapidity of the development team in an agile driven organization. Rapidity is defined as how many story points the team is capable to deliver in a given sprint cycle. If the team is not capable to deliver the agreed story points then the process is decreasing apart. The factors that affect process quality in the three phases such as: design, construction, and operation of the life cycle of a developing project. The management commitment to continuous quality improvement, management leadership to promoting high process quality, quality training of all employees, efficient teamwork to promote quality issues at the business level, and effective cooperation between parties taking element in the project are generic factors that affect process quality [43].

The quality of a product depends on quality of a process is an acknowledged truth. Process quality is the control of product within defined limits, even of those limits develops a product which doesn't work. Process quality is building a product reliably to some standards. It does not mean that those standards will produce a useful and quality product [44]. Process quality keeps the software organization owner from unnecessary costs in software product repair, maintenance, and rejects. These process guidelines come in the form of commands and instructions that describe each step and give details of how to differentiate a well-performed action from a bugs, problems, and errors.

So, there is a huge difference between product quality and process quality i.e. even when you follow the process with all quality features it may or may not produce a quality product, on the other hand product quality comprises of all the features required to create it acceptable to user and hence solve all the pain points. In order to develop high quality software products, the process also must be capable which involves the use of methods, tools, and practices. A software metric is a standard to measure the degree to which a software, system, and process possesses some property. Since quantitative measurements are vital in all science and engineering, there is a constant effort by computer science practitioners, researches, and theoreticians to bring same approaches to software development. The goal is obtaining reproducible, objective, and quantifiable measurements which may have many valuable applications in planning, time, cost, cost estimation, testing, quality assurance, software performance optimization, software debugging, and optimal personnel task assignments. Software metrics are a significant tool that provides continuous insight into products and processes, and helps develop required software in mission-critical environments [45-46]. Software metric is a measure of software characteristics which are quantifiable and countable. Software metrics are significant for various reasons such as: planning work items, measuring software performance, measuring productivity, and many other uses. Within the software development process there are many software metrics that are all related to each other. Software metrics are interrelated to the four functions of management such as: planning, organization, control, and improvement.

4.7 MEASUREMENT OF PROCESS AND PRODUCT QUALITY

Software measurement is the continuous process of defining, collecting, and analyzing data on the software development process and its products in order to understand, control, and optimized the software process and its products. Measurement can be used in both a software development process and a software product. It can be used to evaluate quality characteristics of a particular software product or to determine effects of a certain software process. Basically there are five types of measurement scales such as: nominal, ordinal, interval, ratio, and absolute. These scale types have growing levels of richness which means that the scales become additional complicated and give stronger statistical analysis methods.

Nominal measurement consists of assigning objects to categories or groups. No quantitative information is conveyed and there is no ordering of the objects is implied. Variables measured on a nominal scale are regularly referred to as categorical or qualitative variables. The ordinal scale is related to the nominal scale but has the added characteristic that the categories can also be ranked in that single class is better or upper than another class. The ranges between two classes have no importance for the ordinal scale. The interval scale is an ordinal scale with equivalent differences between the classes. Interval scales do not have a proper zero point and consequently it is not feasible to make statements about how many times higher one score is than another. Ratio variables are exceptionally similar to interval variables; in addition to all the properties of interval variables, they feature an individual absolute zero point, and allowing statements such as x is two times more than y. The measurement for an absolute scale is basically made by counting the number of fundamentals that needs to be measured. The absolute scale resembles the ratio scale robustly [47]. The difference is that in absolute scale the scale is unique. The absolute scale is the generally restricted scale.

A fractional solution to the unknown relationship between process and product quality is provided by software process measurement, as it supports in the innovation of the impact of process actions on product quality in a particular context. Through measurement and analysis of the software development process and its success in achieving the proposed product qualities, an assessment instrument becomes accessible. As data collection and analysis of software engineering process is infrequently executed in present practice. Many researchers and authors have insisted on extended application of measurement during software development process, however software industry still has not adopted this sufficiently. By monitoring the overall performance of the software development process, it is promising to provide an outline on definite results of that process, and to acquire corrective action based on these results. Software measurement is a tremendous mechanism for control of a software process. Numerous methods are available that illustrate how to carry out software process measurement such as: the goal/question/metric (GQM) approach.

Software product measurements can be divided into external and internal product measurements. External product measurement measures the external (dynamic) behavior of the software product such as: number of screens, function response times, time to add a feature, and mean time to failure, etc. Internal product measurement measures the internal (static) structure of the software product such as: cyclomatic complexity, measurements of lines of code, test defects found, and number of comment lines, etc. Both types of measurement are used in practice. External measurement is more motivating from a user point of view while internal measurement is more attractive from a developer point of view. External product measurement can be used to evaluate conformance of a software product to user needs and requirement. This approach supposes that user needs are known and made explicit. These methods are based on the software specification of product quality and evaluating the finish product to this software specification. The additional objective and existing specifications are, the better it can be evaluated. Software specification of product quality in measurable terms is consequently recommended. In the existing methods for internal product measurement, it is mostly the static attributes of a software product that are measured. These measurements can be used to evaluate internal product measurements to (organization specific) norms and standards to accomplish a confident level of quality. Software measurement related literature contains many internal software metrics. These metrics are simple to measure due to the availability of automated calculation tools for these metrics. The major advantage of both internal and external product measurement is that purpose numbers are provided to distinguish a software product. The similar stands for software process measurements. Even though the limits of metrics are still questionable, they are in various way objective. This level of impartiality is critical for co-ordination using standardizing processes or products, their benefits for control of software product quality are known and clear [48]. Present practice has inadequately adopted this kind of software measurement working to achieve software quality products which is a huge shortcoming in the existing state of practice.

4.8 SUMMARY

The quality of software involves product and process quality. Process oriented methods deal with the institution of principles, standards, rules, guidelines, process definitions, manuals, and improvement of software quality processes. The high quality software product is resulted from high quality software development process. The lack of process and product quality increase struggle during software development project. Process and product quality are process driven approaches with exact steps to accomplish software development goals. The process and product quality consider requirement, design, development, and production. Software project measures are the discipline that ensures that the software project is stay in control. Measurements of a software project apply to people, processes, and products.

4.9 REFERENCES

1. Singh, B.; Gautam, S. (2016). Situational Factors Affecting the Software Process: A Systematic Literature Review. International Conference on Advanced Computing and Software Engineering, ICACSE Oct'16, pp. 229-237.
2. Zahra, K. et al. (2017). Success Factors of Organizational Change in Software Process Improvement: A Systematic Literature Review. In Proceedings of the ACM 5th International Conference on Information and Education Technology, pp. 155-160.
3. Kuilboer, J. P.; Ashrafi, N. (2000). Software process and product improvement: an empirical assessment. Information and software technology, Vol. 42, Issue 1, pp. 27-34.
4. García-Mireles, G. A.; Moraga, M. A.; García, F.; Piattini, M. (2015). Approaches to promote product quality within software process improvement initiatives: A mapping study. Journal of Systems and Software, Vol. 103, pp.150-166.
5. Kobyliński, A. (2013). The Relationships between Software Development Processes and Software Product Quality. In International Conference on Business Informatics Research, Springer Berlin Heidelberg, pp. 161-169.
6. Kenett, R. S.; Baker, E. (1999). *Software process quality: management and control*. CRC Press.
7. Wagner, S. (2013). *Software product quality control*. Springer Berlin.
8. Paulk, M. C. (Ed.) (1995). *The capability maturity model: Guidelines for improving the software process*. Addison-Wesley Professional.
9. Herbsleb, J.; Zubrow, D.; Goldenson, D.; Hayes, W.; Paulk, M. (1997). Software quality and the capability maturity model. Communications of the ACM, Vol. 40, No. 6, pp.30-40.
10. Paulk, M. (1993). Comparing ISO 9001 and the capability maturity model for software. Software Quality Journal, Vol. 2, Issue 4, pp.245-256.
11. West, J. (2000). Quality management principles. Quality Progress, Vol. 33, Issue 3, p. 79.
12. Mesquida, A. L.; Mas, A.; Amengual, E.; Calvo-Manzano, J. A. (2012). IT Service Management Process Improvement based on ISO/IEC 15504: A systematic review. Information and Software Technology, Vol. 54, Issue 3, pp.239-247.

13. Ehsan, N.; Perwaiz, A.; Arif, J.; Mirza, E.; Ishaque, A. (2010). CMMI/SPICE based process improvement. In IEEE International Conference on Management of Innovation and Technology (ICMIT), pp. 859-862.
14. Xenos, M.; Christodoulakis, D. (1995). Software quality: The user's point of view. In Software Quality and Productivity, Springer US, pp. 266-272.
15. Shankar, G.; Maurya, E. L. S. (2017). Software Quality Models: A Comparative Study. SRMS Journal of Mathematical Sciences, Vol. 1, Issue 1, 180-190.
16. Dromey, R. G. (1995). A model for software product quality. IEEE Transactions on software engineering, Vol. 21, No. 2, pp.146-162.
17. Boehm, B.; Brown, J. R.; Kaspar, H. (1978). *Characteristics of software quality*. North Holland Publishing Co., NY.
18. Al-Kilidar, H.; Cox, K.; Kitchenham, B. (2005). The use and usefulness of the ISO/IEC 9126 quality standard. In IEEE International Symposium on Empirical Software Engineering, p. 7.
19. Zhao, Y.; Jiayu, G.; Yun Hu, H.; Zhenyu, L.; Lizhi, C. (2017). Analysis of quality evaluation based on ISO/IEC SQuaRE series standards and its considerations. In IEEE/ACIS 16th International Conference on Computer and Information Science (ICIS), pp. 245-250.
20. Chrissis, M. B.; Konrad, M.; Shrum, S. (2003). *CMMI guidlines for process integration and product improvement*. Addison-Wesley Longman Publishing Co., Inc.
21. Miller, R. H. (2002). Evaluation of Software Systems. Encyclopedia of Library and Information Science, Vol. 72, Supplement 35.
22. Gediga, G.; Hamborg, K. C.; Düntsch, I. (2002). Evaluation of software systems. Encyclopedia of computer science and technology, Vol. 45, Supplement 30, pp. 127-53.
23. Gediga, G.; Hamborg, K. C.; Düntsch, I. (1999). The IsoMetrics usability inventory: an operationalization of ISO 9241-10 supporting summative and formative evaluation of software systems. Behaviour & Information Technology, Vol. 18, No. 3, pp.151-164.
24. Azarian, A.; Siadat, A. (2011). A Synthesis of Software Evaluation Methodologies and the Proposal of a New Practical Approach. JSW, Vol. 6, Issue 11, pp. 2271-2281.
25. Coutaz, J. (1994). Evaluation techniques: Exploring the intersection of HCI and software engineering. In Workshop on Software Engineering and Human-Computer Interaction, Springer, Berlin, Heidelberg, pp. 35-48.
26. Helander, M. G.; ed. (2014). *Handbook of human-computer interaction*. Elsevier.
27. Harter, D. E.; Krishnan, M. S.; Slaughter, S. A. (2000). Effects of process maturity on quality, cycle time, and effort in software product development. Management Science, Vol. 46, No. 4, pp. 451-466.
28. Zhang, Y.; Chen, H. (2006). Predicting for MTBF failure data series of software reliability by genetic programming algorithm. In IEEE Sixth International Conference on Intelligent Systems Design and Applications, ISDA'06, Vol. 1, pp. 666-670.
29. Barney, S.; Wohlin, C. (2009). Software product quality: Ensuring a common goal. In International Conference on Software Process, Springer, Berlin, Heidelberg, pp. 256-267.
30. Al-Qutaish, R. E. (2009). Measuring the software product quality during the software development life-cycle: An international organization for standardization standards perspective. Journal of Computer Science, Vol. 5, Issue 5, pp. 392-397.
31. Hendriks, R.; Veenendaal, E. van; Vonderen, R. van. (2002). Measuring Software Product Quality. Software Quality Professional, Vol. 5, No. 1.
32. McCall, J. A.; Richards, P. K.; Walters, C. F. (1977). Factors in Software Quality: Concept and Definitions of Software Quality. Final Technical Report.
33. Cavano, J. P.; McCall, J. A. (1978). A Framework For the Measurement of Software Quality. In ACM SIGMETRICS Performance Evaluation Review, Vol. 7, No. 3-4, pp. 133-139.
34. Kitchenham, B.; Peeger, S. (1996). Software quality: The elusive target. IEEE Software, Vol. 13, No. 1, pp. 12-21.
35. ISO, International Organization for Standardization, ISO 9126-1:2001, Software engineering - product quality, Part 1: quality model, 2001.

36. Jung, H. W.; Kim, S. G.; Chung, C. S. (2004). Measuring software product quality: A survey of ISO/IEC 9126. IEEE software, Vol. 21, No. , pp.88-92.
37. ISO/IEC 25010:2011, Systems and software engineering - Systems and software Quality Requirements and Evaluation (SQuaRE) - System and software quality models, 2011.
38. Suryn, W.; Abran, A.; April, A. (2003). ISO/IEC SQuaRE. The second generation of standards for software product quality.
39. Ortega, M.; Pérez, M.; Rojas, T. (2003). Construction of a systemic quality model for evaluating a software product. Software Quality Journal, VOl. 11, Issue 3, pp. 219-242.
40. Ortega, M.; Pérez, M.; Rojas, T. (2000). A model for software Product Quality with a Systemic Focus. In 4th World Multi conference on Systemics, Cybernetics and Informatics SCI 2000 and the 6th International Conference on Information Systems, Analysis and Synthesis ISAS 2000, International Institute of Informatics and Systemics, Orlando, pp. 395-401.
41. Kitchenham, B.; Linkman, S.; Pasquini, A.; Nanni, V. (1997). The SQUID approach to defining a quality model. Software Quality Journal, Vol. 6, Issue 3, pp.211-233.
42. Guceglioglu, A. S.; Demirors, O. (2005). Using software quality characteristics to measure business process quality. In International Conference on Business Process Management, Springer, Berlin, Heidelberg, pp. 374-379.
43. Arditi, D.; Gunaydin, H. M. (1998). Factors that affect process quality in the life cycle of building projects. Journal of Construction Engineering and Management, Vol. 124, No. 3, pp.194-203.
44. Hwang, S. M. (2009). Process quality levels of ISO/IEC 15504, CMMI and K-model. International Journal of Software Engineering and Its Applications, Vol. 3, Issue 1, pp. 33-42.
45. Baharom, F.; Yahaya, J.; Deraman, A.; Hamdan, A. R. (2011). SPQF: software process quality factor. In IEEE International Conference on Electrical Engineering and Informatics ICEEI 2011, pp. 1-7.
46. Thiruvathukal, G. K.; Hayward, N. J.; Läufer, K. (2018). Metrics Dashboard: A Hosted Platform for Software Quality Metrics. arXiv preprint arXiv:1804.02053.
47. Norman, F. E.; Pfleeger, S. L. (1997). *Software metrics: a rigorous and practical approach*. PWS Publishing Company, Second Edition.
48. Humphrey, W. S. (1989). *Managing the software process*. SEI series in software engineering, Addison-Wesley.

Chapter - 5

SOFTWARE PROCESS MODELS

This chapter discuss the software process models as a means of instructing software organization on how to achieve software development, business objective, and improvement goals. This chapter is structured into four main parts. These sections are described and discussed with respect to their importance and issue. Section I describes the introduction of this chapter. Section II describes the prescriptive process models. Section III describes the descriptive process models. Section IV describes the process modelling notations and tools. Section V describes the process improvement. Section VI describes the summary of this chapter.

5.1 INTRODUCTION

A software process model (process model) is a model of a software process. A software process model is an explanation of a software process. Process models are often used as a resource for problem solving. The requirement of the enactment of a software process by a process model is comparable to the specification of baking a cake using a recipe. Process models can be represented by using different notations (e.g. natural language, graphical, and machine-readable notations). A process model can illustrate a process on different levels of abstraction (e.g. engineering process level, lifecycle process level, and atomic step level).

Basically there are two types of process models such as: prescriptive and descriptive process models. Both types may exist in parallel in a software organization often in numerous instances. They do not automatically differ in content but quite in their proposed purpose. Prescriptive process models tell us how something should be completed, whereas descriptive process models illustrate how something is finished in reality. This constitutes a most important difference. Whereas a descriptive process model is formed by observing the process actually performed, a prescriptive process model normally aims to tackle all related issues of developing a piece of software. It is consequently often based on an ultimate model of software development. For example: a set of best practices that must be applied in software projects in order to yield their benefits.

5.2 PRESCRIPTIVE PROCESS MODELS

The aim of prescriptive modelling is to describe the required or recommended resources of executing the process. Prescriptive models are the outputs of prescriptive modelling. The software process models that go by the name of prescriptive models answer the question, "how should the software be developed?" and "how to develop source code". McChesney [1] divides prescriptive models into two categories (manual prescriptive models and automated prescriptive models) which is depend on the objective of the prescriptions.

Manual prescriptive models can be standards, methodologies, and methods centred on development, management, evaluation, software development life cycle and organisational life cycle support process. The models belong to this category include:

- Traditional structured methodologies.
- Object oriented methodologies.
- Knowledge engineering methodologies.
- Organisational design methodologies which are considered as several of the most representative in the area of socio-cultural systems.
- Software development life cycle process standards (ISO 12207-1995 revised in 2017 and IEEE 1074-1991 revised in 2006).
- Software process evaluation standard or model-based methods.

The automated prescriptive models perform activities related to support, assistance, management, and computer-assisted software development techniques. The models belong to this category include: ALF [2], MARVEL [3], SPADE [4], GRAPPLE [5], MERLIN [6], and unified software development process (unified process) [7]. These models are computerised specifications of software process standards. Main objective is to act as a guide for the software modelling process. They are focussed at aiding process agents by automatically interpreting software process models.

ALF research project (1987-1992) intends to provide a computerized facility to support software development activities. It is also intended to make an environment an active role during software development. ALF applies knowledge based systems and advanced information systems techniques to software engineering environments (SEE). Its main objectives were to movement on the ways of building customizable SEE and to move from a mechanist point of view of SEE to a further one which is more oriented toward user's activities, objective, and goals [8].

MARVEL's object management system is quick-and-dirty: The object-oriented data model is rather easy and does not support methods excluding for the special case of tools, every relationships with objects are maintained in memory, objects are mapped directly to the UNIX file system in the understandable manner, and object clustering is by alphabetical order. The long term goal of the MARVEL project is to examine rule-based process modelling as the foundation for architecture, design, SEE, and mainly software development environments [9].

SPADE project aims are developing an environment to define, analyze, and execute process models which is defined in the SLANG process modelling language. It illustrates the services provided by SLANG to sustain process evolution and shows how the SPADE-1 architecture supports integration of the process interpreter with peripheral tools available in the environment [10]. GRAPPLE is not tied to a particular software environment rather it accepts direct streams transcribed from real terminal sessions or made-up for experimental uses.

MERLIN is a prototype of process centred software development environment (PSDE). This prototype uses a rule based technique to illustrate and enact a software process. Users of Merlin are either software developers or managers who are involved in a software development process to develop a software product. The main benefit of using an environment like Merlin for software development is refined team support i.e. support for coordinating access to collective information on diverse levels of granularity, committed message servers for broadcasting information about project states, (urgent) tasks to do, and getting response of completed work packages etc.

The automated prescriptive models can be divided into two categories which are depending on the guiding standard selected to tackle software process modelling such as: a) activity-oriented models and b) people-oriented models. Activity-oriented models centre on the functions, activities and information concerning parts of the development, management, and software development life cycle support process such as: Marvel and SPADE. People-oriented models focus on the specifications of the people concerned in the software process and their relationships such as: ALF and unified process. Individuals are the least formalised factor in existing software process models. Their importance is patent; their behaviour is subjective, non-deterministic and has a critical impact on the results of software development which is an essentially a rational and social activity [11]. Moreover the non-specification of person resources means that the process does not imitate the real status of the software process of the modelled organisation, having the added threat of process not appropriate to the capability of the organisation's person resources being executed.

A prescriptive process model tells individual what work to do and perhaps also how to do it. It is built deductively, possibly drawing from an exterior standard and documentation from other software projects. A prescriptive process is a standard procedure or process accompanied by an authorization to follow it exactly. Prescriptive process become further appropriate as the work becomes more cyclic. A prescriptive model suggests individual to do things in a different way which means that they require changing their behaviour. Getting persons to change their behaviour is one of the trickiest tasks in software engineering and software development. So the successful execution of a prescriptive process model is complex, too. Deploying a prescriptive process model means changing person's behaviour. Prescriptive process models were initially proposed to fetch order to the chaos of software development. History indicated that these conventional models have brought a certain amount of useful structure to software engineering work and have provided a practically effective road map for software development and teams. However, software engineering work and the software product that it developed remain on the edge of chaos. It is called prescriptive because they impose a set of process elements framework activities, software engineering actions, work products, tasks, quality assurance, and change manage mechanisms for each software project [12]. Every process model also prescribes a process flow (work flow) that is the method in which the process elements are interconnected to one another.

Prescriptive process model divided into two classes of process models such as: lifecycle models and engineering models. Lifecycle process models capture the whole lifecycle of a software product. Usually, they abstract from a number of details and instead offer a broader view on the process (concentrate on what, not on how). There are various lifecycle models such as: waterfall model, iterative enhancement model, prototyping model, and spiral model. Another example lifecycle models frequently found in industry that adopt these general concepts: unified process and IBM clean room process. Engineering process models depict a portion of the total software development lifecycle process such as: specific type of inspection. Engineering process models can be extremely detailed regularly not only describing what to do, but also explaining how to do it [13]. There are various engineering process models such as: a process model for statistical testing, a process model for hybrid cost estimation, and extreme programming process model.

Lifecycle process models as contrasting to engineering process models characteristically cover the entire lifecycle of a software product or a large portion of it. This means that they often cover relatively a lot of ground which in turn commonly leads to a high level of abstraction. One of the best known lifecycle models is the waterfall model. It was first formally described by Winston Royce in 1970, suggests the implementation steps to develop a small computer program for internal operations and large computer program for delivery to a user. The waterfall model is used when software requirements are very well known and fixed. The idea behind the waterfall model is the chronological creation of products on diverse levels of abstraction (e.g. precede design by requirements and precede code by design) and integration in reverse direction. The stringent sequence can be weakened by controlled iterations [14]. The requirements themselves must be constant and one must acquire high capabilities for effort estimation.

Importance: Naturally waterfall projects face simply few problems during integration. Version and configuration management is also simplified.

Issue: Unexpected software requirements or circumstance changes pose high risks. Gaining experience and learning during a software project is difficult. Fixed delivery deadline are hazardous for waterfall projects. The documentation created is often huge and heavyweight. The waterfall approach does not extent very well for large projects and long cycle times [15]. The waterfall process model is often referenced but hardly applied in its strict form.

Basili and Turner proposed an iterative model to develop a software product in 1975. Iterative model helps when software requirements are not completely clear. Iterative approach maintains efficient learning. The iterative enhancement model proposes to initial implement a (accurately chosen) part of the entire system and then adds functionality in a number of iterations which all collectively finally form the whole system. Every iterations is completed in a waterfall style, i.e. a requirements analysis for the individual iteration is followed by designing the system part of the iteration, which is again followed by its implementation and so on. The result of each iteration is integrated with the previously existing system. The focus of iteration may be on introducing latest functionality but it might also be on refinement, improvement or architectural consolidation. The iterative enhancement model is based on three iterations. The first iteration (checkerboard pattern) develops the centre part of the complete system, the second iteration (vertical stripes), and the third iteration (horizontal stripes) mutually add functionality. A prerequisite for applying the iterative enhancement model is that the problem permits incremental development [16]. For example, requirements to be considered in later iterations should not require an entire redesign of the software architecture.

Importance: Iterative projects support efficient learning. With iterations designed properly, the core of the final software product is available very timely, thus featuring necessary properties of the entire software product. This allows for early customer participation and feedback. Iteratively developing the requirements helps when requirements are not completely clear or still unstable. Integration testing is supported due to comparatively small increments being added at a time. In case of fixed delivery time, incremental development helps to ensure that the most vital functionality can actually be delivered and the customer can choose what is most important to him.

Issue: Since the software product design is based on the existing set of requirements, there is a risk that requirements showing up later may be expensive to fulfil, due to software design decisions made earlier. Good, comprehensive version, and configuration management is necessary to distinguish increments. Integration may become increasingly complicated with the number of iterations, depending on how well requirements may be partitioned and on the software architecture. Many recent software process models such as: the unified process, extreme programming, or scrum follow an iterative pattern [17].

Another common lifecycle model is the prototyping model. Prototyping concentrates on the software development of an executable version of the software system that fulfils a limited number of requirements. Specifically, user interfaces are often simplified, performance, and throughput requirements are reduced or even ignored. In other cases e.g. for performance-critical components, diverse designs may be tested for their capabilities. The principal goal of building prototypes is to expand initial experience (e.g. concerning unclear requirements or complex design aspects). Getting an initial version of the final software product should usually not be an objective. In fact, when all questions have been answered by the prototype, it should be thrown away and a system fulfilling all requirements should be developed based on one of the other lifecycle models.

Importance: A prototype can be developed when the final properties are not completely clear. The direct contact between user and developer reduces misunderstandings. Inconsistent requirements are exposed earlier, either by the developer or by the user who evaluates the prototype. Complete version and configuration management is not essential because the code produced is not supposed to exist for a long time. The risk of software project failure (e.g. developing a product that does not satisfy the user requirements) is reduced due to the early involvement of the customer. In some cases, prototypes may even be used for evaluating software models before a lot of money is spent on developing the actual software.

Issue: There is an inherent risk that side effects are not adequately considered, especially non-functional requirements such as usability, reliability, maintainability, and safety. Another risk is that the user (or the developer) considers the prototype as the primary version of the software and that the system will be evolved from the prototype, potentially foremost to inadequately documented, and badly architected systems. The prototyping phase of a software project may also induce higher costs compared to a non-prototyped approach [18]. A prototype is an enormous way to clarify uncertain or unknown requirements; however, don't ever confuse it with the primary version of the actual software.

The spiral model was first published by Barry Boehm in 1986. It represents a risk driven approach, i.e. the assessment of risks determines the next software project phase. The spiral model combines aspects of the waterfall model, the iterative model, and prototyping. The primary step of each spiral cycle identifies the objectives of the software product element being elaborated (e.g. performance, reliability, and functionality), the diverse alternatives for implementing the software product part (e.g. diverse designs or reuse of existing components), and the constraints for each of the identified alternatives (e.g., cost or time). The next evaluates the identified alternatives, identifies and resolves risks that come with the different alternatives. The third step is determined by the risks that remain after the second step. During the third step the development approach that is suited most excellent for the risks is chosen. This may be the development of a prototype or the application of a strict waterfall process for implementing a smoothly defined feature. Finally, the next phases are planned, and the entire cycle is reviewed by the stakeholders [19-20]. Obviously a requirement for applying the spiral model is the ability to identify and assess risks, as well as a project context that allows for changing the process model during project runtime.

Importance: The third step accommodates features of other lifecycle process models as required. The spiral model is consequently very flexible. The explicit thoughtfulness of risks avoids numerous of the difficulties of other process models. Similarly, unpleasant alternatives are identified and removed early. The spiral model forms a single approach for software development and maintenance, while other models often focus on one or the other.

Issue: For bond software, the spiral model is tricky to apply because individual project phases are not wholly determined during project setup. It relies deeply on the organization's capability with respect to risk assessment. Therefore a bad risk assessment may guide to the selection of bad alternatives or development approaches. Boehm's spiral model embraces risk as its central aspect and chooses the paramount approach for all iteration [21-22]. This distinguishes it from other lifecycle models which consider risks but do not alter the basic model accordingly.

5.3 DESCRIPTIVE PROCESS MODELS

A descriptive process model attempts to collect tacit knowledge regarding how work is really done. It tries to explain key features of the 'as is' reality. It is built inductively. A descriptive process model describes what people do all day which means that it reflects existing practices. Software process models that is descriptive models answer the questions "how is software currently developed?" and "how has software been developed?". A descriptive model can expose many previously unknown problems in the organisation's software development process and knowing what is really being done is the primary step to be capable to improve it. McChesney divides this group of models into two main categories: informal descriptive models and formal descriptive models, depending on the objective of the descriptions. The goal of informal descriptive models is basically to provide an informal and qualitative model. Examples of informal models are the process cycle model [23], Bendifallah & Scacchi's model [24], and the stratified behaviour model [25].

Process cycle is a short and integrated view of engineering, management, performance, and improvement of software processes. The process cycle bounded by the three sectors A (engineering process models), B (managing software processes), and C (performing software processes) which defines the span of the total set of process steps essential for the development and evolution of software processes. In addition the cycle defines the key roles played by human beings, goals, categories of tools used, policies governing the processes, and inter-relationships and feedback among the different roles. The feed-front and feedback links across the three sectors complete a cycle of process engineering, management, performance, and improvement hence it is called process cycle.

Formal descriptive models of the software process can be related to process assessment, process improvement, and process prediction. This category includes: the systems dynamics based model [26], the PRISM change model [27], the FUNSOFT network based model [28], and the TAME model [29].

Systems dynamics based model offers a quantitative and deterministic explanation of the organization's human resources from the viewpoint of the human resource recruitment, training, inclusion, and transfer experience level and not as regards the logic of competencies of the human resources and roles. The models that relation for people does not provide a defined set of stages or activities for performing the modelling process [30].

Prism project is a model of changes which support two change related environment infrastructures with the unique key features: (1) separation of changes to the described items as of the changes to the environmental facilities encapsulating these items, (2) a facility called dependency structure for describing various items and their interdependencies, and for identifying the items affected by a given change, (3) a facility called change structure for classifying, recording, and analyzing change related data, and for assembly qualitative judgments of the consequences of a change (4) identification of the numerous separate properties of a change, and (5) a built in mechanism for providing feedback [31].

TAME (Tailoring A Measurement Environment) project at the University of Maryland an improvement oriented software engineering process model was developed that uses the goal/question/metric (GQM) paradigm to incorporate the productive and systematic aspects of software development process. The model provides a method for formalizing the characterization and planning tasks, controlling and getting better projects based on quantitative analysis, learning in a deeper and more systematic way about the software process and product, and feeding the suitable experience back into the existing and future projects. The TAME system is an instantiation of the TAME software engineering process model as an ISEE (integrated software engineering environment) [32].

5.8 Systems and Software Process

The actual process present in an organization, those that are used in practice to produce and deliver the organization's products and services, can regularly be counted as one of its major assets. Normally these processes encompass an excellent deal of the knowledge and experience that make an organization successful. For this reason managing them appropriately can develop into a critical activity for an organization's long-time continued existence. The discipline behind this type of process management is called descriptive process modelling. As its name implies, this discipline is worried with producing an explicit and correct representation of an organization's actual processes, for rationale such as documentation, analysis, dissemination, and improvement [33].

Software process can be exceptionally complex activity, involving large numbers of interrelated activities. For every software development process, these activities must be suspiciously orchestrated, so that dependencies are respected and resources are accessible where and when they are needed. This complexity makes it very tough to guarantee that processes are constantly executed in the similar way. In a large number of individual activities and their difficult interdependencies, there is a high risk that various activities are not planned or are skipped because of exterior pressure. Also, there is forever the risk that people will modify the process over time without even noticing it or maybe eliminating valuable activities. A documented process helps to ease these risks. With a proper process explanation at hand, both project planners and project performers can make sure that they are performing all significant and expected activities, and that they are taking into account all related interrelations between those activities. Also, a documented process makes it easier to initiate changes in a controlled manner, which leads understanding of the process. Understanding the process is essential for managing this risk [34]. An explicit process demonstration makes it greatly easier to evaluate the overall impact of process changes thus allowing improved management of process changes.

A universal problem in large organizations is that dissimilar development groups in the organization track different processes for the similar task. Sometimes these process differences are interrelated to the truth that diverse groups may have different needs. However it frequently happens that they work differently simply because it is tricky for a group to accurately tell what other groups are doing. By explicitly describing the processes followed by a variety of organizational units, it is easier to accomplish a unified process. Differences between groups can still survive, but they are the result of aware decisions based on analyzing each group's fastidious needs [35].

As an organization reaches a superior degree of process maturity, it becomes increasingly vital to compute process performance. Aspects that are generally measured in a process are its productivity (amount of output produced vs. resources spend into producing it), its capability to deliver results on time, and its ability to identify and acceptable product defects as early as possible, between many others. The reasons for measuring diverge widely from one organization to another but frequently incorporate the desire to get better planning and enhance the quality of software products by removing the number of defects exist in them when they are delivered [36].

In the long time, one of the chief reasons for explicitly describing an organization's development process is to be capable to describe goals for the process and work thoroughly on achieving them. Doing this involves at least several of the descriptive process modeling goals such as: process understanding and process measurement [37]. Long-period goals for a process are often related to improvement (e.g. increase product quality, reduce time dedicated to product rework, and product quality) but may be associated to other process aspects.

It happens reasonably often that one time a process has been properly described and stabilized, opportunities for supporting components of it with automated tools develop into evident. Analyzing a comprehensive process explanation is possibly one of the best ways to establish the exact requirements of new process supporting tools. Usually, descriptive process modeling is performed in an iterative fashion, with the elicitation and modeling phases being interwoven. Process engineers work to increase some preliminary knowledge of the modeled process, explain this knowledge using a suitable process notation and then discuss this original description with the process stakeholders. This conversation often leads to extra knowledge being collected which can be included into the depiction and validated once more with the stakeholders [38]. This cycle can be repetitive until satisfactory levels of aspect and accuracy of the process explanation have been achieved.

The concluding step of descriptive process modeling is perform a process analysis, i.e. to utilize the process model to track or evaluate process performance, depending on the process modeling objectives stated in step 1. One opportunity is to track the process by asking process performers to log. For example, begin and end times for their activities. Another option is to make snapshots by asking people regarding their existing activities at usual intervals. The resulting data can be used for qualitative or for quantitative analyses. Qualitative analyses intend at identifying weaknesses of the processes, e.g. when moreover many responsibilities are pinned to a particular role or when too many roles are assigned to a sole person or when feedback loops become so extreme that they consume more time than productive project work. Quantitative analyses try at identifying correlations between process attributes e.g., when the number of requirements is more than 30% higher than average, the number of defects rises by 70%. Both kinds of analysis results can then be used to adjust the process (in order to eliminate the weaknesses), or the process model (in case it does not robust the process after all), or to influence project planning (e.g. assign more resources to requirements reviews when the number of requirements is more than 30% higher than average) [39].

Typically, descriptive and prescriptive models track each other in a cycle. Regularly, the present processes are modeled and transferred into a descriptive process model. Using this model, problems are identified and mitigated by changes to the process model frequently using supplementary external knowledge such as best practices. This model then becomes a prescriptive process model, effectively instructing group to do things differently than before. When the altered process is fully incorporated into people's daily work, the process model becomes descriptive once again, and an additional improvement cycle can establish. Figure 5.1 shows the relationship between descriptive and prescriptive models as reported by Abramowicz et al [40].

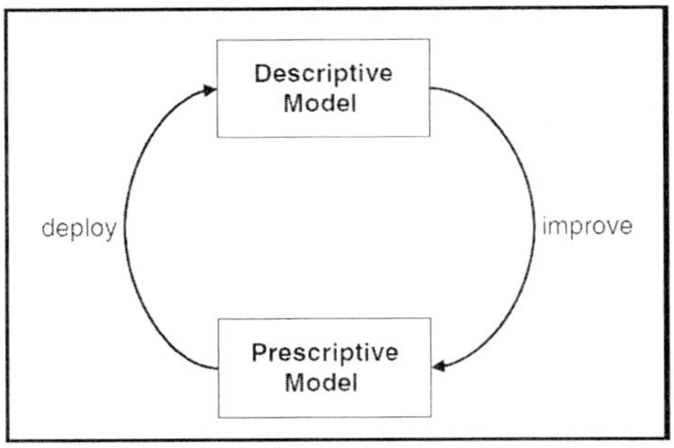

Figure 5.1 The Relationship between Descriptive and Prescriptive Process Models

That figure shows descriptive model improve prescriptive model and prescriptive model deploy descriptive model in cyclic form.

5.4 PROCESS MODELLING NOTATIONS AND TOOLS

Using process modelling notations, we identify an excess of diverse approaches for software development. This is due to the fact that during the chronological development of process modelling notations, dissimilar communities have influenced the discipline of process modelling. In software engineering processes, two main groups influenced the software development. Process modelling notations can be identified to develop software products [41]. The first group was extensively influenced by tool developers and programmers. Within this group, notations for the demonstration of processes were developed or adopted aimed at creating representations that could be interpreted by system and technology. Therefore, this group concentrate on process automation and the notations used were normally not designed to be interpreted by users. The fundamental vision was to generate software development environments where the implementation of software development tools would be affected by a process driven engine. The core centre was on little, low level processes such as the code, compile, test, and fix cycle [42]. As a consequence, this approach determined on processes with a high potential of automation.

The second group has its beginning in the area that was concerned with software process improvement. In this group, the aim was to develop software development more mature by means of introducing most excellent practices and establishing learning cycles. For this reason, the need arose to signify software processes in order to recognize and improve the processes of software development performed by developers. The notation constructs developed in this perspective aimed at describing real world concepts and creating models that people can interpret. This approach represents software engineering processes which focused on advanced level processes and a minor amount of automation. Consequently processes are described in an additional informal and less detailed way and most significantly they provide direction that can be interpreted and enacted by people. In this circumstance process guides based on natural notation became accepted [43]. They focus on providing people with the information essential to properly enact the software process.

Presently, a large quantity of different process modelling notations exists and consequently a strong need for standardization has developed. As an output of this development the Software Process Engineering Metamodel (SPEM) was developed. Its objective is to enable the demonstration of different software engineering concepts and theory.

Software developers, practitioners, and process engineers are in necessitate of software support for process modelling in order to be capable to deal efficiently with process model development and the administration of changes, updations, and modifications. Usual process guidelines are not only widespread but also cross referenced and subsequently changes in definite areas guide to changes in other parts. Support is consequently useful for maintaining stability, consistency, and reliability. Such following software can have dissimilar functionalities. In order to be capable to evaluate different solutions such as: ECMA/NIST Reference Model for Software Engineering Environments which provides a framework for the discussion of different Software Engineering Environments (SEE) this reference model is useful. Next is the Eclipse Process Framework (EPF), mainly of the EPF Composer, as a specific tool for process modelling [44].

The purpose of process modelling is to get better performance by increasing efficiency, performance, and productivity. The tools used are essentially visual aids that can rapidly and visibly tell a story of a process such as: making a bus/train/air ticket reservation, online examination system, completing online orders, and transferring funds from one bank account to another. There are a various common tools and techniques of process modelling such as:

- **Flowcharts:** A flowchart is a diagram that represents a process and can be formed with readily available software and tool. Flowcharts hold a starting and ending point. Typically, symbols such as squares, circles, and diamonds symbolize events or activities, while arrows designate sequential flow and relations between steps [45].
- **Functional Flow Block Diagram:** A functional flow block diagram (FFBD) is used to explain requirements in functional terms. Somewhat than being solution oriented like a flowchart, the FFBD is functionally oriented and illustrates the efficient architecture of a system. In a FFBD, elements such as equipment, training and software are identified and defined, support requirements are fixed to detailed functions and appropriate sequence is established [46].
- **Control Flow Diagram:** A control flow diagram is used to help depict the usual flow of a process with added limits and constraints. Control flow diagrams illustrate how definite conditions such as: additional data or equipment, alternate operations, and alternative inputs affect a process flow [47].
- **Gantt Chart:** A Gantt chart is a different way to visually communicate information. Typically a bar chart, Gantt charts concentrate on activities and tasks involved in a process, depicted sequentially and against a timeline. They also illustrate the dependency relationships among activities. With a Gantt chart, a professional can see at a quick look whether a software project is on ahead or behind time [48].

- **PERT Diagram:** A PERT (Program Evaluation and Review Technique) diagram is used to assist identify the smallest amount of time needed to complete the software project. The idea behind this technique is that assured activities cannot be started until others are completed. The PERT diagram helps avoid idealistically short timelines by estimating the through most likely and longest time every step of a process will acquire [49].
- **IDEF Diagram:** IDEF is a short form for Integration Definition, a generally used technique in business process modelling. IDEF refers to 16 methods, designed to collect a type of information through modelling processes. IDEF methods are used to build diagrams that define system control, data flow, and can graphically characterize a wide diversity of processes with any required level of detail [50].

5.5 PROCESS IMPROVEMENT

This section describes the concepts for improving software process which is structured into four major parts. First, model based improvement approaches in common and additionally introduces two specific model based improvement frameworks such as: CMMI and SPICE. Secondly explain continuous improvement approaches, beginning with a short outline of commonly used continuous improvement approaches and moreover introducing specific software engineering concepts, especially the Quality Improvement Paradigm (QIP) and the Experience Factory (EF). Third is the role of measurement in the context of process improvement. Operationalizing process improvement creates a need for measurement and consequently the goal question metric (GQM) method which is a de-facto standard for goal oriented software measurement. Finally, the organizational and business context of process improvement is explain by introducing the Balanced Scorecard (BSC) and the GQM+ strategies approaches which can be used as means to support specific improvement activities with business goals, objective, and strategies [51].

Mature software development process can be seen as a requirement for high quality software products. In significance lacking good process quality and an understanding of the effects that processes have in a detailed development environment, achievement, and success is not effortlessly repeatable. This is correct in particular for processes that rely exclusively on some sort of high quality process documentation exclusive of adapting and improving the process in the context of the development environment. If processes do not change with the development context and do not hold developers in a sufficient way the resulting product quality frequently remains short. Past knowledge with inadequate process quality has motivated the development of software process improvement (SPI) approaches that tackle these issues [52]. Presently, two types of SPI approaches are being used in tradition: model based SPI approaches (also called as problem oriented approach or top down approach) and continuous SPI approaches (also called as solution oriented approach or bottom up approach).

Model based SPI approaches contrast the recent processes and practices of a development organization against a reference model or a benchmark. They can be used to recognize difficult process areas with respect to the used reference model. With the identified challenging process areas helps to obtain potential improvement area and options. Typically, model based SPI approaches give different so called capability and maturity levels with changed sets of processes and practices. These levels often describe an improvement roadmap. Continuous SPI approaches concentrate on solutions for the mainly essential challenges of a software development organization and usually occupy improvement cycles based on a preliminary baseline. Continuous approaches centre on solving a detailed problem by analyzing the problem, implementing and observing problem focused improvement activities, and measuring the effects of the performance. The explanation of the measurement data is used as input for additional optimization of the solution. Model based and continuous SPI approaches can be seen as being corresponding. Model based approaches can be used to recognize problem areas and possible improvement options. Continuous approaches can be used to resolve respective company specific problems. Continuous approaches can be effectively applied, autonomous of the maturity of an organization, whereas model based approaches generally involve continuous improvement starting a certain point on. The requirement for SPI is being widely acknowledged nowadays. Due to the information that software development processes are usually human based and depend on the development (context including such as: workforce capabilities, domain characteristics, and organizational maturity), changes to these processes usually cause considerable costs and should be measured carefully. Different improvement options should to be evaluated with respect to their development cost and their possible impact on business objective and goals. To tackle these organizational aspects concepts of business configuration will be used by organization. Model based SPI approaches evaluate organizational processes with a reference model and can be used to recognize coarse grained problem areas and impending improvement options [53]. Continuous SPI approaches can be used to develop organization or company specific solutions for significant problems and assess the belongings of improvement activities.

Generally ISO/IEC 15504 and the SEI's CMM are popular frameworks for the evaluation of software processes. There is an elementary difference in the way they help to set improvement goals. While ISO/IEC 15504 deals with processes individually, the CMM organizes the processes in a levelled depiction that defines a practical sequence for improvement. Each level in the CMM groups a set of KPAs, and each KPA group a set of related activities that performed cooperatively, help to achieve business goals that are measured significant for establishing the potential of the software process at the agreed maturity level. For example, if an organisation is at level 2, the CMM recommends that their instant improvement efforts should be focussed towards level 3 KPAs. On the other hand, ISO/IEC 15504 provides a comprehensive snapshot of the existing state of each process throughout process profiles but makes no suggestions as to how to develop on that state. The new integrated version of the CMM is the CMMI, which has been newly released by the SEI aims to merge both approaches and provides two option representations such as: a continuous and a staged representation. The staged representation obviously orders the maturity levels, while the continuous representation concentrates on different process areas and provides success profiles (a list of process areas and their respective capability levels) as an outcome of the assessment. Using the continuous approach, the improvement should be designed as target staging that is a chain of target profiles to be achieved by the organisation. Equivalence has been clear between the maturity levels of the staged representation and several target profiles. Thus the continuous approach allows organisations to set their own improvement goals more openly and flexibly, while staging is an additional guided and restricted approach, although it allows for benchmarking between organisations or projects [54].

Harmonizing to the CMM, the SEI has defined an improvement model called IDEAL that divides the improvement process into five stages such as: initiation, diagnosing, establishing, acting and leveraging. The initiation stage is preceding to software process evaluation. It is the stage at which the senior management first realises the requirement to improve the organisation's software process and sets the context for and makes an assurance to the improvement program. Sustained senior management commitment is essential for the success of the improvement program. The basic infrastructure for improvement is established at this phase including a Software Engineering Process Group (SEPG) and a Management Steering Group (MSG) [55].

The diagnosing stage of the IDEAL model is the one of the most related to the evaluation of the current software processes. The purpose of this stage is to set a baseline for the recent state. This baseline will assist recognize the errors and deficiencies to be conquer and later on, to calculate the degree of improvement that has been improved and achieved. The next step is to establish the objective and goals for the improvement program. The acting stage used to execute and track installation whereas leveraging phase used to document, analyze lessons, and revise organizational approach. The SEPG provides instruction for the process developers and performers, i.e. they offer a helpdesk that persons can contact if they have questions. The MSG should develop a strategic action plan for the next three to five years for setting the short and long term goals, according to the baseline and timeline the organisation's business needs and objective. A tactical action plan is also developed for the first improvement cycle and organisational arrangement is completed including one or more Technical Working Groups (TWG). Each TWG will be in charge of developing and implementing solutions for one of the selected improvement areas. In spite of the numerous approaches to software process improvement, there are no magic actions and recipes. Improvement is a long, complex, rigid effort and any organization that embarks on a SPI program should understand that the major benefits will be supposed in the long term [56].

5.6 SUMMARY

A software process model is an explanation of a software process. Prescriptive process models tell us how something should be finished, whereas descriptive process models describe how something is completed in actuality. A prescriptive model tells developer to do something differently which means that we need to modify their behaviour. Getting individuals to alter their behaviour is one of the most complex tasks in software development so the successful deployment of a prescriptive process model is tricky too: Deploying a prescriptive process model means changing people's behaviour. A descriptive process model describes what developer do every day which means that it reflects current practices of software development. This discipline is concerned with developing a clear and correct illustration of a software organization's real processes for purposes such as analysis, documentation, dissemination, and improvement. The notation constructs developed in the context aimed at relating real world concepts and creating models that people can interpret. Software process improvement used to mature and improve the software process to get a quality product. SPI addresses the software development and quality issues using model-based SPI approaches and continuous SPI approaches.

5.7 REFERENCES

1. McChesney, R. (1995). Toward a classification scheme for software process modelling approaches. Information and Software Technology, Vol. 37, Issue 7, pp. 363-374.
2. Canals, G.; Boudjlida, N.; Derniame, J. C.; Godart, C.; Lonchamp, J. (1994). ALF: A framework for building processcentred software engineering environments. In Software Process Modelling and Technology chapter 7, Research Studies Press, pp. 153-185.
3. Kaiser, G. E.; Feiler, P. H.; Popovich, S. S. (1988). Intelligent assistance for software development and maintenance. IEEE Software, pp. 40-49.
4. Bandinelli, S. C.; Fuggetta, A.; Ghezzi, C. (1993). Software process model evolution in the SPADE environment, IEEE transactions on software engineering. Vol. 19, No. 12, pp. 1128-1144.
5. Huff, K. E.; Lesser, V. R. (1988). A plan-based intelligent assistant that supports the software development process. ACM SIGSOFT Software Engineering Notes, Vol. 13, No. 5, pp. 97-106.
6. Junkermann, G.; Peuschel, B.; Schäfer, W.; Wolf, S. (1994). MERLIN: Supporting cooperation in software development through a knowledge-based environment. In Software Process Modelling and Technology, chapter 5, Research Studies Press, pp. 103-129.
7. Jacobson, I.; Booch, G.; Rumbaugh, J. (1999). *The Unified Software Development Process*. Addison Wesley.
8. Boudjlida, N.; Derniame, J. C.; Godart, C.; Lonchamp, J. (1993). A Short Tour Through ALF.
9. Kaiser, G. E.; Barghouti, N. S.; Sokolsky, M. H. (1990). Preliminary experience with process modeling in the Marvel software development environment kernel. In Proceedings of the IEEE Twenty-Third Annual Hawaii International Conference on System Sciences, Vol. 2, pp. 131-140.
10. Bandinelli, S. C.; Fuggetta, A.; Ghezzi, C.; Lavazza, L. (1994). SPADE: An environment for software process analysis, design, and enactment. In Software Process Modelling and Technology, pp. 223-247.
11. De Leoni, M.; Marrella, A. (2017). Aligning real process executions and prescriptive process models through automated planning. Expert Systems with Applications, Vol. 82, pp. 162-183.
12. Gericke, K.; Blessing, L. (2011). Comparisons of design methodologies and process models across domains: a literature review. In DS 68-1: Proceedings of the 18th International Conference on Engineering Design (ICED 11), Impacting Society through Engineering Design, Vol. 1: Design Processes, Lyngby/Copenhagen, Denmark.
13. De Leoni, M.; Lanciano, G.; Marrella, A. (2017). A tool for aligning event logs and prescriptive process models through automated planning. In CEUR Workshop Proceedings, Vol. 1920.
14. Royce, W. W. (1970). Managing the development of large software systems. In: Proceedings of IEEE WESCON, Los Angeles, CA, USA, pp. 1-9.
15. Simões, D.; Antunes, P.; Carriço, L. (2018). Eliciting and Modeling Business Process Stories. Business & Information Systems Engineering, Vol. 60, No. 2, pp. 115-132.
16. Basili, V. R.; Turner, A. J. (1975). Iterative enhancement: a practical technique for software development. IEEE T Software Eng, Vol. 1, Issue 4, pp. 390-396.
17. Mapikou, G. L. M.; Etoundi, R. A. (2016). A process mining oriented approach to improve process models analysis in developing countries. In IEEE/ACS 13th International Conference of Computer Systems and Applications (AICCSA 2016), pp. 1-8.
18. Sommerville, I. (2011). *Software engineering*. 9th ed., Pearson Education, Inc.

19. Boehm, B. W. (1986). A spiral model of software development and enhancement. ACM Sigsoft Software Eng Notes, Vol. 11, N0. 4, pp. 22-42.
20. Boehm, B. W.; Lane, J. A. (2008). Guide for using the Incremental Commitment Model (ICM) for systems engineering of DoD projects. Center for Systems and Software Engineering, University of Southern California, Los Angeles, CA.
21. Boehm, B. W. (2011). Some future software engineering opportunities and challenges. In: Nanz S. (ed) The future of software engineering, Springer, Heidelberg.
22. Boehm, B. W. (2009). The incremental commitment model (ICM), with ground systems applications. In: Ground systems architectures workshop (GSAW), Torrance, CA, USA.
23. Madhavji, N. H. (1991). The process cycle. Software Engineering Journal, Vol. 6, No. 5, pp. 234-242.
24. Bendifallah, S.; Scacchi, W. (1989). Work structures and shifts: An empirical analysis of software specification teamwork. Proceedings of the 11th International Conference on Software Engineering, pp. 260-270.
25. Curtis, B.; Krasner, H.; Iscoe, N. (1988). A field study of the software design process for large systems. Communications of the ACM, Vol. 31, No. 11, pp. 1268-1287.
26. Abdel-Hamid, T.; Madnick, S. (1991). *Software Project Dynamics: An Integrated Approach*. Prentice-Hall.
27. Madhavji, N. H. (1991). The prism model of changes. In Proceedings of the IEEE 13th international conference on software engineering, pp. 166-177.
28. Deiters, W.; Gruhn, V. (1991). Software process analysis based on FUNSOFT nets. Systems Analysis Modelling Simulation, Vol. 8, No. 4-5, pp. 315-325.
29. Basili, V. R.; Rombach, H. D. (1988). The TAME project: Towards improvement-oriented software environments. IEEE Transactions on Software Engineering, Vol. 14, Issue 6, pp. 758-773.
30. Acuña, S. T.; Juristo, N. (2004). Assigning people to roles in software projects. Software: Practice and Experience, Vol. 34, No. 7, pp. 675-696.
31. Madhavji, N. H. (1992). Environment evolution: the Prism model of changes. IEEE Transaction on Software Engineering, Vol. 18, No. 5, pp. 380-392.
32. Oivo, M.; Basili, V. R. (1992). Representing software engineering models: The TAME goal oriented approach. IEEE Transactions on Software Engineering, Vol. 18, No. 10, pp. 886-898.
33. Eckert, C. M.; Stacey, M. K. (2010). What is a process model? Reflections on the epistemology of design process models. In Modelling and Management of Engineering Processes, Springer, London, pp. 3-14.
34. Browning, T. R.; Fricke, E.; Negele, H. (2006). Key concepts in modeling product development processes. Systems Engineering, Vol. 9, No. 2, pp. 104-128.
35. Acuna, S. T.; De Antonio, A.; Ferre, X.; Maté, L.; López, M. (2001). The software process: Modeling, evaluation and improvement. In Handbook of Software Engineering and Knowledge Engineering, Vol. I: Fundamentals, pp. 193-237.
36. Salviano, C. F. (2011). A Modeling View of Process Improvement. In International Conference on Software Process Improvement and Capability Determination, Springer, Berlin, Heidelberg, pp. 16-27.
37. Gong, H.; Zhang, H.; Yu, D.; Liu, B. (2017). A systematic map on verifying and validating software process simulation models. In Proceedings of the ACM International Conference on Software and System Process, pp. 50-59.
38. West, J. (2000). Quality management principles. Quality Progress, Vol. 33, Issue 3, p. 79.
39. Blaukopf, S.; Mendling, J. (2017). An Organizational Routines Perspective on Process Requirements. In International Conference on Business Process Management, Springer, Cham, pp. 617-622.
40. White, S. A. (2004). Process modeling notations and workflow patterns. Workflow handbook, pp. 265-294.
41. Abramowicz, W.; Filipowska, A.; Kaczmarek, M.; Kaczmarek, T. (2012). Semantically enhanced business process modeling notation. In Semantic Technologies for Business and Information Systems Engineering: Concepts and Applications, IGI Global, pp. 259-275.

42. Shankar, G.; Maurya, E. L. S. (2017). Software Quality Models: A Comparative Study. SRMS Journal of Mathematical Sciences, Vol. 1, No. 1, pp. 180-190.
43. Rosemann, M.; Recker, J.; Indulska, M.; Green, P. (2006). A study of the evolution of the representational capabilities of process modeling grammars. In International Conference on Advanced Information Systems Engineering, Springer, Berlin, Heidelberg, pp. 447-461.
44. Figl, K. (2017). Comprehension of procedural visual business process models. Business & Information Systems Engineering, 59(1), 41-67.
45. Dumas, M.; Pfahl, D. (2016). Modeling Software Processes Using BPMN: When and When Not? In Managing Software Process Evolution, Springer, Cham, pp. 165-183.
46. Kalenkova, A. A.; van der Aalst, W. M.; Lomazova, I. A.; Rubin, V. A. (2017). Process mining using BPMN: relating event logs and process models. Software & Systems Modeling, Vol. 16, No. 4, pp. 1019-1048.
47. Polančič, G.; Jošt, G. (2016). The impact of the representatives of three types of process modeling tools on modeler's perceptions and performance. Journal of Software: Evolution and Process, Vol. 28, Issue 1, pp. 27-56.
48. Hinkelmann, K. (2016). Business Process Flexibility and Decision-Aware Modeling - The Knowledge Work Designer. In Domain-Specific Conceptual Modeling, Springer, Cham, pp. 397-414.
49. Fellmann, M.; Metzger, D.; Jannaber, S.; Zarvic, N.; Thomas, O. (2018). Process Modeling Recommender Systems. Business & Information Systems Engineering, Vol. 60, No. 1, pp. 21-38.
50. Khider, H.; Benna, A.; Meziane, A.; Hammoudi, S. (2017). Toward an approach to improve business process models reuse based on linkedin social network. In World Conference on Information Systems and Technologies, Springer, Cham, pp. 179-189.
51. Larrucea, X.; O'Connor, R. V.; Colomo-Palacios, R.; Laporte, C. Y. (2016). Software process improvement in very small organizations. IEEE Software, Vol. 33, No. 2, pp. 85-89.
52. Lee, J. C.; Shiue, Y. C.; Chen, C. Y. (2016). Examining the impacts of organizational culture and top management support of knowledge sharing on the success of software process improvement. Computers in Human Behavior, Vol. 54, pp. 462-474.
53. Afzal, W.; Alone, S.; Glocksien, K.; Torkar, R. (2016). Software test process improvement approaches: A systematic literature review and an industrial case study. Journal of Systems and Software, Vol. 111, pp. 1-33.
54. Khan, A. A.; Keung, J. (2016). Systematic review of success factors and barriers for software process improvement in global software development. IET software, Vol. 10, No. 5, pp. 125-135.
55. Niazi, M.; Wilson, D.; Zowghi, D. (2005). A maturity model for the implementation of software process improvement: an empirical study. Journal of systems and software, Vol. 74, Issue 2, pp. 155-172.
56. Behari, S. (2018). IT service management: process capability, process performance, and business performance. Doctoral dissertation, University of Southern Queensland.

Chapter - 6

SITUATIONAL FACTORS AFFECTING THE SOFTWARE PROCESS

Various methods, standards, and models are used to develop a quality product with the help of software process. Software development process is a large and complex activity in which technical and managerial tool are used to develop software. To achieve quality in finished product well defined process is required because quality of product directly related to quality of process. Situational factors are important to consider during development process to develop a quality product. This chapter provides the overview of situational factors that affect software process and analyze how situational factors affect software process. In this chapter situational factors affecting the software process are analyzed based on literature review, categorize each factor according situation domain and analyzed how these factors affect software process. Section I describes the introduction of software process and situational factors. Section II describes the software development models and standards. Section III describes the software development cost estimation, Section IV describes the software development risk factors, Section V describes the software development success factors, Section VI describes the software development environment factors, and Section VII describes the summary of this chapter.

6.1 INTRODUCTION

A process is an activity or a function that is performed for some specific business reason. A process accepts input data needed for the process to be carried out and produce software. Software process provides a way in which software development is organized, measured, managed, and improved. Software process is used to solve technical, managerial, and environmental issues during software development [1].

In last four decades various kind of software development process is develop as per need of time, technology, environment, and structure such as: waterfall model [2], iterative model [3], prototyping model [4], evolutionary model [5], spiral model [6], V Model [7], agile model [8], and hybrid spiral model [9]. Software process has been used in models to develop software product as well as for software evaluation [10].

There are various activities and relationship among the software products which is use at the time of software development [11]. Software process can be categorize or characterize as software development models, standards and maturity framework [12]. In the past decade, software processes has evolved significantly and there are various critical issues and challenges too. Most relevant directions in software process are:

social aspects, empirical and global software engineering, model-driven engineering, and application lifecycle management (ALM) [13-14]. There are various challenges involve in software process because there are various situational factors which affect software process [15].

Software process has its own role and importance in software development. In this chapter various situational factors are analyzed which affect software process and its performance. To develop a quality product various situational factors are involve which affect software process. These situational factors is important and necessary for consider at the time of development.

The objective of this chapter is to identify the situational factors which affect software process and its impact on software quality. Also analyze how situational factors affect the software process. There are main two research questions that are reported in this chapter:

RQ1: What are the situational factors which affect software process?

RQ2: How situational factor affect the software process.

Situational factors are analyzed and categorized as: I) Software development models and standards, II) Software development cost estimation, III) Software development risk factors, IV) Software development success factors, and V) Software development environment factors, which affect the software process.

First situational factor consider as software development models and standards which is used to develop a quality product at minimum time, cost, and effort [16-19]. Second situational factor consider as software development cost estimation model which is used for early prediction of cost, effort, and time to develop a quality product [20-24]. Third situational factor consider as software development risk factors are used to identify the risk which is involved at the time of development. These risk factors are used to identify an undesired event or condition associated with a software project [25-30].

Fourth situational factor consider as software development success factors based upon the success of project and satisfied customer. These success factors show that quality of product and performance of software is same as per customer or user requirement in given cost and time [31-35]. Fifth and last situational factor consider as Software development environment factors describes the organizational impact on software process, product, and quality. There are organizational, environmental, behavioral, team knowledge, decision making, and performance factors which affect the software process [36-44].

6.2 SOFTWARE DEVELOPMENT MODELS AND STANDARDS

There are various software development processes which are used to develop a software product such as: waterfall model, V model, iterative model, spiral model, prototype model, and agile model [45]. All theses process uses well recognized and established standards such as Capability Maturity Model (CMM), ISO/IEC 12207, ISO/IEC 15504, and Capability Maturity Model Integration (CMMI) [46].

When software development process fails to produce a quality product then Software Engineering Institute (SEI) release CMM to assist the software organization to maintain and develop a quality product. The Capability Maturity Model for Software provides software organizations with guidance on how to gain control of their processes for developing and maintaining software and how to evolve toward a culture of software engineering and management excellence. The CMM was planned to guide software organizations to select process improvement strategies by finding current process maturity and determining the issues which is most critical to process improvement and software quality. Using limited set of activities and working assertively an organization can progressively improve its organization extensive software process to facilitate continuous and lifelong gains in software process capability and maturity. There are five levels of software process maturity such as: Initial, Repeatable, Defined, Managed, and Optimizing. There are various situational factors which affect the CMM such as: Maturity level, Key Process Areas (KPA), Common features, and Key practices [47-49].

ISO/IEC 12207 provides a common framework for developing and maintaining software products. The standard defines a set of process from conception to retirement. These set of processes are defined in terms of activities and these activates are divided into set of tasks. The processes are defined in three parts: Primary life cycle processes, supporting life cycle processes, and organizational life cycle processes. Primary life cycle processes categories as: acquisition process, supply process, development process, operation process, and maintenance process. Supporting life cycle processes divided into audit process, configuration management, joint review process, documentation process, quality assurance process, problem solving process, verification process, and validation process. Organizational processes are based on management process, infrastructure process, improvement process, and training process. There are various situational factors which affect ISO/IEC 12207 such as: Nature of tasks, Applicability to organizations and projects, Software metrics, and Documentation of outputs [50-52].

ISO/IEC 15504 is the International Standard also known as Software Process Improvement and Capability dEtermination (SPICE) for process assessment. ISO/IEC 15504 initially was derived from process lifecycle standard ISO/IEC 12207 and from maturity models like CMM. ISO/IEC 15504 is a reference model. The reference model defines process dimension and capability dimension. The process dimension divided into five process categories such as: customer/supplier, engineering, supporting, management, and organization. Capability level define as: optimizing process, predictable process, established process, managed process, performed process, and incomplete process The SPICE standard is designed for software process to help software developer, organization, and customer too. There are various situational factors which affect ISO/IEC 15504 such as: Process management, Process improvement, Process assessment, and Compatibility with other standards [53-56].

To solve the issue when multiple CMMs are used CMMI is developed. CMMI provides an opportunity to minimize these barriers using integrated model that exceed disciplines. Five principal ideas which inspired CMMI model are:

1. Planning, tracking, and schedule management
2. Requirements definition and configuration control
3. Process assessment
4. Quality measurement and continuous improvement

5. Evolutionary improvement

Like CMM there are five levels of software process maturity such as: Initial, Managed, Defined, Quantitatively Managed, and Optimizing. CMMI provide guidance to use when developing process. There are various situational factors which affect the CMMI such as: Maturity level, KPA, Specific and generic goals, and Specific and generic practices [57-59].

Situational factors that affect software development models and standards are shown in Table 6.1. Table 6.1 shows the summary of situational factors that affect software development models and standards.

As per requirement of software project and organization various models and standards have been used as specified in Table 6.1 which is affected by various situational factors. CMM is used for mature software process, ISO/IEC 12207 is used for software life cycle, ISO/IEC 15504 is used for software process improvement, and CMMI is used process maturity and improvement both. So CMMI is more important and effective model than other models and standards.

Table 6.1 Situational Factors that Affect Software Development Models and Standards

Software Development Models and Standards	Situational Factors
CMM	Maturity level, KPA, Common features, and Key practices
ISO/IEC 12207	Nature of tasks, Applicability to organizations and projects, Software metrics, and Documentation of outputs
ISO/IEC 15504	Process management, Process improvement, Process assessment, and Compatibility with other standards
CMMI	Maturity level, KPA, Specific and generic goals, and Specific and generic practices

6.3 SOFTWARE DEVELOPMENT COST ESTIMATION

Software development cost estimation is used to estimate cost and effort which is important for software development. It projects the cost of software development. Early cost estimation is helpful to plan and develop a quality product within estimated time and cost [60-61].

The function point measurement method was developed by Albrecht [62] which is an alternative to estimating Source Lines of Code (SLOC). Function Point Analysis (FPA) is the most commonly used approach for functional size measurement. It is promising to estimate them early in the life cycle such as: requirements definition which is a significant benefit to estimate the effort which is required for software development. There are two steps concerned in counting Function Points: 1) counting the user functions, and 2) adjusting for processing complexity. There are presently five user function categories such as: external input types, external output types, logical internal file types, external interface file types, and external inquiry types [63]. Today function point-based method called as Common Software Measurement International Consortium (COSMIC) [64]. There are various situational factors which affect the FPA such as: User functions, Processing complexity, and Environment [65].

There are number of approaches exist for cost estimation which is divided into three categories such as: i) algorithmic models, ii) expert judgement, and iii) analogy principal [66-68].

Kitchenham [69] divided algorithmic models into two categories such as: i) empirical factor models, and ii) constraint models. Empirical factor models provide an estimate value of a cost. The Constructive Cost Model (COCOMO) and COCOMO II are most accepted and used empirical factor models. COCOMO is used for estimating cost, effort, and schedule for software projects. COCOMO II is a successor of COCOMO which is better suitable for estimating recent software projects. It provides additional support for current software development process. The requirement for a new model came when software development technology shift from mainframe and batch processing to desktop development, code reusability, and use of off the shelf software components. There are various situational factors which affect the COCOMO such as: Number of lines of source code, Product complexity, Language, and Process maturity [70-71].

Constraint models show the relationship between the various cost factors such as effort, cost, and time. The Software Lifecycle Methodology (SLIM) is most commonly used constraint models. The SLIM method is a theory based technique to estimate software development effort and schedule. The SLIM method is based on Putman's life cycle analysis of the Rayleigh distribution. The Rayleigh distribution is used to estimate how many people should be allocated to the development and maintenance cost during the development process. The SLIM model uses two equations such as: software productivity level equation and manpower equation. The software productivity level equation defines development effort in terms of project size and development time. The manpower equation suggests the buildup of manpower as a function of time. There are various situational factors which affect the SLIM such as: Manpower Buildup Index (MBI), and Technology constant or Productivity Factor (PF) [72-73].

Expert judgement predicts cost based on the skill and experience of experts such as: Delphi Technique and Program (or project) Evaluation and Review Technique (PERT), and analogy predict cost using comparison of one or more completed projects with new project to estimate cost and time [74]. All these cost estimation models such as: SLIM, FPA, and COCOMO are used for cost estimation but there is emerging requirement for cost estimation using machine learning methods such as: Case Based Reasoning (CBR) and Artificial Neural Networks (ANN) based cost estimation model [75].

CBR means using old experience identify a solution of new problems. CBR have a spontaneous application for software effort estimation. It has the potential to model the expert estimation as well as illuminate the reasoning apply to adapt past cases. CBR is problem solving method proposed by Bisio and Malabocchia [76] which is based on the reuse of past experiences and knowledge by using analogy paradigm. It is promising to find past cases whether a predefined cost model can consistently estimate a new project or whether analogy can be chosen. There are various situational factors which affect the CBR such as: Complexity of the system, Experts experience or knowledge, Similarity check, and Risk factors [77-79].

ANN based estimation model proposed by Hakkarainen, Laamanen, and Rask [80] estimate the software size at specification level. A multilayer feed forward network is trained using back propagation algorithm. The training and testing data consist of randomly generated structured analysis descriptions as input data and corresponding algorithm based size metric values as output data. The size metrics used in the experiments are Albrecht's function points, Symons's Mark II Function Points, and DeMarco's Function Bang metric. ANN model estimate development effort using trained historical data to produce better results. ANNs have the capability to model complex non linear relationships and approximating several measurable functions. There are various situational factors which affect the ANN such as: Layout of neurons, Connections between network nodes, Learning speed, Historical project data inputs, Actual known values, and Delta value [81-83].

Situational factors that affect software development cost estimation are shown in Table 6.2. Table 6.2 shows the summary of situational factors that affect software development cost estimation.

Table 6.2 Situational Factors that Affect Software Development Cost Estimation

Software Development Cost Estimation	Situational Factors
FPA	User functions, Processing complexity, and Environment
COCOMO	Number of lines of source code, Product complexity, Language, and Process maturity
SLIM	Manpower Buildup Index (MBI), and Technology constant or Productivity Factor (PF)
CBR	Complexity of the system, Experts experience or knowledge, Similarity check, and Risk factors
ANN	Layout of neurons, Connections between network nodes, Learning speed, Historical project data inputs, Actual known values, and Delta value

As per requirement of software development various models or methods have been used for cost estimation as specified in Table 6.2 which is affected by various situational factors. FPA is used to calculate functional size measurement. COCOMO is used to estimate value of a cost. SLIM is used to estimate various cost factors. CBR is used estimate cost by reusing knowledge and experience. ANN is used to cost estimation by trained data. So ANN based cost estimation model is more important and effective model than other models.

6.4 SOFTWARE DEVELOPMENT RISK FACTORS

Software development is a large and complex process in which various risk involves in development of software such as: budget overrun, late delivery, and low quality product. To maintain balance relationship between cost, time, and quality; well and accurate risk management are used by software organization. By using that organizations prevent the occurrence of risks that might affect a project [84-85].

Numbers of research papers have been publish to identify risk in software development known as: i) Risk Framework, ii) Risk Assessment, iii) Risk Management, and iv) Risk Model [86-88].

Risk framework is used to identify the risk factors which are associated with projects. A risk categorization framework which is divided into four quadrants such as: Customer mandate, Scope and requirements, Execution, and Environment. There are

53 project risk factors which are mapped into the each quadrant to categorize the risk factor where top five risk factors are:

1. Lack of top management commitment to the project
2. Failure to gain user commitment
3. Misunderstanding the requirements
4. Lack of adequate user involvement
5. Failure to manage end user expectations

There are various situational factors which affect the risk framework such as: Lack of top management commitment, Lack of adequate user involvement, Execution risks, Undefined project success criteria, and Poor project planning [89-90].

Risk assessment tools such as: checklists or surveys are a simplest form to quickly identify and assess risk. Boehm describes top ten risk checklists which is necessary to use and assess several continuing challenges such as: Achieve commitment of all key stakeholders (developers, users, maintainers, and others) to a risk management approach. Establish an evolving knowledge base of risk management experience and expertise which is organized for easy and mutual use by all stakeholders. Define and circulate mature guidelines on when and how to avoid, check, transfer, manage and accept risk. Develop metrics and tools for reasoning about risk management's return on investment issues, guidelines for deciding how much of a risk reduction activity is enough. Tools that might be used in this way include specifying, risk focused prototyping, formal verification and validation, testing, configuration management, and quality assurance [91]. There are various situational factors which affect risk assessment such as: Unrealistic schedules and budgets, Wrong functions and properties, Corporate environmental risk, and Decision making problem [92-94].

Risk management used to manage risk which is related to development success and reduce the chance of project failure. Basically there are six components of software development risk such as: scheduling and timing risks, system functionality risks, subcontracting risks, requirement management risks, resource usage and performance risks, and personnel management risks. To manage risks through tailoring processes Octopus Situational Method Engineering (SME) Risk Management Approach (OSRiMA), implemented by using Analytic Hierarchy Process (AHP) technique. OSRiMA considers method fragments which are retrieved from method base according to the project risks and prioritized by to project. There are various situational factors which affect risk management such as: Environmental factors, Requirement management, Technology, and External dependencies [95-97].

Risk model developed by Wallace, Keil, and Rai [98] find out the weak area for correction and improve the process to recover from risk. They also suggest the six dimensions of risk:

1. Organizational environment risk
2. User risk
3. Requirements risk
4. Project complexity risk
5. Planning and control risk
6. Team risk

Main development risks related to developer turnover and requirement implementation which affect the software project. There are various situational factors which affect risk model such as: Project planning, Project complexity, Constraints of process, and Maintenance policies [99-100].

Software development cost overrun is another risk factor for software development. Appari and Benaroch [101] have developed a method for pricing the risk factors for software development in 2010. They also identified a list of software development risk factors and reduced the 13 risk factors into four macro level orthogonal risk factors such as:

1. Application Task (complexity, reliability, database size, storage, and execution time)
2. Process Maturity (modern development practices, and use of software development tools)
3. Personnel Capability (platform volatility, platform experience, language & tools experience)
4. Technology Platform (analyst capability, programmer capability, and application experience)

There are various situational factors which affect software development cost such as: Extra cost incurred per unit exposure, Project sensitivity, and Variability in project cost [102].

Situational factors that affect software development risk are shown in Table 6.3. Table 6.3 shows the summary of situational factors that affect software development risk.

Table 6.3 Situational Factors that Affect Software Development Risk

Software Development Risk	**Situational Factors**
Risk Framework	Lack of top management commitment, Lack of adequate user involvement, Execution risks, Undefined project success criteria, and Poor project planning
Risk Assessment	Unrealistic schedules and budgets, Wrong functions and properties, Corporate environmental risk, and Decision making problem
Risk Management	Environmental factors, Requirement management, Technology, and External dependencies
Risk Model	Project planning, Project complexity, Constraints of process, and Maintenance policies
Software Development Cost	Extra cost incurred per unit exposure, Project sensitivity, and Variability in project cost

There are various risk factors for software development specified in Table 6.3 which is affected by various situational factors. Risk framework is used to identify and categorize the risk factors. Risk assessment is used to assess the risk factors. Risk management is used to manage the risk involve in software development and risk model is used for correction in process. Software development cost is used to develop a software product. So risk management is most important than other because using effective and efficient risk management we can minimize the risk which is associated with software projects.

6.5 SOFTWARE DEVELOPMENT SUCCESS FACTORS

Software development success based on the development of good software consistently and this goodness is multidimensional which is divided into user and developer perspective [103]. The success of any project is based on the Iron Triangle such as cost, time, and quality. The criteria for success are: The delivery stage (process) doing it right and Post delivery stage (system) getting it right [104] which is now shifted as: quality, scope, time, and cost; and there are main five dimension for success: i) People, ii) Process, iii) Technology, iv) Project, and v) Organization [105].

For successful software development user and developer play an important role because user provides an idea and image of software product which is developed by developer. This chapter considers successful software development on developer point of view only.

Software development is a complex activity that required a technical and managerial tool to manage the software project successfully. There are five main essential factors which affect the successful software project: i) Start on the right foot means plan properly, ii) Maintain momentum used to maintain or rebuild team effort, iii) Track progress using a conceptual model a blueprint buildup that people can see, iv) Make smart decisions means make good decisions to fulfill user requirements, and v) Institutionalize post-mortem analyses used to learn from their past mistakes [106].

Project manager is attentive to control working of project and plan a strategy for project success. There are technical and non-technical factors which affect project success. There are three types of critical success factors (CSFs) of software project such as: People (7 CSFs), Process (16 CSFs), and Technology (3 CSFs).

List of CSFs for people:

1. Effective project management skills/methodologies (project manager)
2. Support from top management
3. User/client involvement
4. Skilled and sufficient staffs
5. Good leadership
6. Committed and motivated team
7. Good performance by vendors/contractors/ consultants

List of CSFs for process:

1. Clear requirements and specifications
2. Clear objective/goal/scope
3. Realistic schedule
4. Effective communication and feedback
5. Realistic budget
6. Frozen requirement
7. Proper planning
8. Appropriate development processes/methodologies (process)
9. Up-to-date progress reporting
10. Effective monitoring and control
11. Adequate resources
12. Risk management
13. Effective change and configuration management
14. Good quality management
15. Clear assignment of roles and responsibilities
16. End-user training provision

List of CSFs for technology:

1. Familiar with technology/development methodology
2. Complexity, project size, duration, number of organizations involved
3. Supporting tools and good infrastructure

There are various situational factors which affect project management such as: Clear requirements and objectives, Effective project management skills, User involvement, Proper planning, and Good quality management [107].

Effective risk management used to implement a successful project by using time, cost, and effort. The success of any project can be measure by resource utilization. The reason of the projects failure can be directly related to the area of risk management undertaken. Further the point of risk management process undertaken during a project impacts directly on the success of the project. Effective risk management should be always undertaken during the project lifecycle to increase project success. There are various situational factors which affect risk management for project success such as: Effective and efficient utilization of resources, Definition of clear goals, Project team performance, and Acceptance by the customer [108].

Existing models of CSFs of software projects have less concentration on communication, team, and product related factors. A conceptual model of CSFs for software projects categorize the success factors as:

List of CSFs for communication factors:

1. Communication in project
2. Leadership
3. Relationship between users and IS staff
4. Reduce ambiguity
5. Maximize stability

List of CSFs for technical factors:

1. Technical tasks
2. Trouble shooting
3. Technical uncertainty
4. Technical implementation problems
5. Integration of the system

List of CSFs for organizational factors:

1. Top management support
2. Realistic expectations
3. Organizational politics
4. Financial support
5. Power

List of CSFs for environmental factors:

1. User involvement
2. Customer involvement
3. Vendor partnership
4. External environment events
5. Client acceptance

List of CSFs for Product factors:

1. Accuracy of output
2. Reliability of output
3. Timeliness of output
4. Quality control
5. Documentation of systems and procedures

List of CSFs for team factors:

1. Team capability/competence
2. Teamwork
3. Select right project team
4. Project team coordination
5. Task orientation

List of CSFs for project management factors:

1. Project planning
2. Project control mechanisms
3. Project schedule
4. Project manager's competence
5. Clear project goal

There are various situational factors which affect conceptual model such as: Communication skills, Ability to manage people well, Company morale, Business process reengineering, Team capability, and Leadership [109].

A contingency fit model is developed by Ahimbisibwe, Cavana, and Daellenbach [110] compare traditional plan based and agile methodologies. They identify good decision making lead to project success and identify or categorize CSFs. There are total 37 CSFs for software development success such as:

1. Top-level management support
2. User/client participation
3. Project team commitment
4. Organizational culture
5. Level of project planning
6. Leadership characteristics
7. Vision and mission
8. Monitoring and controlling
9. Change management skills
10. Internal project communication
11. User support
12. Technological uncertainty
13. Development processes/methodologies
14. Technical complexity
15. Project team empowerment
16. Project team's composition
17. Customer training and education
18. Customer (client) experience
19. Project team's expertise with the task
20. Project team's general expertise
21. Lack of development team skill
22. Urgency/duration

23. Relative project size
24. Specification/requirement changes
25. Project team's experience with SDM
26. Project criticality
27. Lack of end user experience
28. Requirements and specifications
29. Good performance by vendors/contractors
30. Supporting tools and good infrastructure
31. Realistic schedule
32. Adequate resources
33. Risk management
34. Realistic budget
35. Good quality management
36. Up-to-date progress reporting
37. Clear assignment of roles and responsibilities

There are various situational factors which affect contingency fit model such as: Organizational culture, Vision and mission, Relative project size, Realistic schedule, Up-to-date progress reporting, and clear assignment of roles and responsibilities.

After product development project success is analyzed. Main indicator for any project success is: Project and its environment which affect productivity.

Factors in the project:

1. Co-location of the project team members
2. Customer collaboration, commitment and involvement
3. Experience level, personnel skills, and team maturity
4. Procedural empowerment of the project team
5. Project cost
6. Project criticality
7. Project duration
8. Project size (man hours)
9. Requirements uncertainty/requirements stability
10. Proportion of the organization affected
11. Team size
12. Technological uncertainty
13. Urgency

Factors in the project's environment:

1. Compliance and governance factors
2. Culture (entrepreneurship, risk taking willingness)
3. Instability of organizational environment
4. Nature of the contract
5. Power distance, national culture
6. Top management support for one approach
7. Training

There are various situational factors which affect project success analyses such as: Time, Budget, Functionality, Quality, Need, Product, User satisfaction, and Team satisfaction. To achieve project success a balance relationship established and implemented in software projects [111].

New product success (NPS) holds importance for every company. Over the past decades, Meta analytical method has been used to summarize empirical findings on NPS factors. There are lots of factors such as: Product, Strategy, Process, Marketplace, and Organizational characteristics [112].

Situational factors that affect software development success are shown in Table 6.4. Table 6.4 shows the summary of situational factors that affect software development success.

Table 6.4 Situational Factors that Affect Software Development Success

Software Development Success	Situational Factors
Project Management	Clear requirements and objectives, Effective project management skills, User involvement, Proper planning, and Good quality management
Risk Management	Effective and efficient utilization of resources, Definition of clear goals, Project team performance, and Acceptance by the customer
Models Conceptual Model	Communication skills, Ability to manage people well, Company morale, Business process reengineering, Team capability, and Leadership
Contingency Fit Model	Organizational culture, Vision and mission, Relative project size, Realistic schedule, Up-to-date progress reporting, and Clear assignment of roles and responsibilities
Project Success Analyses	Time, Budget, Functionality, Quality, Need, Product, User satisfaction, and Team satisfaction

There are various success factors for software development specified in Table 6.4 which is affected by various situational factors. Project management is used to manage the project successfully. Risk management suggests how to utilize the resources to perform effectively. A model (conceptual and contingency fit model) is used to find out the CSFs and shows the importance of these factors for project success. A project success analyses is used to analyze project success. So project management is most important than other because using accurate project management we can develop a successful software product.

6.6 SOFTWARE DEVELOPMENT ENVIRONMENT FACTORS

To be successful in software industry and market every software organizations must be managed, organized, and executed to perform effectively, respond immediately in dynamic environment [113]. For desired project outcome four dimensions are required such as: i) development process, ii) people and action, iii) project content, and iv) institutional context [114]. So, software development environment is important to develop software products.

In 1996, Jones and Kavanagh [115], examined three situational variables: i) quality of the work experience, ii) peer influences, and iii) managerial influences, and three individual variables: i) locus of control, ii) Machiavellianism, and iii) gender. Organizational environment (professional bureaucracy) when combined with developers and tools than create or build a correct software products. Individuals and organization both impact on software product because both are involve to development of software. There are various situational factors which affect individuals and organizations such as: Behavioral intentions, Social desirability, Organization structure and learning, Process maturity, Risk management, Entrepreneurial intentions (job satisfaction and self-efficacy), Innovative orientation and climate, and Technical excellence incentives [116-117].

For software development team is assigned to do their job positively. There are various internal boundaries (geographic, time zones, languages, and culture) to get the job done. Knowledge sharing is an important key component for successful team work and project success. There are various situational factors which affect development team such as: Teamwork, Distance and time separation, Coordination, Task programming, Communication, Team knowledge, Team cognition, Shared task knowledge, Team awareness, Team diversity, Team size, Team climate, Team innovation, Team leader behavior, and Top management support [118-122].

To survive in competitive and changing environment every organization adopts an up to date technological changes to maintain a well recognized credit in the market. Dynamic capabilities and strategic management provide a valuable path to maintain that. There are various situational factors which affect technology and development culture such as: Technological skills, Technology transfer, Dynamic capabilities, System support, System service support, Technical competences, Information and communication technology (ICT), Self-service technologies (SSTs), and Technology readiness dimensions [123-126].

Situational factors that affect software development environment are shown in Table 6.5. Table 6.5 shows the summary of situational factors that affect software development environment.

Table 6.5 Situational Factors that Affect Software Development Environment

Software Development Environment	Situational Factors
Individuals and Organizations	Behavioral intentions, Social desirability, Organization structure and learning, Process maturity, Risk management, Entrepreneurial intentions (job satisfaction and self-efficacy), Innovative orientation and climate, and Technical excellence incentives
Development Team	Teamwork, Distance and time separation, Coordination, Task programming, Communication, Team knowledge, Team cognition, Shared task knowledge, Team awareness, Team diversity, Team size, Team climate, Team innovation, Team leader behavior, and Top management support
Technology and Development Culture	Technological skills, Technology transfer, Dynamic capabilities, System support, System service support, Technical competences, ICT, SSTs, and Technology readiness dimensions

There are various environmental factors for software development specified in Table 6.5 which is affected by various situational factors. Individuals and organizations provide the social and organizational factor that affect software development. Development team is assign to develop the software products. By using technology and development culture software is develop. So development team is most important than other because using right people we develop a right product.

6.7 SUMMARY

During development process there are various managerial and technical issues. Software process is used to solve these issues and develop a quality product. There are various situational factors which affect software process. This chapter identified the situational factors based on literature review and analyzed five main important situational factors.

First situational factor consider as software development models and standards provide a way to develop a quality product with the help of CMM, ISO/IEC 12207, ISO/IEC 15504, and CMMI. CMMI is more important and effective model than other models because CMMI is used process maturity and improvement both. Second situational factor is software development cost estimation which is used to estimate the total cost of development. To predict and estimate cost FPA, COCOMO, SLIM, CBR, and ANN based cost estimation model is used. ANN based cost estimation model is more useful model than other models because ANN uses trained data for cost estimation.

Third situational factor consider as software development risk factors to identify risk in software development by using risk framework, risk assessment, risk management, risk model, and software development cost. Risk management is more important and helpful than other because using effective and efficient risk management we can minimize the risk which is associated with software projects. Fourth situational factor is software development success factors to find out important success factors in software development using project management, risk management, models, and project success analyses. Project management is more important than other for successful development because using accurate project management we can develop a successful software product. Fifth situational factor is software development environmental factors to identify affect of environment on software process by using individuals and organizations, development team, and technology and development culture. Development team is more important than other factors to produce quality product because using right people we develop a right product.

So all five factors are important to consider at the time of development but risk management is more important factor to consider because to develop a quality product risk management is useful. Risk management manages resource effectively, minimize risk, and develop quality product within time and cost.

6.8 REFERENCES

1. Boehm, B. W. (2006). A view of 20th and 21st century software engineering. In Proceedings of the ACM 28th International Conference on Software Engineering, New York, pp. 12-29.
2. Royce, W. W. (1970). Managing the development of large software systems: Concepts and techniques. In Proceedings of Western Electric Show and Convention Technical Papers, IEEE WESCON, Los Angeles, CA, USA, Vol. 26, No. 8, pp. 328-338.
3. Basili V. R.; Turner, A. J. (1975). Iterative enhancement: a practical technique for software development. IEEE Trans. on Software Engineering, Vol. 1, Issue 4, pp. 390-396.
4. Boehm, B. W.; Gray, T. E.; Seewaldt, T. (1984). Prototyping versus Specifying: A Multiproject Experiment. IEEE Trans. on Software Engineering, Vol. 10, Issue 3, pp. 290-303.
5. Lehman, M. M.; Belady, L. A. (1985). *Program Evolution: Processes of Software Change*. Academic Press, London.
6. Boehm, B. W. (1988). A Spiral Model for Software Development and Enhancement. IEEE Computer, Vol. 21, Issue 5, pp.61-72.

7. Bucanac, C. (1999). The V-Model. University of Karlskrona/Ronneby.
8. Cockburn, A. (2001). *Agile Software Development*. Addison-Wesley.
9. Singh, B.; Gautam, S. (2016). Hybrid Spiral Model to Improve Software Quality Using Knowledge Management. International Journal of Performability Engineering, Vol. 12, No. 4, pp. 341-352.
10. Humphrey, W. S. (1989). *Managing the Software Process*. Addison-Wesley.
11. Feiler, P. H.; Humphrey, W. S. (1992). Software Process Development and Enactment: Concepts and Definitions. Technical Report CMU/SEI-92-TR-04.
12. Humphrey, W. S. (1987). Characterizing the Software Process: A Maturity Framework. Technical Report CMU/SEI-87-TR-11.
13. Fuggetta, A. (2000). Software Process: A Roadmap. In Proceedings of the ACM Conference on the Future of Software Engineering, ACM, pp. 25-34.
14. Fuggetta, A. (2014). Software Process. In Proceedings of the ACM Conference on the Future of Software Engineering, FOSE'14, Hyderabad, India, pp. 1-12.
15. Bekkers, W.; van de Weerd, I.; Brinkkemper, S.; Mahieu, A. (2008). The Influence of Situational Factors in Software Product Management: An Empirical Study. In Second IEEE International Workshop on Software Product Management, IWSPM'08, pp. 41-48.
16. Harter, D. E.; Krishnan, M. S.; Slaughter, S. A. (2000). Effects of Process Maturity on Quality, Cycle Time, and Effort in Software Product Development. Management Science, Vol. 46, No. 4, Information Technology Industry, pp. 451-466.
17. Liu, J. Y.-C.; Chen, V. J.; Chan, C.-L.; Lie, T. (2008). The impact of software process standardization on software flexibility and project management performance: Control theory perspective. Information and Software Technology, Vol. 50, Issue 9, pp. 889-896.
18. Trendowicz, A.; Münch, J. (2009). Factors Influencing Software Development Productivity- State of the Art and Industrial Experiences. Advances in Computers, Vol. 77, pp. 185-241.
19. dos Santos, R. P.; de Oliveira, K. M.; da Silva, W. P. (2009). Evaluating the service quality of software providers appraised in CMM/CMMI. Software Qual J, Vol. 17, Issue 3, pp. 283-301.
20. Jack, R.; Mannion, M. (1995). Improving the software cost estimation process. WIT Transactions on Information and Communications Technologies, Vol. 11, pp. 245-256.
21. Jørgensen, M.; Shepperd, M. (2007). A Systematic Review of Software Development Cost Estimation Studies. IEEE Transactions on Software Engineering, Vol. 33, No. 1, pp. 33-53.
22. Muzaffar, Z.; Ahmed, M. A. (2010). Software development effort prediction: A study on the factors impacting the accuracy of fuzzy logic systems. Information and Software Technology, Vol. 52, Issue 1, pp. 92-109.
23. Wallshein, C. C.; Loerch, A. G. (2015). Software cost estimating for CMMI Level 5 developers. The Journal of Systems and Software, Vol. 105, pp. 72-78.
24. Whigham, P. A.; Owen, C. A.; MacDonell, S. G. (2015). A baseline model for software effort estimation. ACM Trans. Softw. Eng. Methodol, Vol. 24, Issue 3, Article 20, pp. 20:1-20:11.
25. Barki, H.; Rivard, S.; Talbot, J. (1993). Toward an Assessment of Software Development Risk. Journal of Management Information Systems, Vol. 10, No. 2, pp. 203-225.
26. Lyytinen, K.; Mathiassen, L.; Ropponen, J. (1996). A Framework for Software Risk Management. Journal of Information Systems, Vol. 8, Issue. 1, pp. 53-68.
27. Wallace, L.; Keil, M.; Rai, A. (2004). Understanding software project risk: a cluster analysis. Information & Management, Vol. 42, Issue 1, pp. 115–125.
28. Bannerman, P. L. (2008). Risk and risk management in software projects: A reassessment. The Journal of Systems and Software, Vol. 81, Issue 12, pp. 2118-2133.
29. Nakatsu, R. T.; Iacovou, C. L. (2009). A comparative study of important risk factors involved in offshore and domestic outsourcing of software development projects: A two-panel Delphi study. Information & Management, Vol. 46, Issue 1, pp. 57-68.
30. Jain, R.; Dey, S. (2004). A life-cycle taxonomy for assessing software development risks. In Conference on Systems Engineering Research, University of Southern California.

31. Baccarini, D. (1999). The Logical Framework Method for Defining Project Success. Project Management Journal, Vol. 30, No. 4, pp. 25-32.
32. Dybå, T. (2000). An Instrument for Measuring the Key Factors of Success in Software Process Improvement. Empirical Software Engineering, Vol. 5, Issue 4, pp. 357-390.
33. Savolainen, P.; Ahonen, J. J.; Richardson, I. (2012). Software development project success and failure from the supplier's perspective: A systematic literature review. International Journal of Project Management, Vol. 30, Issue 4, pp. 458-469.
34. Khan, S. U.; Niazi, M.; Ahmad, R. (2012). Empirical investigation of success factors for offshore software development outsourcing vendors. IET Software, Vol. 6, Issue 1, pp. 1-15.
35. Khan, A. W.; Khan, S. U. (2013). Critical success factors for offshore software outsourcing contract management from vendors' perspective: an exploratory study using a systematic literature review. IET Software, Vol. 7, Issue. 6, pp. 327-338.
36. Lord, R. G.; Smith, J. E. (1983). Theoretical, Information Processing, and Situational Factors Affecting Attribution Theory Models of Organizational Behavior. The Academy of Management Review, Vol. 8, No. 1, pp. 50-60.
37. Oldham, G. R.; Kulik, C. T.; Stepina, L. P.; Ambrose, M. L. (1986). Relations between Situational Factors and the Comparative Referents Used by Employees. The Academy of Management Journal, Vol. 29, No. 3, pp. 599-608.
38. Hawk, S. R.; Santos, B. L. D. (1991). Successful System Development: The Effect of Situational Factors on Alternate User Roles. IEEE Transactions on Engineering Management, Vol. 38, No. 4, pp. 316-327.
39. Harrison, A. W.; Rainer, R. K. (1992). The Influence of Individual Differences on Skill in End-User Computing. Journal of Management Information Systems, Vol. 9, No. 1, pp.93-111.
40. Mathieu, J. E.; Heffner, T. S.; Goodwin, G. F.; Salas, E.; Cannon-Bowers, J. A. (2000). The Influence of Shared Mental Models on Team Process and Performance. Journal of Applied Psychology, Vol. 85, No. 2, pp. 273-283.
41. Stelzer, D.; Mellis, W. (1998). Success Factors of Organizational Change in Software Process Improvement. Software Process Improvement and Practice, Vol. 4, Issue 4, pp. 227-250.
42. de Waal, A. (2003). Behavioral Factors Important for the Successful Implementation and Use of Performance Management Systems. Management Decision, Vol. 41, Issue 8, pp. 688-697.
43. Brodbeck, F. C. (2001). Communication and performance in software development projects, European Journal of Work and Organizational Psychology. Vol. 10, Issue 1, pp. 73-94.
44. Clarke, P.; et al. (2016). An Investigation of Software Development Process Terminology. In International Conference on Software Process Improvement and Capability Determination, SPICE 2016, Springer International Publishing, pp. 351-361.
45. Wang, Y. (2008). *Software Engineering Foundations a Software Science Perspective.* Auerbach Publications, Taylor & Francis Group.
46. Goodman, F. A. (2006). *Defining and deploying software processes.* Auerbach Publications, Taylor & Francis Group.
47. Paulk, M. C.; Curtis, B.; Chrissis, M. B.; Weber, C. V. (1993). Capability Maturity Model[SM] for Software. Version 1.1, Technical Report, CMU/SEI-93-TR-024.
48. Paulk, M. C.; Curtis, B.; Chrissis, M. B.; Weber, C. V. (1993). Capability Maturity Model for Software. Encyclopedia of Software Engineering, pp. 1-26.
49. Raynus, J. (1999). *Software Process Improvement with CMM.* Artech House, Inc.
50. ISO/IEC 12207:1995, Information Technology - Software life cycle processes, 1995.
51. Singh, R. (1996). International Standard ISO/IEC 12207 Software Life Cycle Processes. Software Process Improvement and Practice, Vol. 2, Issue 1, pp. 35-50.
52. Ramasubbu, N.; Balan, R. K. (2009). The Impact of Process Choice in High Maturity Environments: An Empirical Analysis. In Proceedings of the 31st International Conference on Software Engineering, pp. 529-539.

53. ISO/IEC 15504:1998, Information Technology - Software Process Assessment, Part 1, 1998.
54. Rout, T. P. (2003). ISO/IEC 15504 - Evolution to an International Standard, Software Process: Improvement and Practice. Vol. 8, Issue 1, pp. 27-40.
55. Pyhäjärvi, M. (2004). SPICE - International Standard for Software Process Assessment. Seminar on Quality Models for Software Engineering, Helsinki.
56. Rout, T. P.; El Emam, K.; Fusani, M.; Goldenson, D.; Jung, H.-W. (2007). SPICE in retrospect: Developing a standard for process assessment. The Journal of Systems and Software, Vol. 80, Issue 9, pp. 1483-1493.
57. CMMI Product Team, Capability Maturity Model® Integration (CMMISM). Version 1.1, CMU/SEI-2002-TR-029, Aug 2002.
58. Clarke, P.; O'Connor, R. V.; Leavy, B.; Yilmaz, M. (2015). Exploring the Relationship between Software Process Adaptive Capability and Organisational Performance. IEEE Transactions on Software Engineering, Vol. 41, Issue 12, pp. 1169-1183.
59. Chrissis, M. B.; Konrad, M.; Shrum, S. (2011). CMMI® for Development, Guidelines for Process Integration and Product Improvement. Third Edition, Pearson Education, Inc.
60. Trendowicz, A.; Jeffery, R. (2014). Software Project Effort Estimation Foundations and Best Practice Guidelines for Success. Springer International Publishing.
61. Trendowicz, A. (2013). Software Cost Estimation, Benchmarking, and Risk Assessment The Software Decision-Makers' Guide to Predictable Software Development. Springer-Verlag Berlin Heidelberg.
62. Albrecht, A. J. (1979). Measuring Application Development Productivity, In Proceedings of the Joint SHARE/GUIDE/IBM Application Development Symposium. Vol. 10, pp. 83-92.
63. Kemerer, C. F. (1987). An Empirical Valuation of Software Cost Estimation Models. Communications of the ACM, Vol 30, No. 5, pp. 416-429.
64. Lagerstrom, R.; von Wurtemberg, L. M.; Holm, H.; Luczak, O. (2012). Identifying factors affecting software development cost and productivity. Software Qual J, Vol. 20, Issue 2, pp. 395-417.
65. Low, G. C.; Jeffery, D. R. (1990). Function Points in the Estimation and Evaluation of the Software Process. IEEE Transactions on Software Engineering, Vol. 16, No. I, pp. 64-71.
66. Jørgensen, M. (2007). Forecasting of software development work effort: Evidence on expert judgement and formal models. International Journal of Forecasting, Vol. 23, Issue 3, pp. 449-462.
67. Boehm, B. W.; Abts, C.; Chulani, S. (2000). Software development cost estimation approaches - A survey. Annals of Software Engineering, Vol. 10, Issue (1-4), pp. 177-205.
68. Lagerstrom, R.; von Wurtemberg, L. M.; Holm, H.; Luczak, O. (2010). Identifying Factors Affecting Software Development Cost. In the Fourth International Workshop on Software Quality and Maintainability (SQM), Madrid, Spain.
69. Kitchenham, B. (1997). Making process predictions, in: N. E. Fenton (Ed.), Software Metrics: A Rigorous Approach. PWS Publishing Company.
70. Boehm, B. W. (1981). Software Engineering Economics. Prentice Hall PTR, Upper Saddle River, NJ, USA.
71. Boehm, B. W.; Clark, B.; Horowitz, E.; Westland, C.; Madachy, R.; Selby, R. (1995). Cost models for future software life cycle processes: COCOMO 2.0. Annals of Software Engineering, Vol. 1, Issue 1, pp. 57-94.
72. Pendharkar, P. C.; Subramanian, G. H.; Rodger, J. A. (2005). A Probabilistic Model for Predicting Software Development Effort. IEEE Transactions on Software Engineering, Vol. 31, No. 7, pp. 615-624.
73. Nassif, A. B.; Ho, D.; Capretz, L. F. (2013). Towards an early software estimation using log-linear regression and a multilayer perceptron model. The Journal of Systems and Software, Vol. 86, Issue 1, pp. 144- 160.
74. Leung, H.; Fan, Z. (2002). Software Cost Estimation, Handbook of Software Engineering. Hong Kong Polytechnic University, pp. 1-14.

75. Clarke, P.; O'Connor, R. V. (2012). The situational factors that affect the software development process: Towards a comprehensive reference framework. Information and Software Technology, Vol. 54, Issue 5, pp. 433-447.
76. Bisio, R.; Malabocchia, F. (1995). Software Projects through Case Base Reasoning. In Proceedings of First International Conference Case Base Reasoning, ICCBR-95 Sesimbra, Portugal, Springer, pp. 11-22.
77. Finnie, G. R.; Wittig, G. E.; Desharnais, J-M. (1997). Software Development Effort with Case-Based Reasoning. In International Conference Case Base Reasoning, Springer Berlin Heidelberg, pp. 13-22.
78. An, S.-H.; Kim, G.-H.; Kang, K.-I. (2007). A case-based reasoning cost estimating model using experience by analytic hierarchy process. Building and Environment, Vol. 42, Issue 7, pp. 2573-2579.
79. Delany, S. J.; Cunningham, P. (2000). The Application of Case-Based Reasoning to Early Software Project Cost Estimation and Risk Assessment. Department of Computer Science, Trinity College Dublin, Dublin, Ireland.
80. Hakkarainen, J.; Laamanen, P.; Rask, R. (1993). Neural Networks in Specification Level Software Size Estimation. IEEE Proceeding of the Twenty-Sixth Hawaii International Conference on System Sciences, Vol. 4, pp. 626-634.
81. Finnie, G.; Wittig, G. E. (1996). AI Tools for Software Development Effort Estimation. IEEE Proceedings of 1996 International Conference on Software Engineering: Education and Practice (SE: E&P'96) pp. 336-353.
82. Wen, J.; Li, S.; Lin, Z.; Hu, Y.; Huang, C. (2012). Systematic literature review of machine learning based software development effort estimation models. Information and Software Technology, Vol. 54, Issue 1, pp. 41-59.
83. Nassif, A. B.; Capretz, L. F.; Ho, D. (2012). Software Effort Estimation in the Early Stages of the Software Life Cycle Using a Cascade Correlation Neural Network Model. IEEE 13th ACIS International Conference on Software Engineering, Artificial Intelligence, Networking and Parallel/Distributed Computing, SNPD'12, pp. 589-594.
84. McManus, J. (2004). Risk Management in Software Development Projects. Elsevier.
85. Pandian, C. R. (2007). *Applied software risk management: a guide for software project managers*. Auerbach Publications, Taylor & Francis Group.
86. Boehm, B. W.; DeMarco, T. (1997). Software Risk Management. IEEE Softw., Vol. 14, No. 3, pp. 17-19.
87. Arnuphaptrairong, T. (2011). Top Ten Lists of Software Project Risks: Evidence from the Literature Survey. In proceeding of the International MultiConference of Engineers and Computer Scientists, IMECS 2011, Hong Kong, Vol. 1, pp. 1-6.
88. Weiler, M.; Engel, A. (2012). Risk Management Processes - A Quantitative Model for Fusing Cost and Time of Risks and Risk Response Actions. Bernard M. Gordon Center (BMGC), The Technion - Israel Institute of Technology, pp. 1-36.
89. Keil, M.; Cule, P. E.; Lyytinen, K.; Schmidt, R. (1998). A framework for identifying software project risks. Commun. of the ACM, Vol. 41, No. 11, pp. 76-83.
90. Wallace, L.; Keil, M. (2004). Software Project Risks and Their Effect on Outcomes. Communications of the ACM, Vol. 47, No. 4, pp. 68-73.
91. Boehm, B. W. (1991). Software Risk Management: Principles and Practices. IEEE Softw., Vol. 8, No. 1, pp. 32-41.
92. Keil, M.; Wallace, L.; Turk, D.; Dixon-Randall, G.; Nulden, U. (2000). An investigation of risk perception and risk propensity on the decision to continue a software development project. The Journal of Systems and Software, Vol. 53, Issue 2, pp. 145-157.
93. Keil, M.; Li, L.; Mathiassen, L.; Zheng, G. (2006). The Influence of Checklists and Roles on Software Practitioner Risk Perception and Decision-Making. IEEE Proceedings of the 39th Hawaii International Conference on System Sciences - 2006, pp. 1-12.
94. Li, L. (2013). The Impact of Risk Checklists on Project Manager's Risk Perception and Decision-Making Process. Proceedings of the Southern Association for Information Systems Conference, Savannah, GA, USA, pp. 106-110.

95. Ropponen, J.; Lyytinen, K. (2000). Components of Software Development Risk: How to Address Them? A Project Manager Survey, IEEE Transactions on Software Engineering, Vol. 26, No. 2, pp. 98-112.
96. Schmidt, R.; Lyytinen, K.; Keil, M.; Cule, P. (2001). Identifying Software Project Risks: An International Delphi Study. Journal of Management Information Systems, Vol. 17, No. 4, pp. 5-36.
97. Pereira, G. V.; Severo, F.; Fontoura, L. (2012). A Risk Management Approach Based on Situational Method Engineering. In CIbSE 2012, Springer-Verlag Berlin Heidelberg, pp. 230-235.
98. Wallace, L.; Keil, M.; Rai, A. (2004). How software project risk affects project performance: an investigation of the dimensions of risk and an exploratory model. Decision Sciences, Vol. 35, Issue 2, pp. 289-321.
99. Li, J.; Li, M.; Wua, D.; Song, H. (2012). An integrated risk measurement and optimization model for trustworthy software process management. Information Sciences, Vol. 191, pp. 47-60.
100. Melo, A.; Tavares, E. A. G.; Sousa, E.; Nogueira, B. C. e S.; Marinho, M. (2015). Dependability approach for evaluating software development risks. IET Software, Vol. 9, Issue 1, pp. 17-27.
101. Appari, A.; Benaroch, M. (2010). Monetary pricing of software development risks: a method and empirical illustration. Journal of Systems and Software, Vol. 83, Issue 11, pp. 2098-2107.
102. Benaroch, M.; Appari, A. (2010). Financial pricing of software development risk factors. IEEE Software, Vol. 27, Issue 5, pp. 65-73.
103. Donaldson, S. E.; Siegel, S. G. (2001). *Successful Software Development*. Second Edition, Prentice Hall.
104. Atkinson, R. (1999). Project management: cost, time and quality, two best guesses and a phenomenon, its time to accept other success criteria. International Journal of Project Management, Vol. 17, No. 6, pp. 337-342.
105. Chow, T.; Cao, D-B (2008). A survey study of critical success factors in agile software projects. The Journal of Systems and Software, Vol. 81, Issue 6, pp. 961-971.
106. Reel, J. S. (1999). Critical Success Factors in Software Projects. IEEE Software, Vol. 6, Issue 10, pp. 18-23.
107. Nasir, M. H. N.; Sahibuddin, S. (2011). Critical success factors for software projects: A comparative study. Scientific Research and Essays, Vol. 6, Issue 10, pp. 2174-2186.
108. Kishk, M.; Ukaga, C. (2008). The impact of effective risk management on project success. In: Dainty, A (Ed.) Procs 24th Annual ARCOM Conference, Cardiff, UK, Association of Researchers in Construction Management, pp. 799-808.
109. Sudhakar, G. P. (2012). A model of critical success factors for software projects. Journal of Enterprise Information Management, Vol. 25, Issue 6, pp. 537-558.
110. Ahimbisibwe, A.; Cavana, R. Y.; Daellenbach, U. (2015). A contingency fit model of critical success factors for software development projects. Journal of Enterprise Information Management, Vol. 28, Issue 1, pp. 7-33.
111. Sheffield, J.; Lemétayer, J. (2013). Factors associated with the software development agility of successful projects. International Journal of Project Management, Vol. 31, Issue 3, pp. 459-472.
112. Evanschitzky, H.; Eisend, M.; Calantone, R. J.; Jiang, Y. (2012). Success Factors of Product Innovation: An Updated Meta-Analysis. J. Prod. Innov. Manag, Vol. 29, Issue S1, pp. 21-37.
113. Dede, B.; Lioufko, I. (2010). Situational Factors Affecting Software Development Process Selection. Master of Science, University of Gothenburg.
114. McLeod, L.; MacDonell, S. G. (2011). Factors that affect software systems development project outcomes: A survey of research. ACM Computing Surveys (CSUR), Vol. 43, Issue 4, pp. 24-76.

115. Jones, G. E.; Kavanagh, M. J. (1996). An Experimental Examination of the Effects of Individual and Situational Factors on Unethical Behavioral Intentions in the Workplace. Journal of Business Ethics, Vol. 15, No. 5, pp. 511-523.
116. Baker, E. W. (2011). Why situational method engineering is useful to information systems development. Info Systems J, Vol. 21, Issue 2, pp. 155-174.
117. Lee, L.; Wong, P. K.; Foo, M. D.; Leung, A. (2011). Entrepreneurial intentions: The influence of organizational and individual factors. Journal of Business Venturing, Vol. 26, Issue 1, pp. 124-136.
118. Nambisan, S.; Wilemon, D. (2000). Software Development and New Product Development: Potentials for Cross-Domain Knowledge Sharing. IEEE Transactions on Engineering Management, Vol. 47, No. 2, pp. 211-220.
119. Espinosa, J. A.; Pickering, C. (2006). The Effect of Time Separation on Coordination Processes and Outcomes: A Case Study. IEEE In proceedings of the 39th Hawaii International Conference on System Sciences (HICSS'06), Vol. 1.
120. Espinosa, J. A.; Slaughter, S. A.; Kraut, R. E.; Herbsleb, J. D. (2007). Team Knowledge and Coordination in Geographically Distributed Software Development. Journal of Management Information Systems, Vol. 24, No. 1, pp.135-169.
121. He, J.; Butler, B. S.; King, W. R. (2007). Team Cognition: Development and Evolution in Software Project Teams. Journal of Management Information Systems, Vol. 24, No. 2, pp. 261-292.
122. Lalsing, V.; Kishnah, S.; Pudaruth, S. (2012). People Factors in Agile Software Development and Project Management. International Journal of Software Engineering & Applications (IJSEA), Vol. 3, No. 1.
123. Teece, D. J.; Pisano, G.; Shuen, A. (1997). Dynamic Capabilities and Strategic Management. Strategic Management Journal, Vol. 18, No. 7, pp. 509-533.
124. Cho, V.; Cheng, T.C.E.; Hung, H. (2009). Continued usage of technology versus situational factors: An empirical analysis. Journal of Engineering and Technology Management, Vol. 26, Issue 4, pp. 264-284.
125. C.-Lumbreras, C.; C.-Palacios, R.; S.-Acosta, P.; Misra, S. (2011). Culture dimensions in software development industry: The effects of mentoring. Scientific Research and Essays, Vol. 6, Issue 11, pp. 2403-2412.
126. Gelderman, C. J.; Ghijsen, P. W. Th.; Diemen, R. (2011). Choosing self-service technologies or interpersonal services - the impact of situational factors and technology-related attitudes. Journal of Retailing and Consumer Services, Vol. 18, Issue 5, pp. 414-421.

Chapter - 7

THE IMPACT OF SOFTWARE DEVELOPMENT PROCESS ON SOFTWARE QUALITY

Quality of software products depends upon various phase of software development process. Process of software development is used to create and achieve quality in software products. Software development process uses four main phases which have its own importance for development. Software quality is a conformance to requirements which is divided into functional and non-functional requirements. The objective of this chapter is to present an overview of the impact of software development process on software quality. In this chapter various quality attributes have been considered during analysis of development process to see the impact on software quality. The activities in software development process are analyzed to see the impact of individual phase on various well defined attributes of quality. Every phase has an individual impact on software quality attributes. Section I describes the introduction of software process. Section II describes the software development process. Section III describes impact of software requirement on software quality. Section IV describes impact of software design on software quality. Section V describes impact of software coding/implementation on software quality. Section VI describes impact of software testing on software quality. Section VII provides the summary of this chapter.

7.1 INTRODUCTION

Software process can be defined as "a set of activities, methods, practices, and transformations that people use to develop and maintain software and the associated products" [1]. According to IEEE software development process is a process by which user needs are translated into a software product. The process involves translating user needs into software requirements, transforming the software requirements into design, implementing the design in code, testing the code, and sometimes, installing and checking out the software for operational use [2]. The main objective of the development of a system is its efficient integration in real life situations. Various software development methods have been adopted to develop the software products such as: waterfall model, iterative and incremental model, spiral model, V model, rapid application development, prototyping model, agile model, and hybrid spiral model [3-6]. Some of the most commonly used are waterfall, spiral, V model, and agile model. Software development organizations have realized that adherence to a suitable well defined life cycle model helps to produce good quality products [7-8].

Mainly there are four phase of software development; software requirement, software design, software coding/implementation and software testing, which have been used in various models. Each and every phase have an individual impact on software quality attributes. These phase play an important role to improve quality in finished products [9-10]. A suitable life cycle model can possible be selected based on an analysis of issues such as: characteristics of the software to be developed, characteristics of development team, and characteristics of customer.

There are various issues in traditional software development process. The failure of many software projects in terms of not meeting user/business requirements, prone to errors etc has led to software quality becoming one of the key issues from all stakeholders' perspective [11]. In a competitive environment quality based product is basic need for any product success. To achieve quality, efficient process is required.

The objective of this chapter is to present an overview on the impact of software development process on software quality. Research objective of this chapter is to analyze the impact of software development process on software quality.

Software requirement analysis used to collect needs or requirement of software. Requirement analysis is the first step which involve to the quality because this step used to capture all functional and non-functional requirements to be implemented in final product. Software design is next step to derive quality to create complete structure or architecture of software which is stated into requirement specification. Design provides not only to find out how the software product is going to be appear, but also allows both software users and developers to realize how it's going to function. Because design is the only way to completely translate a requirements into a finished product.

After software design software coding/implementation phase is used for implementing the software. Software implementation is based upon programming language. This phase also play an important role because using coding an executable version of software is created. Programming language can impact not only the coding process, but also the properties of the resulting product and its quality. Software testing is conducted when executable software exists. Testing used to find out errors and fix them to support software quality. Testing check what all functions software supposed to do and also check that software is not doing what he not supposed to do.

7.2 SOFTWARE DEVELOPMENT PROCESS

Software process is an organized set of activities required to develop a software product. Software development is the process of taking a set of requirements from a user, analyzing them, designing a solution to the problem, and then implementing that solution on a computer [12]. The international standard for describing the method of selecting, implementing and monitoring the life cycle for software is ISO/IEC 12207 [13]. There are many approaches to software development, known as software development life cycle models, methodologies, or processes. The waterfall model is a traditional version, and agile software development is a newest version for development. There are many different software processes but every process includes four activities: Requirement, Design, Coding, and Testing. After that maintenance is required for further changes [14].

Presently software organization shifting its focus from product issues to process issues [15]. Software quality is important concern for software industry and organization [16]. A general process model for software encompasses a set of framework and umbrella activities, actions, and work tasks. Process can be used to solve common problems that are occurred as part of the software process. Every model provides a different process flow, but all perform the same set of activities: communication, planning, modeling, construction, and deployment [17]. Sequential process models, such as waterfall [18] and V models [19] are linear process models. This is applicable in situations where requirements are well defined and stable [20].

Incremental process models are iterative in nature and develop working versions of software simply [21]. Evolutionary process models use the iterative, incremental nature to implement software product. Evolutionary models, such as prototyping [22] and spiral model [23], produce incremental work products quickly. These process models can be used from development to long-term system maintenance. Agile [24] is a new model for software development which uses iterative/incremental development, less documentation, lightweight and fewer process controls. It was targeted at small to medium-size software projects and smaller teams of developers and develops complete software quickly [25].

Special models include the component-based model [26] that incorporate component reuse and assembly; the formal methods model that encompasses a mathematical based approach to software development; and the aspect-oriented model [27] which uses crosscutting concerns spanning for system architecture. The Unified Process [28] is a "use case driven, architecture-centric, iterative and incremental" software process designed for UML [29] methods and tools. Personal [30] and team [31] models for the software process have been developed. Both provide measurement, planning, and self-direction as key ingredients for a successful software process [4]. Quality of process affects quality of product. Process is important because quality is derived by well defined process. Selection of most appropriate process model is important concern to achieve quality [32].

7.3 IMPACT OF SOFTWARE REQUIREMENT ON SOFTWARE QUALITY

Software requirements are defined during the early stages of a software development as a specification of what should be implemented. They are descriptions of how the system should behave, or of a system property or attribute. [33]. IEEE defines requirement analysis is a process of studying user needs to arrive at a definition of system, hardware, or software requirements [2].

Requirements analysis is important to the success of a systems or software project [34]. The requirements should be documented, actionable, measurable, testable, traceable, related to identified business needs and defined to a level of detail sufficient for system design. Software requirements include business requirements, user requirements, system requirements, external interface requirement, functional requirements, and non-functional requirements. In requirement statements every individual business, user, functional, and nonfunctional requirement would exhibit the qualities. Characteristics of requirement statements are complete, correct, feasible, necessary, prioritized, unambiguous, and verifiable. But it's not enough to have excellent individual requirement statements. So requirements collections are used to collect a set of requirement or group of requirement [35].

Requirements should state what to do and the design should describe how it does this. Characteristics of requirements are complete, consistent, modifiable, and traceable which reflect the software quality. Quality is heavily dependent on functional or non functional requirements. In this section various parameters of requirement are analyzed i.e. consistency, completeness or correctness, and modifiable.

Consistency means providing predictable, maintainable, and reliable results to the customer. Consistency refers to situations where a specification contains no internal contradictions, whereas completeness refers to situations where a specification entails everything that is known to be true in a certain context. It contains all necessary information to avoid ambiguity and need no amplification to enable proper implementation and verification. Correctness is usually meant to be the combination of consistency and completeness. Correctness is often more pragmatically defined as satisfaction of certain business goals [36].

Modifiable refers to each requirement be uniquely labeled and expressed separately from others so you can refer to it unambiguously. If its structure and style are changes to the requirement can be made easily, completely, and consistently [35]. Table 7.1 depicts the relationship between the software requirement characteristics and software quality attributes.

Using consistent requirement traceability, tailorability, interoperability, usability, scalability, and reliability improve [36, 37, 40, 41].

Table 7.1 Relationship between the Software Requirement Characteristics and Software Quality Attributes

Requirement Characteristics	Quality Attributes
Consistency [36, 37, 40, 41]	Traceability, Tailorability, Interoperability, Usability, Scalability, and Reliability
Completeness or Correctness [38, 40]	Maintainability, Functionality, and Reliability
Modifiable [39, 42]	Performance

Using completeness or correctness of requirement maintainability, functionality, and reliability enhance [38, 40]. Modification in requirement engineering (RE) process contributes to improved business performance [39, 42]. Based on Table 7.1 the impact of software requirement on various attributes of software quality has been shown in Figure 7.1.

Figure 7.1 also shows that all quality attribute individually dependent on requirement characteristics but reliability dependent on all three Cs of requirement consistency, completeness and correctness.

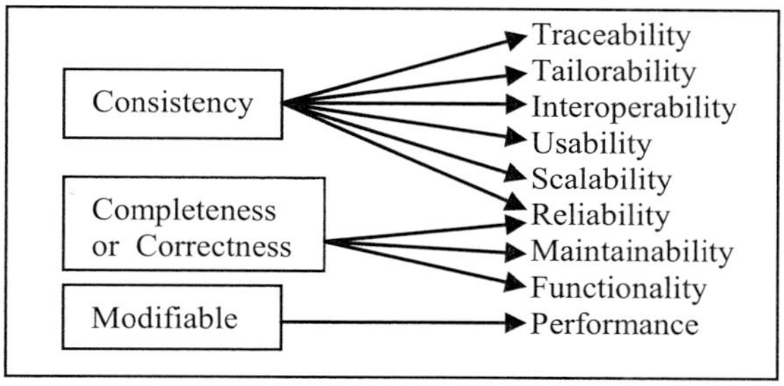

Figure 7.1 Impact of Software Requirement on Software Quality

7.4 IMPACT OF SOFTWARE DESIGN ON SOFTWARE QUALITY

The importance of software design can be stated with a single word- Quality. Design is the place where quality is achieved. Design provides representations of software that can be assessed for quality. Design is the only way to accurately translate a customer's requirements into a finished software product [4]. Software design is the process by which an agent creates a specification of a software artifact, intended to accomplish goals, using a set of primitive components and subject to constraints [43].

IEEE defines design is a process of defining the architecture, components, interfaces, and other characteristics of a system or component [2]. Software design usually involves problem solving and planning a software solution. This includes both low level design and high level design. Fundamental concept of software design has evolved to modularity, software architecture, design patterns, and software component design. Table 7.2 depicts the relationship between the software design attributes and software quality attributes.

Table 7.2 Relationship between the Software Design Attributes and Software Quality Attributes

Design Attributes	Quality Attributes
Software Architecture [47, 48, 49, 50, 51, 52, 53]	Adaptability, Functionality, Security, Efficiency, Portability, Interoperability, Reusability, Maintainability, Reliability, Usability, Performance, and Evolvability
Modularity [54, 55]	Changeability and Adaptability
Design Pattern [56, 57, 58, 59]	Reusability, Composability, Completeness, Maintainability, Stability, Understandability, Flexibility, Simplicity, and Expandability
Software Component Design [60, 61]	Accountability, Stability, and Modularity

The design process is concerned with describing how a requirement is to be met by the design product [44]; there are various design attribute but software architecture, pattern, abstraction, and modularity play an important role.

Quality captured as nonfunctional requirements in the early steps of software development, influence greatly the software systems architecture. Software architecture is a coordination tool among the different phases of software development. Architectural design decisions directly impact software quality [45].

Software architecture provides abstract representation of overall structure of software [46]. Software architecture improves functionality, security, efficiency, portability, interoperability, reusability, maintainability, reliability, usability, performance, evolvability, and adaptability [47-53]. Software architecture includes modularity; software is divided into separately named and addressable components, called modules that are integrated to satisfy problem requirements. Modularity enhances changeability and adaptability [54-55].

Based on Table 7.2 the impact of software design on various attributes of software quality has been shown in Figure 7.2.

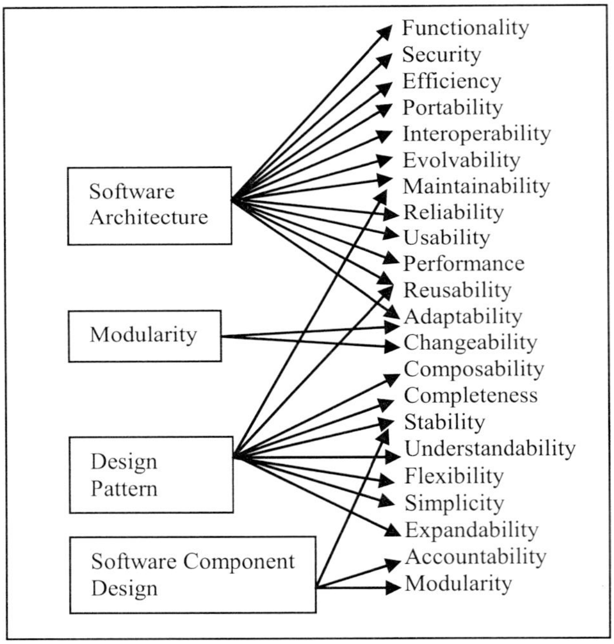

Figure 7.2 Impact of Software Design on Software Quality

Design pattern is a reusable solution of problem occurring in software design. Design pattern is not a finished design that directly transformed into source code. It is a description or template for how to solve a problem that can be used in many different situations. Using design pattern reusability, composability, completeness, maintainability, stability, understandability, flexibility, simplicity, and expandability are improved [56-59]. Software component design also affects the software quality attributes because that transforms structural elements of the software architecture into a procedural description of software components and improves accountability, stability, and modularity [60-61].

Figure 7.2 shows that all attributes individually dependent on design attributes except adaptability, maintainability, reusability, and stability. Maintainability and reusability depends upon software architecture and design pattern. Adaptability depends upon software architecture and modularity. Design pattern and software component design reflects the stability.

7.5 IMPACT OF SOFTWARE CODING/IMPLEMENTATION ON SOFTWARE QUALITY

Implementation/coding may involve developing programs in high level or low level programming languages or tailoring and adapting generic, off the shelf systems to meet the specific requirements of an organization. Programming used to create an executable program. IEEE defines coding is a process of expressing a computer program in a programming language [2].

There are various characteristics of different languages that affect software quality. The choice of language for a particular application is a difficult. The selection affects result and quality of software [62]. Most software codes are written by distant teams. Developers may frequently join or leave software companies. It's very important for source codes to be understandable so it can be easily adapted [63]. In a traditional methodology heavy weight coding is involve there is a need to adopt a light weight codling policy.

Software quality is a broadly characteristic of the software life cycle, defined as what software product should be and what it must contain. The quality of software coding is defined by several characteristics: maintainability, traceability, availability, reliability, reusability, testability, and readability. There are various factors which affect the quality, few are: simple and standardized code, programming features, flexible, traceable, and readable code. Table 7.3 depicts the relationship between the software coding factors and software quality attributes.

Software coding affects various quality attributes using programming features, standardized code, and code clone. Programming Features provides overall characteristics of programming code which influences the code readability and maintainability. Code readability is usually connected with comments and naming standards [63-65]. Standardized Code provides desired product and enhances the comparability of results. Using standardized code maintainability and stability achieves. The technical quality of source code is an important feature for software maintainability [66-67].

Table 7.3 Relationship between the Software Coding Factors and Software Quality Attributes

Coding Factors	Quality Attributes
Programming Features [63, 64, 65]	Readability and Maintainability
Standardized Code [66, 67]	Maintainability and Stability
Code Clones [68, 69]	Reliability and Maintainability

Code clone (duplicated code) provides reusability of code and save effort by copying a pre-existing program section. Code clone affects the maintainability and reliability [68-69]. Based on Table 7.3 the impact of software coding/implementation on various attributes of software quality has been shown in Figure 7.3.

Figure 7.3 shows that every attribute individually depend on coding factor but maintainability depend all three factor of coding.

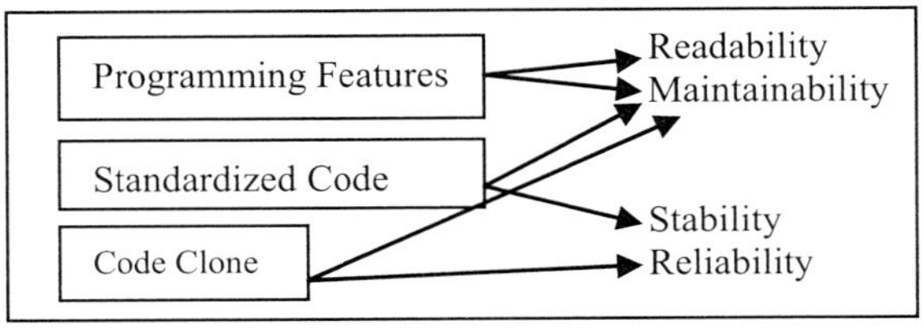

Figure 7.3 Impact of Software Coding on Software Quality

7.6 IMPACT OF SOFTWARE TESTING ON SOFTWARE QUALITY

Software testing is a process of executing a program or application with the intent of finding the errors or quality of the product or service under test. Software testing can be conducted as soon as possible when executable software exists. The overall approach to software development often determines when and how testing is performed. IEEE defines testing is a process of operating a system or component under specified conditions, observing or recording the results, and making an evaluation of some aspect of the system or component [2].

There are many types of software testing. Reviews, walkthroughs, and inspections are referred to as static testing, whereas actually executing programmed code with a given set of test cases is referred to as dynamic testing. Software testing methods are divided into white box and black box testing. There are four levels of tests: unit testing, integration testing, component interface testing, and system testing. Software testing is more difficult because of the vast array of programming languages, operating systems, and hardware platforms that have evolved [70].

Testing is an important factor for software quality. Testing phase support quality by executing software code and find out errors. In testing various important parameters which improve software quality such as: test criteria, coverage, testing effort, and software testability. Table 7.4 depicts the relationship between the software testing parameters and software quality attributes.

Table 7.4 Relationship between the Software Testing Factors and Software Quality Attributes

Testing Parameters	Quality Attributes
Test Criteria [71, 72, 73]	Completeness, Effectiveness, and Maintainability
Testing Effort [74, 75]	User Friendliness, Portability, Reliability, and Maintainability
Software Testability [76, 77]	Reliability, Correctness, Performance, and Effectiveness
Coverage [78]	Reliability

Test criteria focus on measuring the coverage of the test suite upon the structural elements of the program. To test code quality system testing is involve. Code coverage is the most frequently used metric for test code quality assessment and there exist many tools for dynamic code coverage estimation. Test criteria improve completeness, effectiveness, and maintainability [71-73].

To test effort is required; testing effort is collection of run test case and fixes the errors. Testing effort used to fix error, using that user friendliness, portability, reliability, and maintainability enhance [74-75].

Software testability is the tendency of code to reveal existing faults during random testing. This is an analysis for measuring the value provided by a particular software testing scheme, where the value of a scheme can be assessed in different ways. Software testability improves reliability, correctness, performance, and effectiveness [76-77].

Coverage testing helps the tester create a thorough set of tests and gives a measure of test completeness. Coverage analysis of programs during testing not only gives a clear measure of testing quality but also reveals important aspects of software structure. Structure of a program can be a significant component in confident assessment of overall software quality and reliable product [78]. Based on Table 7.4 the impact of software testing on various attributes of software quality has been shown in Figure 7.4.

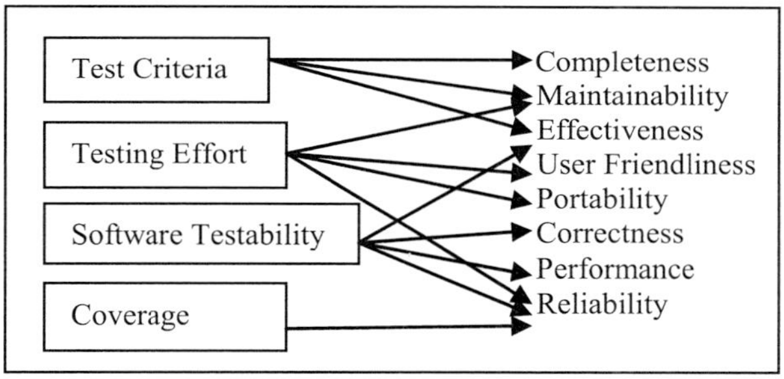

Figure 7.4 Impact of Software Testing on Software Quality

Every attribute individually depend on software testing except maintainability, effectiveness, and reliability as shown in Figure 7.4. Maintainability depends on test criteria and testing effort. Effectiveness depends on test criteria and software testability. Reliability depends on software testability, testing effort, and coverage.

7.7 SUMMARY

Purpose of this analysis is to see the impact of software development process on software quality. The impact of software development process consisting four phase of development process on software quality as shown in Figure 7.5.

Based on analysis every phase of software development is important for software quality however software design is more important than other phase. Software design is used to capture the set of requirements and create architecture of software which realizes the broad class of requirements. Software architecture provides abstract representation of overall structure of software and strategic design for creating or structuring software using the general principles of software design. The impact of software architecture on software quality is shown in Figure 7.6.

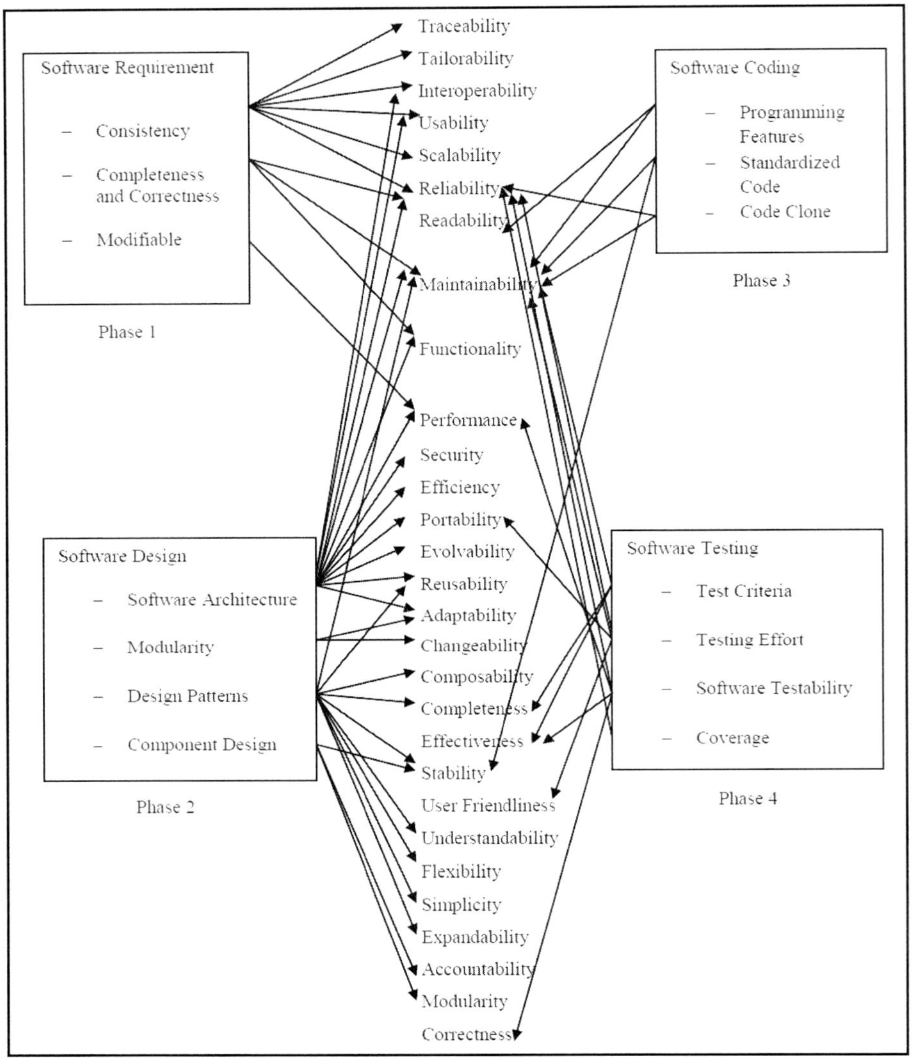

Figure 7.5 Impact of Software Development Process on Software Quality

In literature every quality attributes are important but maintainability is most important attribute for software product because software maintenance is the process of refining the software to make sure it continuous to meet business needs. The importance of maintainability for software development process is shown in Figure 7.7.

Using maintainability software system or component can be modified to correct faults, improve performance or other attributes, or adapt to a changed environment. So maintainability is important for software development process. The main finding of this analysis is software architecture play an important role to achieve quality in software products.

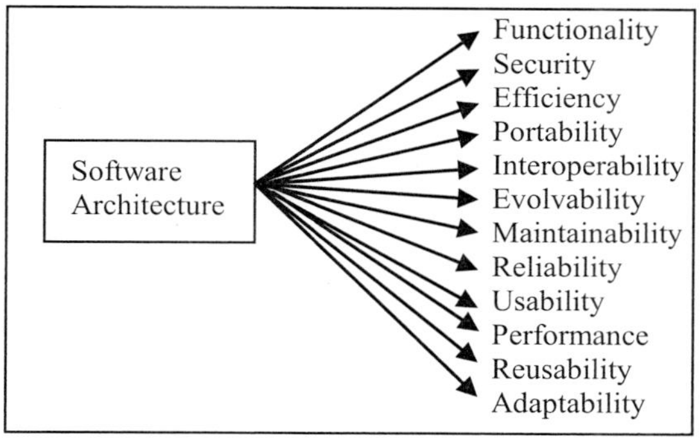

Figure 7.6 Impact of Software Architecture on Software Quality

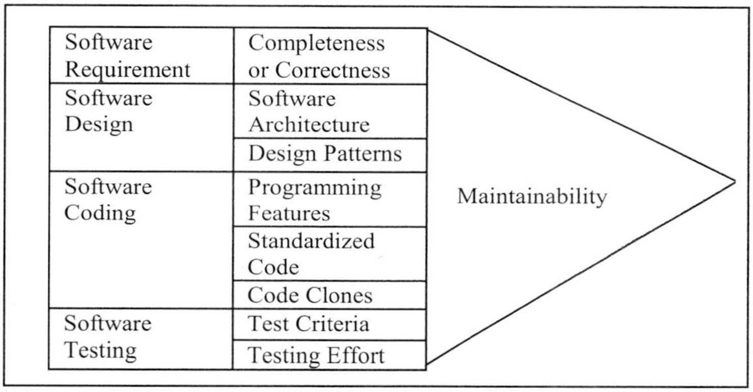

Figure 7.7 Importance of Maintainability for Software Development Process

7.8 REFERENCES

1. Paulk, C.; Weber, C.; Curtis, B.; Chrissis, M. B. (1995). *The Capability Maturity Model: Guidelines for Improving the Software Process* (5th Edition). Addison Wesley Publishing Co.
2. Radatz, J.; Geraci, A.; Katki, F. (1990). *IEEE standard glossary of software engineering terminology*. IEEE Std. 610.12-1990.
3. Aggarwal, K. K.; Singh, Y. (2007). *Software Engineering*. New Age International Publishers.
4. Pressman, R. S. (2010). *Software engineering: a practitioner's approach*. McGraw-Hill.
5. Singh, B.; Kannojia, S. P. (2012). A model for software product quality prediction. Journal of Software Engineering and Applications, Vol. 5, pp. 395-401. doi: 10.4236/jsea.2012.56046.
6. Singh, B.; Gautam, S. (2016). Hybrid Spiral Model to Improve Software Quality Using Knowledge Management. International Journal of Performability Engineering, Vol. 12, Issue 4, pp. 341-352.
7. Isaias, P.; Issa, T. (2015). *High level models and methodologies for information systems*. Springer.
8. Sujatha, P.; Sankar, G. V.; Rao, A. S.; Satyanarayana, T. (2001). The Role of Software Verification and Validation in Software Development Process. IETE Technical Review, Vol. 18, Issue 1, pp. 23-26. doi: 10.1080/02564602.2001.11416938.

9. Godbole, N. S. (2004). *Software quality assurance: Principles and practice*. Alpha Science Int'l Ltd.
10. Singh, B. (2011). *Quality Control and Reliability Analysis*. Khanna Publishers.
11. Land, L. P. W.; Higgs, J. (2007). An Empirical Study of Software Quality Improvement Practices from Multiple Perspectives - An Australian Case Study. In Proceedings of the 11th Pacific-Asia Conference on Information Systems (PACIS), pp. 547-560.
12. Dooley, J. (2011). *Software development and professional practice*. Apress.
13. IEEE Standard 12207-2008, Systems and Software Engineering - Software life cycle processes, IEEE, 2008.
14. Telemaco, U.; Mesquita, R.; Oliveira, T.; Melo, G.; Alencar, P. (2018). A Catalog of Software Development Rules for Agile Methods.
15. Davis, M. J. (1995). Process and Product: Dichotomy or Duality? ACM SIGSOFT Software Engineering Notes, Vol. 20, Issue 2. doi: 10.1145/224155.565634.
16. Cugola, G.; Ghezzi, C. (1998). Software Processes: a Retrospective and a Path to the Future, Software Process: Improvement and Practice. Vol. 4, Issue 3, pp. 101-123. doi: 10.1.1.17.2499.
17. Scacchi, W. (1987). *Models of software evolution: life cycle and process*. Carnegie-Mellon Univ Pittsburgh Pa Software Engineering Inst.
18. Royce, W. W. (1970). Managing the development of large software systems. In Proceedings of IEEE WESCON, Vol. 26, Issue 8, pp. 328-338.
19. Bucanac, C. (1999). The V-Model. University of Karlskrona/Ronneby.
20. Boehm, B. (1976). Software Engineering. IEEE Trans. on Computer, Vol. C-25, Issue 12, pp. 1226-1241. doi: 10.1109/TC.1976.1674590.
21. McDermid, J.; Rook, P. (1991). *Software development process models*. Software Engineer's Reference Book.
22. Jalote, P. (2008). *A concise introduction to software engineering*. Springer.
23. Boehm, B. (1988). A spiral model of software development and enhancement. IEEE Computer, Vol. 21, Issue 5, pp. 61-72. doi: 10.1109/2.59.
24. Cockburn, A. (2006). *Agile software development: the cooperative game*. Pearson Education.
25. Cohn, M. (2010). *Succeeding with agile: software development using Scrum*. Pearson Education.
26. Brown, W.; Wallnan, K. C. (1996). Engineering of component-based systems. In Second IEEE International Conference on Engineering of Complex Computer Systems, pp. 414-422. doi: 10.1109/ICECCS.1996.558485.
27. Clarke, S.; Baniassad, E. (2005). *Aspect-oriented analysis and design*. Addison-Wesley Professional.
28. Arlow, J.; Neustadt, I. (2005). *UML 2 and the unified process: practical object-oriented analysis and design*. Pearson Education.
29. Booch, G.; Rumbaugh, J.; Jacobsen, I. (2005). *The unified modeling language user guide*. Pearson Education India.
30. Humphrey, W. S. (2000). The Personal Software Process-Status and Trends. IEEE Software, Vol. 17, Issue 6, pp. 71-75.
31. Humphrey, W. S. (2000). *Introduction to the team software process*. Addison-Wesley Professional.
32. Ghezzi, C.; Jazayeri, M.; Mandrioli, D. (2002). *Fundamentals of software engineering*. Prentice Hall PTR.
33. Sommerville, I.; Sawyer, P. (1997). *Requirements engineering: a good practice guide*. John Wiley & Sons, Inc.
34. Rehman, T. ur; Khan, M. N.; Riaz, N. (2013). Analysis of requirement engineering processes, tools/techniques and methodologies. International Journal of Information Technology and Computer Science (IJITCS), Vol. 5, Issue 3, pp. 40-48. doi: 10.5815/ijitcs.2013.03.05.
35. Wiegers, K.; Beatty, J. (2013). *Software requirements*. Pearson Education.

36. Zowghi, D.; Gervasi, V. (2002). The Three Cs of requirements: consistency, completeness, and correctness. In International Workshop on Requirements Engineering: Foundations for Software Quality, pp. 155-164.
37. Cimatti, A.; Roveri, M.; Susi, A.; Tonetta, S. (2012). Validation of requirements for hybrid systems: A formal approach. ACM Transactions on Software Engineering and Methodology (TOSEM), Vol. 21, Issue 4, 22. doi: 10.1145/2377656.2377659.
38. Sinnig, D.; Chalin, P.; Khendek, F. (2013). Use case and task models: an integrated development methodology and its formal foundation. ACM Transactions on Software Engineering and Methodology (TOSEM), Vol. 22, Issue 3, 27. doi: 10.1145/2491509.2491521.
39. Sommerville, I.; Ransom, J. (2005). An empirical study of industrial requirements engineering process assessment and improvement. ACM Transactions on Software Engineering and Methodology (TOSEM), Vol. 14, Issue 1, pp. 85-117.
40. Sommerville, I. (2005). Integrated requirements engineering: A tutorial. IEEE software, Vol. 22, Issue 1, pp. 16-23.
41. Alshamari, M.; Mayhew, P. (2009). Technical review: Current issues of usability testing. IETE Technical Review, Vol. 26, Issue 6, pp. 402-406. doi: 10.4103/0256-4602.57825.
42. Liu, H. H. (2011). *Software performance and scalability: a quantitative approach.* John Wiley & Sons.
43. Ralph, P.; Wand, Y. (2013). A Proposal for a Formal Definition of the Design Concept. Annual Review of Policy Design, Vol. 1, Issue 1, pp. 1-35.
44. Budgen, D. (2003). *Software design*. Pearson Education,
45. Mistrík, I.; Bahsoon, R.; Eeles, P.; Roshandel, R.; Stal, M.; editors (2014). *Relating System Quality and Software Architecture*. Morgan Kaufmann.
46. Zhu, H. (2005). *Software design methodology: From principles to architectural styles*. Elsevier.
47. Losavio, F.; Chirinos, L.; Lévy, N.; Ramdane-Cherif, A. (2003). Quality characteristics for software architecture. Journal of Object Technology, Vol. 2, Issue 2, pp. 133-150.
48. Juristo, N.; Moreno, A. M.; Sanchez-Segura, M. I. (2007). Analysing the impact of usability on software design. Journal of Systems and Software, Vol. 80, Issue 9, pp. 1506-1516. doi: 10.1016/j.jss.2007.01.006.
49. Martens, A.; Koziolek, H.; Becker, S.; Reussner, R. (2010). Automatically improve software architecture models for performance, reliability, and cost using evolutionary algorithms. In Proceedings of the First Joint WOSP/SIPEW International Conference on Performance engineering, ACM, pp. 105-116.
50. Breivold, H. P.; Crnkovic, I.; Larsson, M. (2012). A systematic review of software architecture evolution research. Information and Software Technology, Vol. 54, Issue 1, pp. 16-40. doi: 10.1016/j.infsof.2011.06.002.
51. Williams, B. J.; Carver, J. C. (2010). Characterizing software architecture changes: A systematic review. Information and Software Technology, Vol. 52, Issue 1, pp. 31-51. doi: 10.1016/j.infsof.2009.07.002.
52. Karthikeyan, T.; Geetha, J. (2012). A Study and Critical Survey on Service Reusability Metrics. International Journal of Information Technology and Computer Science (IJITCS), Vol. 4, Issue 5, pp. 25-31. doi: 10.5815/ijitcs.2012.05.04.
53. Saxena, S.; Dubey, S. K. (2013). Impact of Software Design Aspects on Usability. International Journal of Computer Applications, Vol. 61, Issue 22, pp. 48-53.
54. Sullivan, J.; Griswold, W. G.; Cai, Y.; Hallen, B. (2001). The structure and value of modularity in software design. In ACM SIGSOFT Software Engineering Notes, Vol. 26, Issue 5, pp. 99-108.
55. Banks, W. (2013). Modularity and adaptability. Lawrence Livermore National Laboratory (LLNL), Livermore.
56. Zhu, H.; Bayley, I. (2015). On the Composability of Design Patterns. IEEE Transactions on Software Engineering, Vol. 41, Issue 11, pp. 1138-1152. doi: 10.1109/TSE.2015.2445341.

57. Ampatzoglou, A.; Chatzigeorgiou, A.; Charalampidou, S.; Avgeriou, P. (2015). The effect of GoF design patterns on stability: a case study. IEEE Transactions on Software Engineering, Vol. 41, Issue 8, pp. 781-802. doi: 10.1109/TSE.2015.2414917.

58. Ampatzoglou, A.; Frantzeskou, G.; Stamelos, I. (2012). A methodology to assess the impact of design patterns on software quality. Information and Software Technology, Vol. 54, Issue 4, pp. 331-46. doi: 10.1016/j.infsof.2011.10.006.

59. Khomh, F.; Gueheneuce, Y. G. (2008). Do design patterns impact software quality positively? In 12th European Conference on Software Maintenance and Reengineering (CSMR), IEEE, pp. 274-278.

60. Benghabrit, W.; Grall, H.; Royer, J. C.; Sellami, M. (2014). Accountability for abstract component design. In 40th EUROMICRO Conference on Software Engineering and Advanced Applications, IEEE, pp. 213-220.

61. Tizzei, P.; Dias, M.; Rubira, C. M.; Garcia, A.; Lee, J. (2011). Components meet aspects: Assessing design stability of a software product line. Information and Software Technology, Vol. 53, Issue 2, pp. 121-36. doi: 10.1016/j.infsof.2010.08.007.

62. Singh, B.; Kannojia, S. P. (2012). Languages and Their Importance in Quality Software. In International Conference on Communication Systems and Network Technologies (CSNT), IEEE, pp. 956-959. doi: 10.1109/CSNT.2012.203.

63. Tashtoush, Y.; Odat, Z.; Alsmadi, I.; Yatim, M. (2013). Impact of programming features on code readability. International Journal of Software Engineering and Its Applications, Vol. 7, Issue 6, pp. 441-458. doi: 10.14257/ijseia.2013.7.6.38.

64. Buse, R. P.; Weimer, W. R. (2008). A metric for software readability. In Proceedings of the 2008 international symposium on Software testing and analysis, ACM, pp. 121-130.

65. Buse, R. P.; Weimer, W. R. (2010). Learning a metric for code readability. IEEE Transactions on Software Engineering, Vol. 36, Issue 4, pp. 546-558.

66. Baggen, R.; Correia, J. P.; Schill, K.; Visser, J. (2012). Standardized code quality benchmarking for improving software maintainability. Software Quality Journal, Vol. 20, Issue 2, pp. 287-307. doi: 10.1007/s11219-011-9144-9.

67. Peercy, E. (1981). A software maintainability evaluation methodology. IEEE Transactions on Software Engineering, Vol. SE-7, Issue 4, pp. 343-351.

68. Monden, A.; Nakae, D.; Kamiya, T.; Sato, S. I.; Matsumoto, K. I. (2002). Software quality analysis by code clones in industrial legacy software. In proceedings of Eighth IEEE Symposium on Software Metrics, IEEE, pp. 87-94.

69. Roy, K.; Cordy, J. R.; Koschke, R. (2009). Comparison and evaluation of code clone detection techniques and tools: A qualitative approach. Science of Computer Programming, Vol. 74, Issue 7, pp. 470-495. doi: 10.1016/j.scico.2009.02.007.

70. Myers, G. J.; Sandler, C.; Badgett, T. (2011). *The art of software testing*. John Wiley & Sons.

71. Athanasiou, D.; Nugroho, A.; Visser, J.; Zaidman, A. (2014). Test code quality and its relation to issue handling performance. IEEE Transactions on Software Engineering, Vol. 40, Issue 11, pp. 1100-1125. doi: 10.1109/TSE.2014.2342227.

72. Hutchins, D.; Foster, H.; Goradia, T.; Ostrand, T. (1994). Experiments of the effectiveness of dataflow-and controlflow-based test adequacy criteria. In proceedings of the IEEE 16th international conference on Software engineering, pp. 191-200.

73. Kapoor, K.; Bowen, J. (2003). Experimental evaluation of the variation in effectiveness for DC, FPC and MC/DC test criteria. In proceedings of the IEEE International Symposium on Empirical Software Engineering (ISESE), pp. 185-194.

74. Yu, T. J.; Dunsmore, H. E. (1985). The Analysis of Software Development and Testing Processes: An Empirical Study. Technical Report Purdue University.

75. Huang, C. Y.; Lo, J. H.; Kuo, S. Y.; Lyu, M. R. (1999). Software reliability modeling and cost estimation incorporating testing-effort and efficiency. In proceedings of 10th International Symposium on Software Reliability Engineering, IEEE, pp. 62-72.

76. Voas, M.; Miller, K. W. (1992). Improving the software development process using testability research. In proceedings of Third International Symposium on Software Reliability Engineering, IEEE, pp. 114-121.
77. Mathur, P. (1991). Performance, effectiveness, and reliability issues in software testing. In Proceedings of the Fifteenth Annual International Conference on Computer Software and Applications (COMPSAC'91), IEEE, pp. 604-605.
78. Horgan, R.; London, S.; Lyu, M. R. (1994). Achieving software quality with testing coverage measures. IEEE Computer, Vol. 27, Issue 9, pp. 60-69.

Chapter - 8

THE EFFECTS OF SOFTWARE PROCESS MATURITY ON SOFTWARE DEVELOPMENT EFFORT

Software product is regularly following time, over cost, non-conforming to software requirements and of poor quality product. Controlling and improving the software process used to develop software has been proposed as a main solution to these problems. The Software Engineering Institute at Carnegie Mellon University has published the Software Capability Maturity Model (SW-CMM) for employ as a set of criteria to evaluate an organization's software process maturity. This model is used as a roadmap to get better a software development process's maturity. The principle of the SW-CMM is that mature development process delivers software products on time, within cost, within software requirements, and high quality software. Software development effort is the main determinant of software development cost and schedule. So this chapter focus on the effects of software process maturity on software development effort. Section I describes the introduction of this chapter. Section II describes the capability maturity model. Section III describes the modelling of effort expenditure. Section IV describes the modelling regression analysis. Section V describes the summary of this chapter.

8.1 INTRODUCTION

Even though growing efforts to improve software development process, recurring concerns regarding software project performance stay largely present. The rate of software development project failure rate has been regularly documented in information systems (IS) research. The management of software development projects is frequently marked by insufficient planning, a poor grasp of the overall software development process, and no obvious management framework, still as the focus in software development shifts from a technology perspective to an additional process centric view. CMM is a methodology used to develop and improve an organization's software development process. The model describes a five level evolutionary path of increasingly organized and systematically more mature software processes. CMM was developed and is promoted by the Software Engineering Institute (SEI), a research and development centre sponsored by the U.S. Department of Defence (DoD). SEI was founded in 1984 to tackle software engineering issues and in a wide sense, to precede software engineering methodologies. More specifically SEI was established to optimize the process of acquiring, developing, and maintaining heavily software reliant systems for the DoD. Because the processes involved are similarly applicable to the software industry as a whole SEI advocates industry wide adoption of the CMM. SEI has developed a Software Capability Maturity Model (SW-CMM) which is used to describe an organization's software process maturity. The inspiration behind the SW-CMM is that a mature software development process will develop the software product on time, within cost, within all requirements, and of high software quality. The model is based on five levels: organizations with ad hoc processes start at level one. To evolution to the next higher level, level two, an organization has to exhibit a repeatable process. To gain a level three rating an organization has to reveal a defined process. In level four organizations has a managed process and a level five organization has an optimizing software development process [1-3].

Here an important question for software industry and government is what is the importance of investing resources to enhance the organization's process maturity? The long term profits of high software process maturity are software delivered on time, within cost, within customer requirements, and of high quality. An essential advantage would be the effect it has on software productivity. It is based on the database productivity index which is consequential from effort, size, and time used to develop the software. It did not divide out other factors that control software productivity, e.g. personnel, product complexity, software technologies, and development practices. It may not be accurate to suppose a relation exists between maturity level and productivity. The integration of software development risk can offer a much needed linkage in the three primary constructs of CMM. From a managerial perspective, in order to promote a better software project performance, IS project leaders and managers should powerfully emphasize devising effective software development risk assessment because a variation of this construct's level might strengthen or weaken the association between software development process maturity and software project performance [4-5].

8.2 CAPABILITY MATURITY MODEL

The Capability Maturity Model was originally funded by military research. The United States air force funded a research at the Carnegie Mellon Software Engineering Institute to develop a model for the military to utilize as an objective assessment of software subcontractors. The outcome was the Capability Maturity Model, published as managing the software process in 1989. CMM was developed by the SEI at Carnegie Mellon University in Pittsburgh. It has been used broadly for avionics software and government projects, in Europe, North America, Asia, South America, Australia, and Africa. Currently some government departments need software development contract organization to accomplish and operate at a level 3 standard [6-7]. The CMM is no longer supported by the SEI and has been outdated by the more widespread Capability Maturity Model Integration (CMMI).

Capability Maturity Model (CMM) mostly refers to a software process improvement approach that is based on a software process model. CMM also refers exclusively to the first model developed by the Software Engineering Institute (SEI) in the mid 1980s, as well as the family of software process models that followed. A software process model is a structured collection of practices that illustrate the characteristics of effective processes and the practices incorporated are those proven by experience to be effective [8].

The CMM was initially intended as a tool to assess the capability of government contractors to execute a contracted software project. Although it comes from the area of software development, it can be, has been, and continues to be broadly applied as a general model of the maturity of processes in IT/IS (and other) organizations. CMM can be used to measure an organization against an extent of five process maturity levels. Each level ranks the software organization according to its standardization of processes in the subject area being assessed. The subject areas can be as different as software engineering, project management, systems engineering, risk management, information technology (IT) services, system acquisition, and personnel management [9].

The model identifies five levels of software process maturity for an organisation. Within each of these maturity levels are KPAs (Key Process Areas) which characterize that level, and for each KPA there are five definitions recognized:

1. Goals
2. Commitment
3. Ability
4. Measurement
5. Verification

The KPAs are not essentially exclusive to CMM, representing as they do the stages that organizations must go throughout on the way to becoming mature. The evaluation is supposed to be led by an authorised lead assessor. One approach in which companies are believed to use the model is first to assess their maturity level and then form a particular plan to get to the next level. Skipping levels is not allowed. So CMM is a way to extend and refine an organization's processes. The first CMM was for the intention of developing and refining software development processes. A maturity model is a planned collection of elements that describe characteristics of effective processes [10]. A maturity model provides:

- A place to start
- The benefit of a community's prior experiences
- A common language and a shared vision
- A framework for prioritizing actions
- A manner to define what improvement means for organization

A maturity model can be used as a standard for assessing different organizations for corresponding comparison. It describes the maturity of the software company based upon the project the company is dealing with and the clients.

Timeline of CMM:

- 1987 SEI-87-TR-24 (SW-CMM questionnaire) released
- 1989 Managing the Software Process is published
- 1991 SW-CMM v1.0 released
- 1993 SW-CMM v1.1 released
- 1997 SW-CMM revisions halted in support for CMMI

- 2000 CMMI v1.02 released
- 2002 CMMI v1.1 released
- 2006 CMMI v1.2 released

Although these models have proved useful to many software organizations, the exercise of multiple models has been problematic. Further, applying multiple models that are not incorporated within and across an organization is expensive in terms of appraisals, training, and improvement activities [11]. The CMM Integration project was created to sort out the problem of using multiple CMMs. The CMMI Product Team's assignment was to merge three source models:

- The Capability Maturity Model for Software (SW-CMM) v2.0 draft C
- The Systems Engineering Capability Model (SECM)
- The Integrated Product Development Capability Maturity Model (IPD-CMM) v0.98

CMMI is the designated descendant of the three source models. The SEI has released a strategy to sunset the Software CMM and previous versions of the CMMI. The equivalent can be said for the SECM and the IPD-CMM; these models were outdated by CMMI. With the release of the CMMI Version 1.2 Product Suite, the active CMMI has been renamed the CMMI for Development (CMMI-DEV), V1.2. Two other versions are being developed, one for services, and the other for acquisitions. In some cases, CMM can be shared with other methodologies. It is generally used in conjunction with the ISO 9001 standard, as well as with the computer programming methodologies of Extreme Programming (XP) and Six Sigma.

There are five levels of the CMM:

Level 1 - Initial

Processes are generally ad hoc and the organization usually does not offer a stable environment. Success in these organizations depends on the capability and heroics of the people in the organization and not on the use of proven processes. In spite of this ad hoc, chaotic environment, maturity level 1 organizations often develop products and services that work; however, they normally exceed the cost and time of their software projects. Organizations are categorized by a tendency to over perform, abandon processes in the time of crisis, and not be capable to replicate their past successes again. Software project success depends on having excellence and quality people [12].

Level 2 - Repeatable

Software development successes are repeatable. The processes may not duplicate for all the software projects in the organization. The organization may use several fundamental software project management techniques to control project cost and time. Process discipline helps ensure that active practices are engaged during times of pressure. When these practices are in position, projects are performed and managed according to their software documented plans. Project position and the delivery of services are observable to management at defined points (for example: at main milestones and at the achievement of main tasks). Basic project management processes are established to track time, cost, and functionality. The smallest process discipline is in place to repeat earlier successes on software projects with related applications and scope. There is still a considerable risk of exceeding cost and time estimate [13].

Level 3 - Defined

The organization's set of standard processes which is the foundation for level 3, is established and enhanced over time. These standard processes are used to create consistency across the organization. Projects establish their defined processes by the organization's set of benchmark processes according to tailoring guidelines. The organization's management establishes process aim based on the organization's set of standard processes and ensures that these objectives are suitably addressed. A critical difference between level 2 and level 3 is the process descriptions, scope of standards, and procedures. At level 2, process descriptions, standards, and procedures may be relatively different in each detailed instance of the process (for example: on a particular software project). At level 3, process descriptions, standards, and procedures for a project are customized from the organization's set of standard processes to suit a particular project or organizational unit [14].

Level 4 - Managed

Using accurate measurements, management can effectively organize the software development effort. In particular, management can recognize ways to adjust and adapt the process to particular projects exclusive of measurable losses of software quality or deviations from software specifications. At this level software organization set a quantitative quality goal for both software process and software maintenance. Sub-processes are selected that considerably contribute to overall process performance. These selected sub-processes are controlled using statistical and other quantitative techniques. A critical difference between maturity level 3 and maturity level 4 is the predictability of process performance. At maturity level 4, the performance of processes is controlled using statistical and other quantitative techniques and is quantitatively predictable. At maturity level 3, processes are simply qualitatively predictable [15].

Level 5 - Optimizing

Main focused on constantly improving process performance using incremental and innovative technological improvements. Quantitative process-improvement objectives for the organization are recognized, continually revised to imitate changing business objectives, and used as criteria in managing process improvement. The effects of deployed process improvements are measured and evaluated beside the quantitative process improvement objectives. Both the defined processes and the organization's set of standard processes are targets of measurable improvement activities. Process improvements to tackle common causes of process dissimilarity and measurably advance the organization's processes are identified, evaluated, and deployed. Optimizing processes that are quick, adaptable, and innovative depends on the contribution of an empowered workforce aligned with the industry values and objectives of the organization. The organization's capability to rapidly respond to changes and opportunities is improved by finding ways to accelerate and contribute to learning [16].

A critical dissimilarity between maturity level 4 and maturity level 5 is the type of process distinction addressed. At maturity level 4, processes are concerned with addressing particular causes of process variation and providing statistical predictability of the results. Though processes may generate predictable results, the results may be inadequate to achieve the established objectives. At maturity level 5, processes are worried with addressing universal causes of process variation and changing the process (that is shifting the mean of the process performance) to improve process performance (while maintaining statistical probability) to accomplish the established quantitative process improvement objectives.

The most valuable elements of CMM Level 2 and 3: Creation of software specifications, stating what is going to be developed, combined with prescribed sign off, an executive sponsor, and approval mechanism. This is not a living document but add-ons are placed in a delayed or out of scope section for later inclusion into the next cycle of software development. A technical specification, stating how exactly the thing specified in the software specifications is to be developed will be used. This is an existing document. Peer review of code (code review) with metrics that allocate developers to walk through an implementation, and to recommend improvements or changes. This is problematic because the code has already been developed and a terrible design cannot be fixed by correction, the code review gives absolute code a formal approval mechanism. Version control, an extremely large number of organizations have no official revision control mechanism or release mechanism in place [17]. The idea that is an accurate way to develop software product is a technical process involving engineering design and that groups of software developers are not around to basically work on the problem du jour.

The CMM is like to ISO 9001, one of the ISO 9000 series of standards specified by the International Organization for Standardization (ISO). The ISO 9000 standards identify an effective quality system for manufacturing and service industries. ISO 9001 deals particularly with software development and maintenance. The main distinction between the two systems lies in their respective purposes: ISO 9001 specifies a minimum acceptable quality level for software processes while the CMM establishes a structure for continuous process improvement and is additional explicit than the ISO standard in defining the means to be employed to that finish [18].

8.3 MODELLING OF EFFORT EXPENDITURE

Effective software project estimation is the one of the most challenging and imperative activities in software development process. Appropriate project planning and control is not probable without a sound and reliable estimate. As a complete, the software industry doesn't estimate software projects well and doesn't utilize estimates appropriately. We experience far additional than we should as an outcome and we require to focus some effort on improving the situation. Under-estimating a software project leads to under-staffing it (resulting in staff burnout), under-scoping the software quality assurance effort (running the risk of low software quality deliverables), and setting excessively short a time or schedule (resulting in loss of trustworthiness as deadlines are missed). For those who outline on avoiding these circumstances by generously filling the estimate, over-estimating a software project can be just concerning as terrible for the software organization. If you give a project extra resources than it actually needs without satisfactory scope controls it will utilize them [19]. The software project is then probable to cost more than it should (a negative impact on the base line), take longer to deliver than essential (resulting in misplaced opportunities), and impediment the use of your resources on the next project.

The four basic steps in software project estimation are:

1. Estimate the size of the development product. This usually ends up in either Lines of Code (LOC) or Function Points (FP), but there are additional possible units of measure.
2. Estimate the effort in person-months (PM) or person-hours.
3. Estimate the schedule or time in calendar months.
4. Estimate the software project cost in rupees or dollars (local currency).

There are modelling various approaches used by models to estimate software development effort. Some are based on analogy, some on theory, and others on statistics for effort expenditure modelling. The most dominant factor in predicting effort in these models is the size of the software product. There are additional factors that also influence effort such as product complexity, the application experience of the software development team, and software development tool support [20].

1. Analogy Models

This method is based on the evaluation of the planned project with previous projects that have same characteristics. This model uses experts or stored project data to find out the effort mandatory to develop a software product. For a novel product it must be determined what subcomponent level is useful for software estimation. There must be an estimate of how lots of components will expected be in the software product. Experts calculate the low, high, and mainly possible estimates for effort required based on the differences among the new and previous software projects. It can offer a detailed estimation of effort depending on how deep into the sub-components the analogies are completed. The model is weak because the amount of resemblance may not be extremely close to the new software project [21].

2. Theoretical Models

A theory based estimation model was put forward by Putnam (1979) and explained in Conte et al. (1986) and Kitchenham (1990). It is based on the probability distribution called the Rayleigh curve. This curve state manpower distribution on a project over time as shown in Figure 8.1. The curve is modelled by the differential equation:

$$\frac{dy}{dt} = 2Kat(e)^{-at^2} \qquad \text{Equation 1}$$

Where dy/dt is the staff build-up rate, t is the elapsed time commencing the start of design to the product replacement, K is the area under the curve and represents total lifecycle effort (including maintenance), and a is a constant that determines the shape of the curve.

Putnam uses productivity to connect the basic Rayleigh manpower distribution model to the software development characteristics of size and technology factors. Productivity in software has been defined as the size of the software product, S, divided by the development effort, E.

$$P = \frac{S}{E} \qquad \text{Equation 2}$$

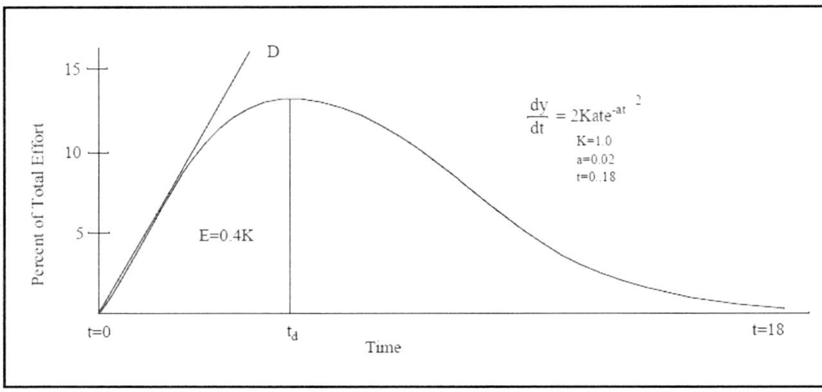

Figure 8.1 Rayleigh manpower distribution model

To find E in the Rayleigh model, Putnam made the assumption that the peak staffing level (peak of the curve) corresponded to the software development time. Through this assumption, the region under the curve represented software development effort, E. E was found to be around 40% of K, the entire lifecycle effort which is the total area under the curve. Putnam observed from project data that the additional productive projects had an early slower staff build-up and the fewer productive projects had an original faster staff build-up [22].

3. Statistical Models

Statistical models use data to obtain the values for model coefficients. Regression analysis is used to create the relationship between model parameters and software development effort. There are two forms of statistical models: linear and non-linear. Linear statistical models have the form:

$$\text{Effort} = b_0 + \sum_{i=1}^{n} b_i x_i \qquad \text{Equation 3}$$

Where x_i is software development factors that are supposed to influence effort and b_i are coefficients. There two reasons that models of this outline do not work fine for estimating software development effort:

1. Empirical evidence represent that the relationship between software development effort and size of the software product is not linear. The more suitable relationship is obvious.

2. As the software product gets better effort exhibits a diseconomy of scale.

The facts of diseconomies of scale linear models are not correct for modelling effort expenditure. This includes the linear model based on a counting metric called Function Points. The original Function Points was proposed by Albrecht in 1979. This metric consists of counting the number of inputs, outputs, interfaces, inquiries, and logical files from the user's perspective and weighting the counts as simple, average, and complex. The total unadjusted function point count was adjusted with 14 complexity characteristics to obtain an adjusted function point count, FP. From data presented in Albrecht and Gaffney (1983) the estimation for effort in person hours was $E = 54 \cdot FP - 13390$. But, when the data points from the article were plotted the effort in person hours was originate to be of the relation $E = 1.1(FP)^{1.49}$. This also supports the conclusion that the relationship between size and effort is nonlinear. Non-linear estimation models have the form:

$$\text{Effort} = A \cdot (\text{Size})^b \qquad \text{Equation 4}$$

Where S is size, A is an arrangement of project factors that affect effort. The exponent b, in non-linear models supports the concept of economies and diseconomies of scale in software development process [23].

Analogy models are not appropriate for software development change research. The model does not provide insight into the possible effects of software development process changes. This would make assessing process maturity's effect on effort an unprofessional estimate. Models based on theory are not practical to predict effort because of the aggregation of input parameters and the reliance of a fundamental theory to describe and predict effort. Many researchers have disagreed with some of the assumptions in these models. It is not obvious how these models account for an iterative software process model where the effort from one build is overlapped with the effort on the subsequently build. Statistical models are simple to understand and use. The effect of the model inputs on effort is complete understandable by observing the arrangement of the inputs in the mathematical model. The model is appropriate to support the research if a technique can be found to standardize the inputs and identify or resolve the correlations between inputs. Statistical models have the drawback of probably producing results that are simply valid for the local environment [24]. Another drawback to a statistical approach is that as the number of model inputs increases, the amount of data required to adjust the model increases (this has to do through the model degrees of freedom).

So once you have an estimate of the size of software product, you can obtain the effort estimate. This conversion from software size to entire project effort can simply be done if you have a definite software development lifecycle and development process that follow to specify, design, develop, and test the software product. A software development project involves far more than basically coding the software, in fact coding is frequently the smallest part of the overall effort. Writing and reviewing documentation, designing the deliverables, implementing prototypes, and reviewing or testing the code acquire the larger portion of overall project effort [25]. The project effort estimate requires to identify and estimate, and then total up all the activities must perform to build a software product of the estimated size. There are two central ways to derive effort from size:

1. The most excellent way is to use organization's own chronological data to find out how much effort previous projects of the estimated size have taken. This, of course assumes (a) organization has been documenting real results from previous projects, (b) you have at least one past project of related size (it is even better if you have numerous projects of similar size as this reinforces that you constantly need a definite level of effort to develop software projects of a specified size), and (c) that you will follow a same development lifecycle, use a related development methodology, use same tools, and use a software team with comparable skills and experience for the new project.
2. If you don't have chronological data from your software organization because you haven't underway collecting it however or because new software project is especially different in one or more aspects, you can utilize a mature and usually accepted algorithmic approach such as COCOMO model or the Putnam methodology to convert a size estimation into an effort estimation. These models have been derived by studying an important number of completed software projects from various organizations to observe how their software project sizes mapped into total project effort. These industry data models may not be as correct as own historical data but they can provide useful approximate effort estimates [26].

8.4 MODELLING REGRESSION ANALYSIS

In statistical modelling, regression analysis is a set of statistical processes for estimating the relationships between variables. It includes numerous techniques for modelling and analyzing various variables, when the centre is on the relationship between a dependent variable (criterion variable) and one or more independent variables (predictors). Specifically, regression analysis helps to recognize how the characteristic value of the dependent variable changes when some one of the independent variables is diverse, while the other independent variables are held permanent. Most frequently, regression analysis estimates the restricted expectation of the dependent variable specified the independent variables that is, the average value of the dependent variable when the independent variables are fixed. Less regularly the focus is on a quantile, or other location parameter of the conditional distribution of the dependent variable specified the independent variables. In all cases, a function of the independent variables called the regression function is to be estimated. In regression analysis, it is also of curiosity to characterize the variation of the dependent variable about the prediction of the regression function using a probability distribution. A related but different approach is Necessary Condition Analysis (NCA) which estimates the maximum (rather than average) value of the dependent variable for a known value of the independent variable (ceiling line rather than central line) in order to categorize what value of the independent variable is essential but not sufficient for a known value of the dependent variable [27]. Regression analysis is extensively used for prediction and forecasting where its use has considerable overlap with the field of machine learning.

Regression analysis is also used to identify with which between the independent variables are linked to the dependent variable and to investigate the forms of these relationships. In restricted conditions, regression analysis can be used to infer causal relationships among the independent and dependent variables. However this can direct to illusions or wrong relationships, so warning is advisable; for example: correlation does not confirm causation. Several techniques for carrying out regression analysis have been developed. Similar methods such as linear regression and ordinary least squares regression are parametric, in that the regression function is distinct in terms of a finite number of unidentified parameters that are estimated from the data. Nonparametric regression refers to techniques that allocate the regression function to lie in a particular set of functions which may be infinite dimensional. The performance of regression analysis methods in tradition depends on the structure of the data generating process and how it relates to the regression approach being used. Since the correct form of the data generating process is usually not known, regression analysis frequently depends to various extents on making assumptions about this process. These assumptions are occasionally testable if a satisfactory quantity of data is accessible. Regression models for prediction are regularly useful even when the assumptions are reasonably violated even though they may not execute optimally. However, in numerous applications, mainly with little effects or questions of causality based on observational data, regression methods can present misleading results. In a narrower sense, regression may refer exclusively to the estimation of continuous response (dependent) variables, as contrasting to the distinct response variables used in classification [28]. The case of a continuous dependent variable may be more exclusively referred to as metric regression to differentiate it from related problems.

As you develop cause & effect diagrams based on data, you may wish to observe the degree of association between variables. A statistical measurement of correlation can be calculated using the least squares method to measure the strength of the relationship between two variables. The output of that calculation is the correlation coefficient, or (r), which ranges between -1 and 1. A value of 1 indicates great positive correlation as one variable increases, the second increases in a linear fashion. Similarly, a value of -1 indicates perfect negative correlation as one variable increases, the second decreases. A value of zero indicates zero correlation. Before calculating the correlation coefficient, the initial step is to create a scatter diagram. Most spreadsheets including Excel, can handle this task. While correlation analysis assumes no fundamental relationship between variables, regression analysis assumes that one variable is dependent upon: A) a different single independent variable (simple regression) and B) multiple independent variables (multiple regression). Regression plots a line of best fit to the data using the least-squares method [29].

Simple linear regression is a linear regression model with a solo explanatory variable. That is, it concerns two-dimensional sample points through one independent variable and one dependent variable (traditionally, the x and y coordinates in a Cartesian coordinate system) and finds a linear function (a non-vertical straight line) that as accurately as feasible, predicts the dependent variable values as a function of the independent variables. The adjective simple refers to the fact that the result variable is related to a single predictor. It is general to make the supplementary stipulation that the ordinary least squares method should be used. The correctness of each predicted value is measured by its squared residual (vertical distance among the point of the data set and the fitted line) and the objective is to make the sum of these squared deviations as small as probable. Other regression methods that can be used in position of ordinary least squares contain least absolute deviations (minimizing the sum of absolute values of residuals) and the Theil-Sen estimator (which chooses a line whose slope is the median of the slopes determined by pairs of sample points). Deming regression (total least squares) as well finds a line that fits a set of two-dimensional sample points, but (contrasting least absolute deviations, ordinary least squares, and median slope regression) it is not actually an instance of simple linear regression because it does not divide the coordinates into one dependent and one independent variable and could potentially return a vertical line as its fit [30].

Multiple regression analysis uses a same methodology as simple regression but includes more than one independent variable. Multiple regression analysis is a statistical technique that can be used to analyze the relationship between a particular response variable and multiple predictor variables. Econometric models are an excellent example where the dependent variable of GNP may be analyzed in conditions of multiple independent variables such as: interest rates, government spending, productivity growth, savings rates, and consumer confidence etc. Many times historical data is used in multiple regression in a challenge to identify the most important inputs to a process. The advantage of this type of analysis is that it can be completed very fast and comparatively simply. However, there are numerous possible pitfalls: The data may be not consistent due to diverse measurement systems, different operators, calibration drift, and recording errors. The range of the variables may be extremely limited and can offer a false indication of low correlation. For example: a process may have temperature controls because temperature has been found in the past to have an effect on the output. Using historical temperature data may therefore specify low significance because the range of temperature is already controlled in tight tolerance. There may be a time lag that influences the relationship, for example: temperature may be much more crucial at an early point in the process than at a later point or vice-versa [31]. There also may be inventory effects that must be taken into account to create sure that all measurements are taken at a reliable point in the process.

It is essential to memorize that correlation is not causality. The regression analysis only tells us what did have an impact on efforts not what could have an impact. If the range of measurements was greater, we might have seen a strong relationship with efforts and extra variability in the output. The regression analysis tool is an advanced tool that can recognize how diverse variables in a process are related. The regression tool will tell you if one or multiple variables are related with a process output. This information can recognize where in the process control is needed or what factors is the most excellent starting point for a process improvement project.

Regression models must satisfy five assumptions to be valid in their results:

1. The independent variables and the dependent variables have a linear relationship.
2. The dependent variable is a continuous random variable and the independent variables are set at various values and are not random.
3. The variances of the dependent variable are evenly distributed given various combinations of the independent variables.
4. Successive observed values of the dependent variable are uncorrelated.

5. The distribution of the sampling error in the regression model is common.

The regression model is used to depict the effects of the independent variables on the dependent variable. If the independent variables are not linearly independent from each other, determining the involvement of each independent variable will be difficult because the effects of the independent variables are mixed. Thus the regression coefficients may be wrongly estimated. Interdependence of independent variables is called multicollinearity [32].

8.5 SUMMARY

Software process maturity was a major factor affecting software development effcrt. Process maturity should be in all cost models. CMM is well defined. A software process model is a structured collection of practices that illustrate the characteristics of effective processes and the practices incorporated are those proven by experience to be effective. It establishes criteria to evaluate processes used to develop software products. It provides an important assessment of the effects of process on development effort. The most dominant factor in predicting effort in these models is the size of the software product. There are extra factors that also influence effort such as product complexity, the application experience of the software development team, and software development tool support. Regression analysis is a set of statistical processes for estimating the relationships between variables. It includes various techniques for modelling and analyzing various variables, when the centre is on the relationship between a dependent variable and one or more independent variables.

8.6 REFERENCES

1. Paulk, M. C.; Curtis, B.; Chrissis, M. B.; Weber, C. V. (1993). Capability maturity model. Version 1.1. IEEE Software, Vol. 10, Issue 4, pp. 18-27.
2. Herbsleb, J.; Zubrow, D.; Goldenson, D.; Hayes, W.; Paulk, M. (1997). Software quality and the capability maturity model. Communications of the ACM, Vol. 40, No. 6, pp. 30-40.
3. http://www.selectbs.com/process-maturity/what-is-the-capability-maturity-model
4. Paulk, M. C. (1993). Comparing ISO 9001 and the capability maturity model for software. Software Quality Journal, Vol. 2, No. 4, pp. 245-256.
5. Kneuper, R. (2018). Software Process Assessment and Improvement. In Software Processes and Life Cycle Models, pp. 211-260, Springer, Cham.
6. Heckman, K. E.; Stech, F. J.; Thomas, R. K., Schmoker, B.; Tsow, A. W. (2015). Capability Maturity Model. In Cyber Denial, Deception and Counter Deception, pp. 127-157, Springer, Cham.
7. O'Regan, G. (2017). Capability Maturity Model Integration. In Concise Guide to Software Engineering, pp. 255-277, Springer, Cham.
8. Paulk, M. C. (2009). A history of the capability maturity model for software. ASQ Software Quality Professional, Vol. 12, No. 1, pp. 5-19.
9. Bamberger, J. (1997). Essence of the capability maturity model. Computer, Vol. 30, No. 6, pp. 112-114.
10. Weber, C. V.; Paulk, M. C.; Wise, C. J.; Withey, J. V. (1991). Key practices of the capability maturity model (No. CMU/SEI-91-TR-25). Carnegie-Mellon Univ Pittsburgh PA Software Engineering Inst.

11. Fraser, P.; Moultrie, J.; Gregory, M. (2002). The use of maturity models/grids as a tool in assessing product development capability. In IEEE International Conference on Engineering Management Conference, IEMC'02. Vol. 1, pp. 244-249.
12. Ferguson, J. R. (2002). Software Acquisition Capability Maturity Model (SA-CMM). Encyclopedia of Software Engineering.
13. Kneuper, R. (2008). *CMMI: Improving Software and Systems Development Processes Using Capability Maturity Model Integration*. Rocky Nook.
14. Niazi, M.; Wilson, D.; Zowghi, D. (2005). A maturity model for the implementation of software process improvement: an empirical study. Journal of systems and software, Vol. 74, Issue 2, pp. 155-172.
15. McGuire, E. G. (1996). Factors affecting the quality of software project management: An empirical study based on the Capability Maturity Model. Software Quality Journal, Vol. 5, No. 4, pp. 305-317.
16. Shang, S. S.; Lin, S. F. (2009). Understanding the effectiveness of Capability Maturity Model Integration by examining the knowledge management of software development processes. Total Quality Management, Vol. 20, No. 5, pp. 509-521.
17. von Wangenheim, C. G., Hauck, J. C. R.; Salviano, C. F.; von Wangenheim, A. (2010). Systematic literature review of software process capability/maturity models. In Proceedings of International Conference on Software Process Improvement and Capabity Determination (SPICE), Pisa, Italy.
18. Padma, P.; Ganesh, L. S.; Rajendran, C. (2008). An exploratory study of the impact of the capability maturity model on the organizational performance of Indian software firms. Quality Management Journal, Vol. 15, Issue 2, pp. 20-34.
19. Software Estimation https://courses.cs.washington.edu/courses/cse403/07sp/assignments/estimationbasics.pdf
20. Shepperd, M.; MacDonell, S. (2012). Evaluating prediction systems in software project estimation. Information and Software Technology, Vol. 54, No. 8, pp. 820-827.
21. Chiu, N. H.; Huang, S. J. (2007). The adjusted analogy-based software effort estimation based on similarity distances. Journal of Systems and Software, Vol. 80, Issue 4, pp. 628-640.
22. Kitchenham, B. (1990). *Software development cost models*. Software Reliability Handbook, P. Rook (Ed.) Elsevier Applied Science, New York.
23. Park, H.; Baek, S. (2008). An empirical validation of a neural network model for software effort estimation. Expert Systems with Applications, Vol. 35, Issue 3, pp. 929-937.
24. Shepperd, M.; Schofield, C. (1997). Estimating software project effort using analogies. IEEE Transactions on software engineering, Vol. 23, No. 11, pp. 736-743.
25. Finnie, G. R.; Wittig, G. E.; Desharnais, J. M. (1997). A comparison of software effort estimation techniques: using function points with neural networks, case-based reasoning and regression models. Journal of systems and software, Vol. 39, Issue 3, pp. 281-289.
26. Prokopova, Z.; Silhavy, P.; Silhavy, R. (2018, September). Influence Analysis of Selected Factors in the Function Point Work Effort Estimation. In Proceedings of the Computational Methods in Systems and Software, pp. 112-124, Springer, Cham.
27. Abdellatif, T. M. (2018). A Comparison Study between Soft Computing and Statistical Regression Techniques for Software Effort Estimation. In 2018 IEEE Canadian Conference on Electrical & Computer Engineering (CCECE), pp. 1-5.
28. de Barcelos Tronto, I. F.; da Silva, J. D. S.; Sant'Anna, N. (2007). Comparison of artificial neural network and regression models in software effort estimation. In IEEE International Joint Conference on Neural Networks IJCNN, pp. 771-776.
29. Jorgensen, M. (1995). Experience with the accuracy of software maintenance task effort prediction models. IEEE Transactions on software engineering, Vol. 21, No. 8, pp. 674-681.
30. Montgomery, D. C.; Peck, E. A.; Vining, G. G. (2012). *Introduction to linear regression analysis*. Vol. 821, John Wiley & Sons.

31. Molokken, K.; Jorgensen, M. (2003). A review of software surveys on software effort estimation. In IEEE International Symposium on Empirical Software Engineering, ISESE, pp. 223-230.
32. Heiat, A. (2002). Comparison of artificial neural network and regression models for estimating software development effort. Information and software Technology, Vol. 44, Issue 15, pp. 911-922.

Chapter - 9

SOFTWARE PROCESS MODEL AND KNOWLEDGE MANAGEMENT

In a fast and dynamic competitive environment it is not easy to survive and maintain a credit in market. It can be possible when user trust on product quality and its performance. This can be possible when software process is well defined because quality of software product directly related to the quality of software process. To develop a mature and effective process lot of resources is required, every organization want to achieve that at a minimum cost and time. That can be achieved when they effectively utilize the knowledge and experience. At present knowledge management is an area where quality is derived. In this chapter hybrid spiral model is discussed to show how knowledge management is used to achieve quality in finished product. Hybrid spiral model is an integration of spiral model and knowledge management which is used to improve software quality using knowledge management. It has been illustrated with example using knowledge flow during process. Section I describes the introduction of this chapter. Section II describes the related work of software process model and knowledge management. Section III describes the hybrid spiral model. Section IV explains the summary of this chapter.

9.1 INTRODUCTION

When experience and knowledge is created, evaluated, engineered, maintained, and disseminated systematically to solve a problem, then this is called knowledge management. Software engineering is knowledge management based activity which requires creativity and experience. When experience and knowledge is created, evaluated, engineered, maintained, and disseminated systematically to solve a problem, then this is called knowledge management [1]. Using software development process software is created. To achieve software quality effective process is required. Knowledge helps organization to perform better, faster and develop quality product within time and budget. Software quality totally based upon the quality of process, which is based on the knowledge and experience of whole team of software development. Well define process is required to achieve quality in finished product.

The knowledge related to a company is known as corporate knowledge. Corporate knowledge is the explicit and tacit knowledge; these are two types of knowledge. Explicit knowledge is tangible knowledge which expressed formally, can be seen, stored and updated. Explicit knowledge found in files, databases, documents, research papers, books, source code, e-mail, messages, any written processes and procedures. Tacit knowledge is richer and more valuable than explicit knowledge, which cannot be easily expressed and communicated. It is about human experience, creativity, feelings, and interpretation. Tacit knowledge is based on mental models and perspectives so that people take them for granted, so making it difficult to articulate. Tacit knowledge is found in the heads of individual person (employees and experts), in the collective sharing of employees, individual person (employees and experts), in the collective sharing of employees, conversations, unwritten processes and procedures and organizational culture [2-5].

There is various quality models [6] are propcsed, to improve and evaluate the quality of software products such as: McCall's Model [7], Boehm Model [8], FURPS Model [9], Dromey Model [10] and ISO 9126 model [11]. McCall's Model and Boehm Model are same and use hierarchical decomposition of quality factors such as maintainability or reliability. After that FURPS model decomposes quality into functionality, usability, reliability, performance and supportability. These quality models become the basis for ISO/IEC 9126 [12].

For survive in competitive environment and perform effectively knowledge management is required because knowledge management (KM) is a quality management (QM). These quality models useful to derive quality but there is a need to merge these quality models to knowledge management for better performance and output. There are few quality models which uses knowledge management [13] such as: Nonaka Model [14], Capability Maturity Model (CMM) [15] and Knowledge Process Quality Model (KPQM) [16].

Organization uses this combination for improvement, enhancement and customization of software development process. Using that transformation we can improve the entire software development process [17]. This trend is continued and lots of new KM model was developed and proposed according the requirement and need. A General Knowledge Management Maturity Models (G-KMMM), objective to assess the maturity of people, process and technology aspects of KM in organizations [18].

9.2 RELATED WORK

For improvement in organization performance effective knowledge management is required because knowledge management is a quality management. Knowledge management is an emerging area which is a dynamic process using that we can create, change and reuse the knowledge to develop high quality product at low cost. Recently academics start to relate knowledge management to quality management [19]. The first study that relate quality and learning, when Fine develop an analytical model and studied the relationship between failure and conformance cost, result show that quality level enhance over time due to learning [20].

Nonaka theory uses knowledge creation as central theme for linking quality to knowledge; it considers both tacit and explicit knowledge. Also suggest knowledge creation in spiral form using four modes of knowledge conversion. Knowledge is created through conversion between tacit and explicit knowledge: 1) from tacit knowledge to tacit knowledge (socialization), 2) from tacit knowledge to explicit knowledge (externalization), 3) from explicit knowledge to explicit knowledge (combination), and 4) from explicit knowledge to tacit knowledge (internalization) which is called SECI model [21].

The research is continued and lot of influences identified on knowledge management. This included culture, technology, leadership, education, measurement, organizational adjustment, administrative knowledge manipulation activities, values and norms, knowledge resources, evaluation of knowledge management activities, employee motivation, knowledge resources, and external factors [22]. KM implementation and its use have rapidly increased to help organization for better productivity. Because software development is changes quickly and many people is involve for development so for improve productivity and quality software organization use KM in software engineering. Using this combination cost and time reduce and quality increase, share and capture knowledge, provides better decision making, know new technology, and accessing domain knowledge [23].

For better management of knowledge experience management also required because knowledge and experience is interrelated. While knowledge management (KM) has received much attention then experience management (EM) to fill this gap a waterfall model for knowledge management and experience management was proposed which integrate experience or knowledge processing and its management [24]. After that a hierarchical spiral model (HSM) for knowledge management was proposed for development of knowledge management (KM) and information systems (IS). This proposed model provides the guidance between the different phases of knowledge management activities [25].

After several researches new direction towards knowledge management (KM) in software requirement engineering (SRE) to join that combination a knowledge management framework in SRE based on the SECI model was proposed. The aim of this research exploit tacit and explicit knowledge related to software requirements in software project. The main part of the proposed framework is a set of four sub systems such as: socialization, externalization, combination, and internalization connected to a couple of domain ontology and a set of knowledge assets [26].

Presently knowledge management becomes more complex and now users demanding more accurate and reliable knowledge management systems (KMS) to enhance quality. There are software quality dimensions, data quality dimensions, information system quality dimensions, and knowledge management system quality dimensions. Using mapping the quality dimensions to the knowledge management processes, the quality of these processes will be improved as a result and assume that the quality of the KMS will be enhanced too [27]. So there is a need for effective knowledge management to develop a quality product.

There are various software development processes such as waterfall model, iterative and incremental model, V model, spiral model and agile model which have its own advantages and disadvantages. These models are selected according the software requirement and projects [28]. As various models have been proposed by various researchers, all of them are based on explicit knowledge. Existing models have used development process, which has been implemented by person who is having explicit knowledge. Hybrid spiral model uses tacit and explicit knowledge in various steps. To explain hybrid spiral model Boehm spiral model is selected because spiral model is a meta-model and accommodate any development process.

9.3 HYBRID SPIRAL MODEL

At present software development process continue change according the need and time. There are many people involve in development process. Software organizations use a close relationship of people, process and technology for software development and quality based product at minimum resources (time, cost, and people). And also want to improve productivity, for that purpose knowledge is required. Using reuse knowledge and experience we can achieve that at a minimum cost [29].

Knowledge Management is being used for efficiency, improvement, performance, maturity, and maintenance of software products. Using integration of technical knowledge with business application domain knowledge software development process increases software development effectiveness and reduces defect density. And also increases software development efficiency and performance [30].

Knowledge Management also used for process improvement. Knowledge management approach adopted in a CMM level 3 to support organizational process tailoring to projects and improvement based on metric data collected from past projects and maintains the project according changes. Using experience and knowledge software process performance improve [31].

Table 9.1 Impact of Knowledge Management in Software Development Process

Concepts	Main Findings	Ref.
Software Process Improvement (SPI)	What knowledge management approach you select in SPI, you need to create both tacit and explicit knowledge. Tacit is necessary to change practice and method while explicit is necessary to create an organizational memory or document.	[32]
	A techno centric approach to SPI may impose unnatural work on an organization and fails to take account of how improvements might occur in practice.	[33]
CMM	Knowledge management is used as development theory in which a set of key process areas supplement to CMM in small or medium sized enterprises (SME) that develop software.	[34]
Unified Process	Iterative approach of unified process effects learning and improves communication and work distribution in the company.	[35]
Software Process	It is possible to define and implement software process in a cost effective way in small organizations. Special considerations must be given to their specific business goals, characteristics, models, and resource limitations.	[36]

We have drawn a Table 9.1 based on various research which impact on software development process using knowledge management in various conditions and environment.

Software development process which are divided into plan-based or traditional development process and agile model. In plan-based or traditional development process there is lack of learning and past experience. In agile model effective use of learning and past experience are involve for high customer satisfaction, better decision making in changing environment and quick delivery of quality product. Table 9.2

show the importance of knowledge and changes in software development process according time and need.

Table 9.2 Software Development Process and Knowledge

Software Development Process	Knowledge
Plan Based or Traditional Development Process	Explicit Knowledge
Agile Methods	Tacit Knowledge

Table 2 shows that in traditional development process such as waterfall model explicit knowledge is used and at present agile methods uses tacit knowledge for development [37].

Spiral model is a risk driven process model uses two main features: 1) Cyclic approach for incrementally growing a system's degree and implementation while decreasing its risk and 2) Anchor Point Milestones for ensuring stakeholder commitment to feasible and mutual satisfactory system solutions [38 -39].

Hybrid spiral model uses knowledge management which is based on knowledge flow during process. Knowledge flow of this model and how the knowledge flows from one phase to another phase as shown in Figure 9.1. Hybrid spiral model which is integration of spiral model and knowledge management are shown in Figure 9.2.

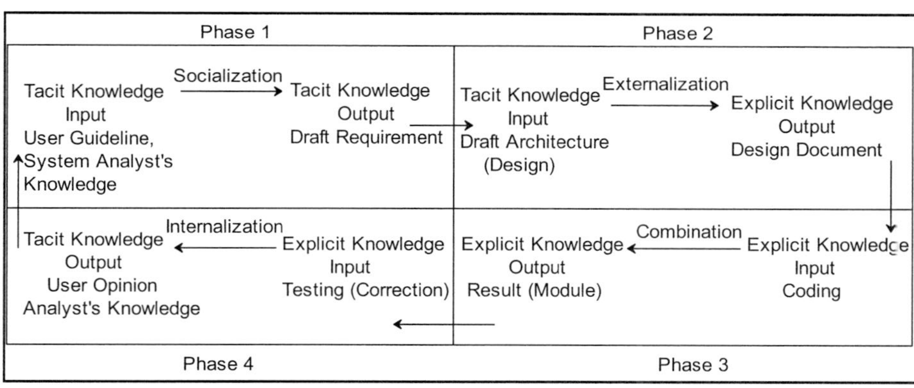

Figure 9.1 Knowledge Flow in Hybrid Spiral Model

Both diagrams are used to describe the how knowledge is used to develop a quality product. Every software development process uses four phase for development such as Requirement, Design, Coding, and Testing. Figure 9.1 shows how these four phases use a knowledge management and flow from one phase to another phase to develop a software product.

PHASE 1: In Figure 9.1 tacit to tacit knowledge conversion which is called socialization is used to show the phase 1 (Requirement). Phase 1 uses tacit knowledge as input and output both. User guideline and system analyst's knowledge used as an input and after requirement gathering draft requirement is created which is an output.

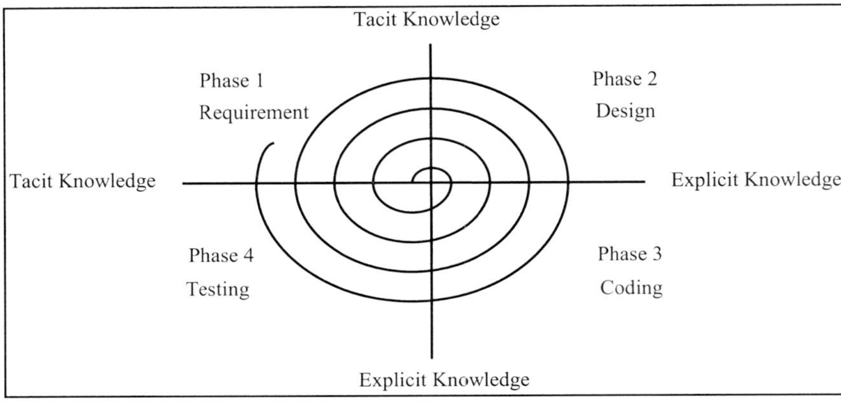

Figure 9.2 Hybrid Spiral Model to Improve Software Quality

PHASE 2: That draft requirement is used as input to phase 2 (Design) in that case after tacit to tacit knowledge conversion draft architecture is build up. That draft architecture is used as input and design document which is an output. Phase 2 uses tacit to explicit knowledge conversion which is called externalization.

PHASE 3: This design document is used an input for phase 3 (Coding) and using explicit to explicit knowledge conversion a code is develop for software. That code is used as input and using knowledge conversion code is run and check by code developer and that module is an output. In phase 3 explicit to explicit knowledge conversion is used which is called combination.

PHASE 4: That module (code) used as input for phase 4 (Testing) and using knowledge conversion module is tested by software testing team and find out errors or bugs and fix it. After correction draft product is created which is used as input and send to user for any suggestions and feedback if any. The user and analyst meeting or gathering used as output. In phase 4 explicit to tacit knowledge conversion is used which is called internalization.

We use a cyclic and iterative flow of knowledge for software development. In phase 4 if user is satisfied that draft product and its quality then we stops further iteration. Otherwise we go to phase 1 and start collecting user guideline and follow the same processing.

The main characteristic of spiral model is its cyclic nature. Each cycle of spiral model uses four stages. Stage 1 is used to determine the objective and alternatives, stage 2 used to evaluate the alternative and find out risks, stage 3 used to develop and identify next level of product and stage 4 are used to review the results and plan for next iteration [40].

Figure 9.2 shows the hybrid spiral model for software development which is combination of spiral model and knowledge management. For requirement tacit to tacit knowledge required, for design tacit to explicit knowledge required, for coding explicit to explicit knowledge required, and for testing explicit to tacit knowledge required.

All four phase have an individual role and importance but requirement and design play a major and important role to derive quality in finished product. So to capture all requirement and quality parameter effective process is required which can be possible to use a balance relationship between development process and knowledge management to improve software quality.

This hybrid spiral model uses simple algorithm which is based upon very simple and few steps:

Step 1: Collect software requirement using tacit knowledge means experienced system analysts are assigning to do that job.

Step 2: When SRS (Software Requirement Specification) documented using design process software architecture is derived with the help of tacit knowledge means experienced system designer or architect are involve to do that phase.

Step 3: In next step with the help of design document software design is translated into source code using explicit knowledge means fresher code developer perform that task.

Step 4: After coding each module is tested using explicit knowledge means fresher software testing team are used to do that work.

Step 5: In a last step when final product is develop then this product goes through the user or customer for using and checking the software. If user satisfied with the product performance and quality then no further processing is required otherwise go to step 1 and start collecting requirement.

This algorithm follows these simple steps and it is based upon iteration and cycle form same as spiral model except step 1 and 2 we use tacit knowledge in place of explicit knowledge. Using hybrid spiral model (Spiral Model and KM) we save time, cost, and resource because using that model we reuse knowledge and experience which can be helpful to develop software in few iteration and quality can be improve with the help of hybrid spiral model.

As per algorithm library management system is developed using Java language and NetBeans IDE 8.1 environment to check to how this hybrid spiral model differs from spiral model and how knowledge management is used to develop a quality product. Two teams is selected which is divided into tacit and explicit knowledge.

In team 1 fresher person are involve which is based upon explicit knowledge and team 2 experience person are involve which is based upon tacit knowledge. Four persons are selected for each team. For team 1 explicit knowledge person (approximate 1-2 years experience) team persons denoted as E1, E2, E3, and E4. For team 2 tacit knowledge persons (approximate 12 years experience) team persons denote as T1 and T2.

First library management system is developed using spiral model then this software is developed using hybrid spiral model. To develop library management system using spiral model team 1 is selected and team 1 and 2 is selected for hybrid spiral model. Spiral model uses explicit knowledge for development and there is more iteration to develop a quality product. Hybrid spiral model uses tacit and explicit knowledge for development and there is less iteration to develop a quality product. Illustration of spiral model and hybrid spiral model are shown in Table 3. Table 3 shows the working of both model and comparison also.

Table 9.3 Illustration of Spiral Model and Hybrid Spiral Model

Spiral Model		**Hybrid Spiral Model**	
Assigned Person	Assigned Job	Assigned Person	Assigned Job
Team 1 E1	For software requirement (**Step 1**) explicit knowledge is used and there are less correctness and consistency	Team 2 T1	For software requirement (**Step 1**) tacit knowledge is used and there are more correctness and consistency
Team 1 E2	For software architecture (**Step 2**) explicit knowledge is used and there are less usability and reliability	Team 2 T2	For software architecture (**Step 2**) tacit knowledge is used and there are more usability and reliability
Team 1 E3	For software coding (**Step 3**) explicit knowledge is used	Team 1 E3	For software coding (**Step 3**) explicit knowledge is used
Team 1 E4	For software testing (**Step 4**) explicit knowledge is used	Team 1 E4	For software testing (**Step 4**) explicit knowledge is used

When we use hybrid spiral model to develop this software same language, environment and processing is used. In step 1 and step 2 tacit knowledge involve means an experienced person are selected to develop the software so we select team 1 and 2 for development. Using that algorithm we develop quality product at minimum cost and time.

For software requirement and design team 2 is assigned to do that job. Team 1 is selected to develop code and perform software testing. After testing final product is used by user and user feedback suggest that all good but more performance is required. Next iteration and same processing is used to develop software after development of final revise product that software used by user and user satisfied with the performance and quality. This illustration shows that hybrid spiral model is effective and efficient than existing spiral model.

Social factors, environment and experience are significant impact upon process of software development [41]. There are environmental standards and management systems for illusion of progress [42]. Knowledge has its own characteristics and importance. Both types of knowledge has important role for development but tacit knowledge more important than explicit knowledge because it is based upon learning, experience and creativity. Using tacit knowledge in requirement and design we achieve quality in finished product because an experience people are involve for development and few chances in failure and overrun of time and cost. Table 4 shows the comparison between existing spiral model and that hybrid spiral model.

Table 9.4 Comparison between Spiral Model and Hybrid Spiral Model

Spiral Model	Hybrid Spiral Model
Explicit Knowledge	Tacit and Explicit Knowledge
Costly model to use	Easy to use and maintain
Risk analysis requires highly specific expertise	No risk problem
Doesn't work well for smaller projects	Used for small projects
Spiral may go infinitely	Few iteration are used
More documentation	Less documentation
More time and cost used for development	Less time and cost used for development
Improve robustness and correctness	Improve consistency, correctness, usability, reliability, and performance

Hybrid spiral model is more efficient based upon various parameters given Table 9.4. It proves that using knowledge management we improve the performance of hybrid spiral model. Knowledge management is required for successful implementation of product and achieves quality in finished product.

9.4 SUMMARY

This chapter focused on software process model and knowledge management. The main aim of this chapter is to design an algorithm for hybrid spiral model which is based upon simple iteration. Hybrid spiral model is a modified version of spiral model which uses knowledge management. Step 1 and 2 of algorithm uses tacit knowledge which improves the software quality of finished product. It has been tested on academic environment to develop a library management system that show that using hybrid spiral model we save time and cost due to reduction of iteration or cycle in comparison of spiral model. So, effective knowledge management is required to drive quality in finished product. Knowledge management is useful to every development phase of software if we use tacit knowledge in requirement and design phase than the output is quality based product

9.5 REFERENCES

1. Schneider, K. (2009). *Experience and knowledge management in software engineering*. Springer-Verlag, Berlin Heidelberg.
2. Meehan, B.; Richardson, I. (2002). Identification of software process knowledge management. Softw. Process Improve. Pract., Vol. 7, No. 2, pp. 47-55.
3. Ebert, C.; Man, J. D. (2008). Effectively utilizing project, product and process knowledge. Information and Software Technology, Vol. 50, Issue 6, pp. 579-594.
4. Dalkir, K. (2005). *Knowledge management in theory and practice*. Elsevier.
5. Bergeron, B. (2003). *Essentials of knowledge management*. John Wiley & Sons Inc.
6. Singh, B.; Kannojia, S. P. (2013). A review on software quality models. IEEE International Conference on Communication Systems and Network Technologies, pp. 801-806.
7. McCall, J. A.; Richards, P. K.; Walters, G. F. (1977). *Factors in software quality*. National Technical Information Service, Springfield.
8. Boehm, B. W.; Brown, J. R.; Kaspar, H.; Lipow, M.; Macleod, G. J.; Merrit, M. J. (1978). *Characteristics of software quality*. North-Holland, Amsterdam.

9. Grady, R. B.; Caswell, D. L. (1987). *Software metrics: establishing a company-wide program*. Prentice Hall, Englewood Cliffs.
10. Dromey, R. G. (1995). A model for software product quality. IEEE Transactions on Software Engineering, Vol. 21, Issue 2, pp. 146-162.
11. ISO/IEC 9126-1:2001: Software engineering - product quality - Part 1: quality model, 2001.
12. Wagner, S. (2013). Software product quality control. Springer-Verlag, Berlin Heidelberg.
13. Haslinda, A.; Sarinah, A. (2009). A review of knowledge management models. The Journal of International Social Research, Vol. 2, No. 9, pp. 187-198.
14. Nonaka, I. (1991). The knowledge creating company. Harvard Business Review, pp. 69-104.
15. Paulk, M. C.; Curtis, B.; Chrissis, M. B.; Weber, C. V.; (1993). Capability Maturity Model for Software Version 1.1, Software Engineering Institute, Pittsburgh, CMU/SEI-93-TR-024.
16. Paulzen, O.; Doumi, M.; Perc, P.; Cereijo-Roibas, A. (2002). A maturity model for quality improvement in knowledge management. 13th Australasian Conference on Information Systems (ACIS), Paper 5, pp. 1-11.
17. Lee, M-C.; Chang, T.; (2005). Applying TQM, CMM and ISO 9001 in knowledge management for software development process improvement. Int. J. Services and Standards, Vol. 2, No. 1, pp. 101-115.
18. Teah, H. Y.; Pee, L. G.; Kankanhalli, A. (2006). Development and application of a general knowledge management maturity model. The Tenth Pacific Asia Conference on Information Systems (PACIS), Paper 12, pp. 401-416.
19. Lindermana, K.; Schroedera, R. G.; Zaheera, S.; Liedtkeb, C.; Choo, A. S. (2004). Integrating quality management practices with knowledge creation processes. Journal of Operations Management, Vol. 22, No. 6, pp. 589-607.
20. Fine, C. H. (1986). Quality improvement and learning in productive systems. Management Science, Vol. 32, No. 10, pp. 1301-1315.
21. Nonaka, I. (1994). A dynamic theory of organizational knowledge creation. Organization Science, Vol. 5, No. 1, pp. 14-37.
22. Holsapple, C. W.; Joshi, K. D. (1999). Description and analysis of existing knowledge management frameworks. 32nd Hawaii International Conference on System Sciences, pp. 1-15.
23. Rus, I.; Lindvall, M. (2002). Knowledge management in software engineering. IEEE Software, Vol. 19, No. 3, pp. 26-38.
24. Sun, Z. (2004). A waterfall model for knowledge management and experience management. IEEE Fourth International Conference on Hybrid Intelligent Systems, pp. 472-475.
25. Sun, Z.; Hao, G. (2006). HSM: a hierarchical spiral model for knowledge management. 2nd International Conference on Information Management and Business, pp. 542-551.
26. Chikh, A. (2011). A knowledge management framework in software requirements engineering based on the SECI model. Journal of Software Engineering and Applications, Vol. 4, No. 12, pp. 718-728.
27. Jabar, M. A.; Alnatsha, A. S. M. (2014). Knowledge management system quality: a survey of knowledge management system quality dimensions. IEEE International Conference on Computer and Information Sciences (ICCOINS), pp. 1-5.
28. Ghezzi, C.; Jazayeri, M.; Mandrioli, D. (2003). *Fundamentals of software engineering*. Second Edition, Prentice Hall of India.
29. Basili, V. R.; Caldiera, G. (1995). Improve software quality by reusing knowledge and experience. MIT Sloan Management Review, Vol. 37, No. 1, pp. 55-64.
30. Tiwana, A. (2004). An empirical study of the effect of knowledge integration on software development performance. Information and Software Technology, Vol. 46, Issue 13, pp. 899-906.

31. Falbo, R. de A.; Borges, L. S. M.; Valente, F. F. R. (2004). Using knowledge management to improve software process performance in a CMM level 3 organization. IEEE Fourth International Conference on Quality Software (QSIC), pp. 162-169.
32. Arent, J.; Nørbjerg, J.; Pedersen, M. H. (2000). Creating organizational knowledge in software process improvement. 2nd Workshop on Learning Software Organizations, pp. 81-92.
33. Segal, J. (2001). Organisational learning and software process improvement: a case study. Springer Third International Workshop on Learning Software Organizations, pp. 68-82.
34. Baskerville, R.; Pries-Heje, J. (1999). Knowledge capability and maturity in software management. The DATA BASE for Advances in Information Systems, Vol. 30, No. 2, pp. 26-43.
35. Folkestad, H.; Pilskog, E.; Tessem, B. (2004). Effects of software process in organization development - a case study. Springer Sixth International Workshop on Learning Software Organizations, pp. 153-164.
36. Wangenheim, C. G. V.; Weber, S.; Hauck, J. C. R.; Trentin, G. (2006). Experiences on establishing software processes in small companies. Information and Software Technology, Vol. 48, Issue 9, pp. 890-900.
37. Bjørnson, F. O.; Dingsøyr, T. (2008). Knowledge management in software engineering: a systematic review of studied concepts, findings and research methods used. Information and Software Technology, Vol. 50, Issue 11, pp. 1055-1068.
38. Boehm, B. (1988). A spiral model of software development and enhancement. IEEE Computer, Vol. 21, Issue 5, pp. 61-72.
39. Boehm, B. (2000). Spiral development: experience, principles, and refinements. Carnegie Mellon Software Engineering Institute, Pittsburgh, PA, CMU/SEI-2000-SR-008.
40. Boehm, B. (1986). A spiral model of software development and enhancement. ACM Sigsoft Software Engineering Notes, Vol. 11, No. 4, pp. 22-42.
41. Adolpha, S.; Kruchtena, P.; Hall, W. (2012). Reconciling perspectives: a grounded theory of how people manage the process of software development. Journal of Systems and Software, Vol. 85, Issue 6, pp. 1269-1286.
42. Yoxon, M.; Sheldon, C. (2008). Environmental standards, management systems and the illusion of progress. International Journal of Performability Engineering, Vol. 4, No. 4, pp. 385-399.

Chapter - 10

DESIGN PATTERN AND SOFTWARE CODING

Software design is the one of the most imperative phase of software process to achieve quality in final product. Design patterns are reusable solutions to general design problems that are estimated to improve various quality attributes. Impact of software design on software quality has been discussed and analyzed how software design is used to develop quality product. Software programming languages are having important role in software quality. It is essential to choose programming language based on functional and non functional requirements of software. C, C++, Java, Python, and Mathematica programming languages analyzed to see the impact of software coding on software quality. Code quality analysis used to explain the importance of programming languages to develop a quality product. Section I describes the introduction of design pattern. Section II describes the related work of design pattern. Section III describes the design pattern assessment method. Section IV describes the assessment of design pattern on software quality. Section V describes the impact of programming languages on software quality. Section VI describes the overview of programming languages. Section VII describes the code quality analysis. Section VIII describes the comparison of programming languages. Section IX describes the impact analysis of software coding on software quality. Section X describes the summary of this chapter.

10.1 INTRODUCTION

Software design patterns provide solutions to common software design problems. Object oriented (OO) design patterns have been proposed in mid 1990s as common solutions of design problems and considered as standard of good software designs. Each design pattern includes class diagrams, explanation, usage information, and real world example. Design patterns offer maintainability, reusability, understandability, and flexible design solution. Design Patterns provides a rapid reference to the original 23 Gang of Four (GoF) design patterns classified according to two main categories. In first category, purpose and motivation of pattern is represented which is grouped into three categories: creational patterns, structural patterns, and behavioral patterns that describe using concepts of aggregation, consultation, and delegation [1].

Creational patterns are used to construct objects which can be decoupled from their implementing system. Creational patterns divided into five categories such as: abstract factory, builder, factory method, prototype, and singleton. Structural patterns are used to outline large object structures between many contrasting objects which is classify as: adapter, bridge, composite, decorator, facade, flyweight, and proxy. Behavioral patterns are used to manage algorithms, relationships, and responsibility between objects which is divided as: chain of responsibility, command, interpreter, iterator, mediator, memento, observer, state, strategy, template method, and visitor [2].

In second category scope defines whether the pattern is applied on object and class level. Object level deals with object relationships which can be modified at runtime. Class level deals with class relationships that can be altered at compile time. Using design pattern reusability, composability, completeness, maintainability, stability, understandability, flexibility, simplicity, and expandability are improved [3].

During past two decades, numbers of researches are concentrated on qualitative evaluation of design patterns and quantitative evaluation of design patterns.

Numerous studies tried to qualitatively evaluate the effects of object oriented design pattern on software quality. In 2001 McNatt and Bieman, observe the concept of pattern coupling to classify how design may contain coupled patterns. They qualitatively evaluate the goodness of pattern coupling as tight and loose which effects on maintainability, factorability, and reusability when patterns are coupled in various ways that provides both benefits and costs [4].

Wendorff presents an industrial case study where inappropriate pattern has lead to rigorous maintainability problems. The reason of inappropriately design patterns classify into two categories: (1) software developers have not understood the basis of design patterns that they have employed and (2) the patterns that have been employed have not met software requirements and suggest the need for documenting pattern usage because pattern removal is extremely costly [5].

Qualitative analysis of software security patterns evaluate security patterns based on how well they track each principle, how well they encounter with probable problems to develop secure software, and which threat categories have to take care to performed. Thirteen security patterns were evaluated based on these three sets of criteria [6].

In 2009 Khomh, Gueheneuc, and Antoniol, presents an empirical descriptive and analytic study on playing roles in design patterns. They investigate the relationship among the quality of the class whether the class plays a major role in design pattern. The results show that there are various differences in quality characteristics of class that involve in patterns [7].

Various studies attempted to quantitative evaluate the effects of design pattern on software quality. Huston demonstrates the effect of three design pattern on metric scores of three different categories. The study methodology is firm since it is based on pattern definitions and mathematical equations but it deals with only one metric per pattern [8].

Hierarchical model for object oriented design quality assessment is an improved model for evaluation of reusability, flexibility, and complexity. Offered a tool QMOOD++ that is flexible, nonintrusive, and applicable to real world projects [9].

In 2007 Ampatzoglou and Chatzigeorgiou, investigate the effect of design pattern in game development using case study. The results of the case study are qualitative and quantitative but the area of the research is limited to games thus results cannot be generalized. The patterns under study are two whereas the considered metric categories are complexity, size, coupling, and cohesion. The results of the case study recommend that pattern enhance coupling, cohesion, and complexity but the size metrics increased [10].

Another quantitative study compare object oriented and aspect oriented implementation of classical patterns with respect to important attributes such as: coupling and cohesion. And found that most aspect oriented solutions improve division of pattern related concerns although some aspect oriented implementation of particular patterns result is higher coupling and more lines of code [11].

Quantitative assessment of design pattern suggests that there is a need to modularize design pattern to achieve reusability and maintainability [12]. In 2008 Hsueh, Chu, and W. Chu, proposed a validation approach to help developers to check design pattern is well designed and measure effectiveness of quality improvement in design pattern that decide which design patterns are applicable to meet functional and non requirements [13].

In 2008, a survey with professional software engineers is performed. The results of empirical study suggest that design patterns do not always impact software quality positively and negative influenced quality issues are recommended to be simplicity, learnability, and understandability [14]. In 2014 another quantitative study suggests that there is a need to be blended (composed) design pattern when they are instantiated in software [15].

10.2 RELATED WORK

Design patterns can speed up the development process by tested and proven development. Design pattern tools have different features for assessment like abstraction, extraction, performance, presentation, scalability, and accuracy but in this section reusability, composability, completeness, maintainability, stability, understandability, flexibility, simplicity, and expandability are consider. Some tools perform experiments on very small examples using patterns and state very high precision and recall rates but using simple examples and suggest they need support of firm conception and scalability. The PTIDEJ system is an automated open source system (OSS) implemented in Java which uses estimated and exact matches to extract different micro architectures, idioms, design patterns, and design defects from source code [16].

Performance of systems, products, and services has always been a concern of designers, operations, and maintenance engineers [17]. If we use tacit knowledge in requirement and design phase then quality product is developed [18]. GoF design patterns present flexible solutions to software development problems. AspectJ implementation of GoF design patterns confirms modularity improvement in 17 of 23 cases for better code locality, composability, reusability, and (un)pluggability [19].

Spine, a pattern specification language allows patterns to define in terms of constraint on their implementation in Java and processed using a proof engine called Hedgehog.

The Hedgehog proof engine reads the Spine definitions and Java source code which attempts to automatically prove whether or not the class correctly realizes the design pattern and answer automatically returned. The functionality has been built into Eclipse giving a menu of design patterns that allow the user to decide if the presently selected Java class meets the design pattern selected [20].

Another method and tool which assist design recovery and program understanding by recognizing instances of design patterns semi automatically. FUJABA supports the definition of UML class, collaboration diagrams, and definition of method behaviors as equivalent graph transformation rules which use the definitions of the class and object diagrams. This environment generates entire and executable Java code from these definitions. The graph transformation rules can be viewed as a subset of UML like collaboration diagrams. It provides a plug in architecture that allows developers to add functionality easily while retaining complete control over their contributions [21-22].

Recovering design patterns able to improve existing source code analysis tools by bringing program to understand at design level. A fully automated pattern detection approach is based on reclassification of the GoF patterns by their pattern intent. GoF pattern catalog classify design patterns in the forward engineering; this reclassification is better suitable for reverse engineering. This approach uses lightweight static program analysis techniques to detain program intent. PINOT implements this new approach. PINOT detects all the GoF patterns that have concrete definitions driven by system behavior or code structure. This tool is more accurate, faster, and targets more patterns than existing pattern detection tools. PINOT has been used successfully in detecting patterns in JHotDraw, Java AWT, Apache Ant, Swing, and many other programs and packages [23].

PatternBox is a design pattern editor for Eclipse which creates Java classes and interfaces that can be customized depending on application needs. Using PatternBox we can easily insert new pattern members whenever required. DPAToolkit help to develop software using design patterns. Design can be visualized using class diagrams. 23 GoF design patterns can be integrated into design easily [24].

To identify set of rules which capture essential information to identify design pattern, JADEPT (JAva Design Pattern deTector) is used. Rules are characterized by weights representing their importance in the detection of design pattern. It captures static and dynamic aspects throughout a dynamic analysis of the software by exploiting the JPDA (Java Platform Debugger Architecture). The collected information is stored in a database. Queries to the database execute the rules defined to recognize the design patterns. The tool has been validated with positive results on several implementations of design patterns [25].

MARPLE (Metrics and Architecture Reconstruction Plug-in for Eclipse) support detection of design patterns detection (DPD) and software architecture reconstruction (SAR) activities using basic elements and metrics that are mechanically extracted from source code. This platform mainly based on the Eclipse framework and plug-ins as well as of different Java libraries for data access, graph management, and visualization [26].

Automated design pattern detection (DPD) is an exigent reengineering task that requires combination of complex structural and behavioral analyses for good results. Still detection quality (precision and recall) of existing tools has so far been unsatisfactory, to make DPD important part of existing IDEs and development practices. DPJF implemented for pattern detectors and provides competitive performance while achieving the best recall of all evaluated tools (with a median of 89%) and 100% precision. This result set the basis for routine application of DPD in program conception and software quality assessment whereas the good performance is achieved by empirically validated simplifications of individual technique [27].

In 2015, FINDER tool detected 22 out of 23 GoF design patterns in Java systems using fine grained static analysis. FINDER extracts static facts from system. These facts are next evaluated against the design pattern detection scripts to produce a list of design pattern candidates. This tool has been evaluated by applying it to various OSS [28].

Design Pattern Advisor tool helps software designers to evaluate design pattern by taking into account the current and upcoming state of software. Basically this tool measure maintainability, design quality (reusability, functionality, understandability, extendibility, effectiveness, and flexibility) and structural quality [29].

At present maintainability, reusability, understandability, and flexibility evaluated by tool and composability, completeness, stability, simplicity, and expandability are measured using mathematical formation. So design pattern assessment method collectively collects all nine quality attributes.

10.3 DESIGN PATTERN ASSESSMENT METHOD

To assess design pattern an analytical method is discussed. Analytical method is a standard process combining the power of the scientific method through the use of formal process to solve any type of problem. This work is motivated by the work of a methodology to assess the impact of design patterns on software quality [30]. This method suggests how to assess the design pattern and collect the required quality attributes in final software product. Framework of design pattern assessment method shows the sequence of execution to assess design pattern in Figure 10.1.

Assume, requirement is well documented and fixed. Main functional requirement is user able to perform login easily. Main non functional requirement is maintainability, reusability, understandability, flexibility, composability, completeness, stability, simplicity, and expandability. This is used as input for software design. Design pattern used for further processing because software design is difficult to apply to a broader range of problems. Design patterns provide general solutions, documented in a format that does not need particulars tied to a particular problem.

In step 1, identify the design problem to solve using design pattern. As initial start firstly identify design problem. Software architect or designer identifies a design problem using visit and by matching software requirement. Design pattern is the reusable structure of a solution to a design problem. Design patterns solve design problems using appropriate objects, object interfaces, object granularity, and object implementations. Design patterns can help to identify few abstractions and objects to capture them.

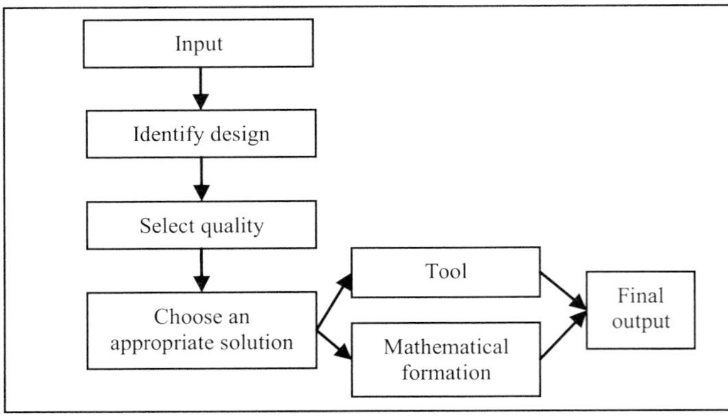

Figure 10.1 Framework of Design Pattern Assessment Method

In step 2, quality attribute are selected to fulfil the design objective. Design objective describe the set of required quality attribute as per requirement. Basically there are nine quality attribute which is required in final product by using design pattern.

In step 3, select an appropriate solution to achieve quality in finished product. If maintainability, reusability, understandability, and flexibility are required then go to step 4 to calculate quality attribute using Percerons Design Pattern Advisor tool. Quantitative analysis used to evaluate these four quality attributes.

In step 5, composability, completeness, stability, simplicity, and expandability are measured using mathematical formation. To evaluate these five quality attributes qualitative analysis is used.

In step 6, after completion combine step 4 and 5 output for required quality product.

Design pattern assessment method asses the design pattern and collect nine quality attribute because design pattern avoid unnecessary complexity, promote quality in finished product, and save time and cost of software development.

10.4 ASSESSMENT OF DESIGN PATTERN ON SOFTWARE QUALITY

For illustration purpose abstract factory pattern is used to describe the design pattern assessment method. Abstract factory pattern known as a creational pattern, provides a method to encapsulate a group of individual factories that have a familiar theme without specifying their concrete classes. In usual case, the client software create a concrete implementation of the abstract factory then using generic interface of the factory to create the concrete objects which is part of the theme. The client doesn't know which concrete objects gets from all of these internal factories, since it uses only the generic interfaces of their products. Abstract products describe interface for every different products of a single product family. Concrete products implement the abstract product interface which is returned by creational methods of concrete factories. Abstract factory defines the interface for creating products which is common to all concrete factories. Concrete factories implement creational methods of the abstract factory and each concrete factory should correspond to a specific products variant as shown in Figure 10.2 [31].

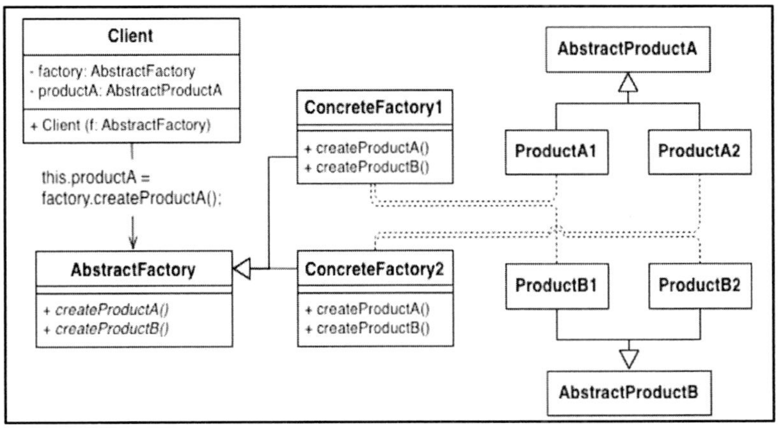

Figure 10.2 Abstract Factory Design Pattern

Purpose of abstract factory design pattern is to provide an interface for creating families of related objects without specifying concrete classes. Figure 10.3 shows the example of abstract factory design pattern for user login where two concrete factory and concrete product used for execution.

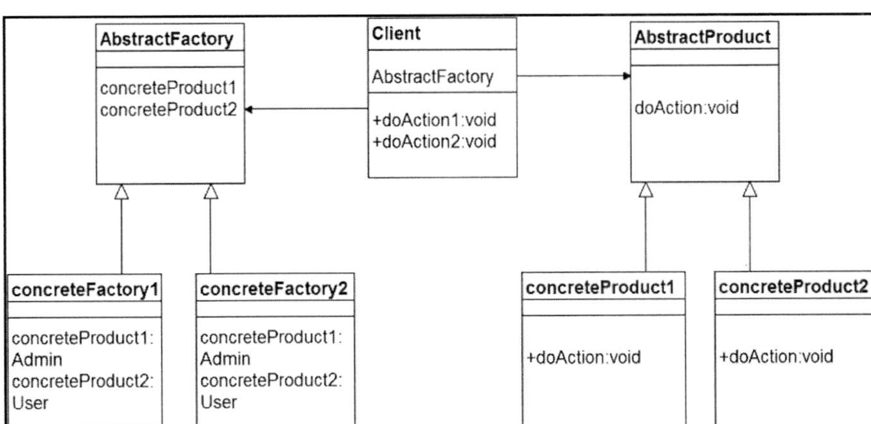

Figure 10.3 Abstract Factory Design Pattern for User Login

Using design pattern assessment method design pattern asses where requirement is well documented and fixed which is used as input. As step 1 firstly identify the design problem using alternative design solutions from literature and experience, and solve using abstract factory design pattern. In step 2 maintainability, reusability, understandability, flexibility, composability, completeness, stability, simplicity, and expandability are selected as design objective.

Using step 3 appropriate solution is selected as step 4 (tool) and step 5 (mathematical formation). In step 4 maintainability, reusability, understandability, and flexibility are calculated using percerons design pattern advisor tool. In step 5 composability, completeness, stability, simplicity, and expandability are measured using mathematical formation. In step 6, combine step 4 and 5 output to get required quality product. Assessment of these nine quality attributes are discussed as:

Maintainability: Use of a design pattern essentially complicates design to usually add abstract classes and additional associations to employ a design pattern. The key advantage of using design pattern is that the resulting software should be easier to adapt, to modify fewer classes, to add functionality to software. So, maintenance programmers should have to use less effort to adapt these changes. Every introduced pattern caused an improvement in different quality attributes [32-33].

In 2014, design pattern guidance tool is developed that retrieve the suitable pattern with respect to software maintainability from a warehouse of patterns. It measures the maintainability of design pattern using some metrics and candidate the more maintainable pattern to the designer. It provides a support for decision making during software design [34].

Maintainability calculated using design pattern advisor tool. This tool requires the value of concrete factory and concrete product, and measure maintainability using 10 metric set. Table 10.1 shows the estimated value of maintainability of illustrated example as shown in Figure 10.3. In Table 10.1 abstract factory design pattern analyze with respect to maintainability. Table shows the estimated mean value of each metric.

Maintainability is a measure of the ease with which a software, system, or component can be modified to correct faults, improve performance, and adapt changing environment. Maintainability index (MI) calculated using several formulae such as: lines of code (LOC) measures, Halstead complexity measures, and McCabe measures [35]. Estimated value of maintainability using LOC is 4.131 which is acceptable.

Table 10.1 Maintainability Assessment

Metric	**Estimated Value**
Depth of Inheritance Tree	0.877
Number of Children Classes	0.619
Message Passing Coupling	0.670
Response for a Class	2.108
Lack of Cohesion of Methods	0.000
Data Abstraction Coupling	0.254
Weighted Methods per Class	1.492
Number of Methods	1.438
Lines of Code	4.131
Number of Properties	1.692

Reusability: Design patterns are reusable micro architectures that add to overall software architecture [36]. Design patterns detain static and dynamic structure and collaborations of components in successful solutions to problems that occur when developing software like business data processing, databases, graphical user interfaces, telecommunications, and distributed communication software. Patterns help development of reusable components and frameworks by using structure and collaboration of participants in software architecture at a level higher than source code and object oriented design models that focus on individual objects and classes. Thus, patterns facilitate reuse of software architecture, even when additional forms of reuse are infeasible [37].

An empirical investigation on reusability of design patterns and software packages, attempts to empirically investigate reusability of design patterns, classes, and software packages to identify the most beneficial starting points for white box reuse, which is pretty popular among OSS. In more than 40% of the cases design pattern based on class selection offers the most reusable starting point for white box reuse. Still there are numerous cases when package based collection might be preferable. The results suggest that every pattern has diverse level of reusability [38].

Design patterns assure novel reuse benefits early in software development process. To gather the benefits of deploying these demonstrated design solutions, define design composition techniques to assemble applications using patterns. These techniques should be supported by adaptable design models [39].

Reusability is calculated using design pattern advisor tool. This tool requires the value of concrete factory and concrete product to calculate reusability. Table 10.2 show the estimated value of reusability of illustrated example as shown in Figure 10.3. Reusability can measure the degree of features or components that are reused to develop similar or different kind of software with minimal change. Various metric set are used to measure reusability such as: arguments per procedure (APP), coupling and cohesion metric, reuse frequency metric, and reusability measure value (RMV) metric [40]. Estimated value of reusability is 4.479 which is good.

Understandability: Design patterns claim to carry understandability. Understandability is one of the main attribute to consider, as it can influence the cost and reliability of software design evolution and maintenance. Understandability is main characteristics for a good model must have accuracy, abstraction, predictiveness, and inexpensiveness [41-42]. Understandability is calculated using design pattern advisor tool. This tool requires the value of concrete factory and concrete product to calculate understandability. Table 10.2 shows the estimated value of understandability of illustrated example as shown in Figure 10.3. It is not easy to calculate understandability because understanding is an internal process of humans. Understandability of software documentation compiled with composition of readability source code (RSC) and quality of documents (QoD) using membership function [43]. Estimated value of understandability is -3.111 which is negative.

Flexibility: Design patterns are used to solve recurring problems in software design and improve maintainability, reusability, comprehensibility, evolvability, and robustness. Design pattern applied when flexibility is actually needed.

Factory + Configuration = Flexibility

Means factory and configuration create flexibility [44-46]. Flexibility calculated using design pattern advisor tool. This tool requires the value of concrete factory and concrete product to calculate flexibility. Table 10.2 show the estimated value of flexibility of illustrated example as shown in Figure 10.3. Typical and existing software design literature suggests that particular implementations are more flexible than others but stops little of suggesting purpose criteria for quantifying such claims. Flexibility can be measured by impact of changes and complexity of software evolution [47]. Estimated value of flexibility is 0.604 which is ok.

Table 10.2 Quality Assessment

Quality Attributes	**Estimated Value**
Reusability	4.479
Understandability	-3.111
Flexibility	0.604

Table 10.2 analyze abstract factory design pattern with respect to reusability, understandability, and flexibility. Table shows the estimated mean value of each quality attributes.

Composability: Design patterns make to assure stringent modularity principles such as: high cohesion, low coupling, conciseness, and separation of concerns. Design pattern assigns roles to their member classes, which define the functionality of the participants in the pattern. The application of design patterns in complex systems is frequently a result of the composition of two or more pattern rather than their instantiations in an isolated manner. Their OO implementations can be composed in several ways ranging from a simple method invocation between the pattern's roles to sharing of one or more classes via pattern roles [48].

There are numerous ways to classify relationship between design patterns according to different purposes. Zimmer proposed a classification of the relationship between the GoF design patterns. The following categories were proposed as: (1) X uses Y; (2) X is similar to Y [49].

Design patterns are generally applied in composed form with each other. It is essential to be able to formally reason about how patterns can be composed and to confirm the properties of composed patterns. A notion of composition of patterns with respect to overlaps is formally defined based on two operations on design patterns, which are the specialization of a pattern by constraints and the stimulating of a pattern with a subset of components as the key. Pattern compositions save the features, semantics, and soundness of the composed patterns [50-51].

To assess composability Zimmer classification is used. Composability of illustrated example as shown in Figure 10.3 is Admin uses User and Admin is similar to User.

Completeness: Design pattern is complete when it is able to perform each and every task that user wants. While amplify completeness would not drastically modify the design dimensions [52-53]. Completeness assessed using performability of user. Completeness of illustrated example as shown in Figure 10.3, users are able to perform login task.

Stability: Stable design patterns are created for semi tangible objects that are always connected with various concepts. The concept either forms their property or represents its relationship to another object. For generating a stable design pattern, firstly recognize the concept that always goes with the object irrespective of the application [54]. The stable design pattern AnyCorrectiveAction is based on the Corrective Action for taking correct action based on evidence, reason, and criteria that govern it [55].

Evidence is an extremely significant concept that deals with augmenting the knowledge about a certain entity. AnyEvidence can be an arrangement of knowledge [56]. Stability of a class diagram indicates its resistance to interclass circulation of changes that the diagram would have when it is modified. Modifications ended to one class can have ripple effects (propagation of changes) on other classes in the diagram [57].

Mathematical models and ideas from mathematical analysis are used to develop a stable design for the state pattern. The design of the state pattern is stable when changes in the state machine result in localized changes in its implementation [58]. Design patterns can offer the estimated protection of certain pattern participating classes against changes, depending on their role in the pattern. Classes that participate in coupled pattern occurrences appear to be the least stable and less stable classes are expected to involve more effort while testing, and advice for refactoring activities that would formulate more resistant to modify propagation [59]. Illustrated example easily adopts the changes when it is modified and stable over time and will not need changes.

Simplicity: Simplicity is a desirable quality attribute for any software. It is the quality or condition of being easy to use or do [60]. Illustrated example is easy to use.

Expandability: It is a degree to which architectural, data, and procedural design of software can be extended. New patterns can be template and integrated into design and support easier understanding for users [61-62]. Illustrated example easily expanded as per user requirement.

After completing assessment of these nine attribute final output is collected as final product. Design patterns do not always improve reusability and understandability but they do improve expandability. Validated with the help of study [14] that shows expandability, simplicity, and reusability perform positively where understandability is negatively impact on software product. As analyze maintainability of illustrated example is acceptable. Reusability is good, understandability is negative, and flexibility is ok. Composability is suitable, completeness as per task, stability is appropriate, simplicity is good, and expandability as per user requirement.

10.5 IMPACT OF PROGRAMMING LANGUAGES ON SOFTWARE QUALITY

Software process is a sequence of well defined phase which is used to create and achieve quality product. Basically there are four phase of software process such as: requirement, design, coding, and testing. Each and every phase have its own role and importance to develop a quality product. This section focused on software coding and its factors to see its impact on software quality. There are various coding factor such as: programming features, standardized code, and code clone. Programming features provide overall characteristics of programming code which affect the code readability and maintainability. Standardized code develop required product and improve the comparability of results, maintainability, and stability of software. Code clone (duplicated code) provides reusability of code and save effort by copying an existing code. Code clone enhance maintainability and reliability of software [3].

According to S. Cass [63], C is a one of the most important programming language for software development as shown in Figure 10.4. Python jumps to no. 1 and Swift enters the top ten ranking as shown in Figure 10.5. Python has sustained its growing path from last year and jumped two places to the no. 1 slot, although the top four- Python, C, Java, and C++ all continue extremely close in popularity. Definitely, C comes out ahead of Python by a good margin. Programming languages trend change very fast, to develop software easiest and efficient way. Using spectrum ranking top 4 programming languages is selected as per Figure 10.6 such as: C, C++, Java, and Python [64-65].

Next Mathematica is selected to compare coding because it is usually less than a third of the length of the similar tasks written in different languages and regularly much better [66]. After selection C, C++, Java, Python, and Mathematica is used to implement various programs and analyze these languages coding factors and quality attributes.

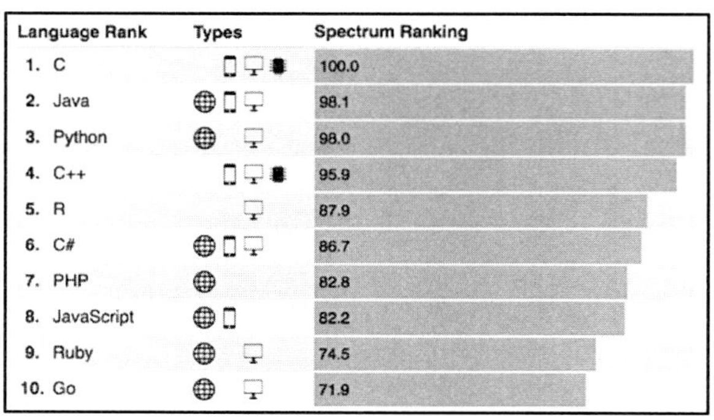

Figure 10.4 Ranking of Languages (IEEE Spectrum, July 2016)

C is a general purpose imperative computer programming language, which support recursion, lexical variable, and structured programming. C was formerly developed by Dennis Ritchie between 1969 and 1973 at Bell Labs which is used to re-implement the Unix operating system. It is designed to promote cross platform programming and is existing on numerous platforms. Today C is a one of the most widely used and popular programming language because of its structure, high level abstraction, and machine independent feature.

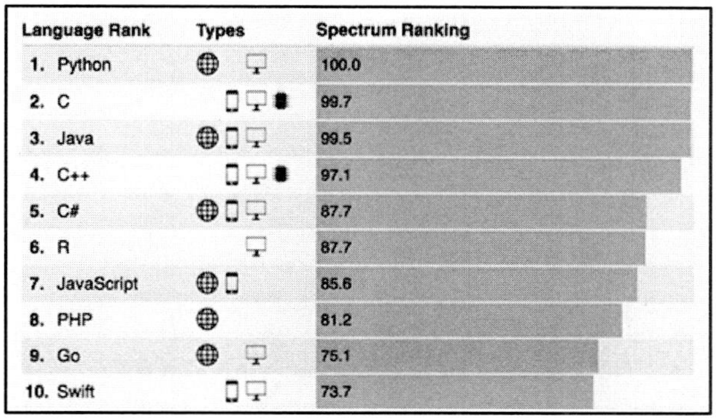

Figure 10.5 Ranking of Languages (IEEE Spectrum, July 2017)

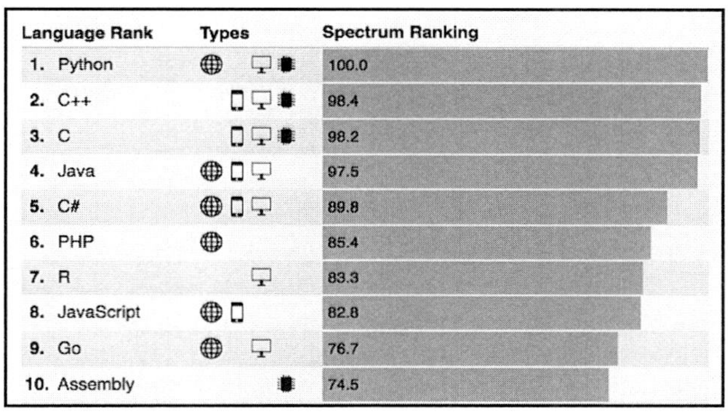

Figure 10.6 Ranking of Languages (IEEE Spectrum, July 2018)

C was adopted as a software development language because it develops code that runs as quick as the code written in assembly language. This language is esteemed for being comprehensible, provide access to hardware, and assembly little binaries. C stands for effectiveness of programming language, excellent style, case sensitive, and sound design. Typically uses a compiler and source code represented in files. These code source files usually a file suffix of .c which called as module or compilation unit. C source code files do not include header field. C has been used effectively for all type of programming problem possible from operating systems to spreadsheets to expert systems. The main measure of C success seems to be based on wholly practical considerations: standard library, portability of the compiler, elegant syntax, powerful and varied repertoire of operators, ready access to the hardware when needed, and ease with which applications can be optimized by furnish coding isolated procedures. C doesn't supply every constructs found in high languages but it provides all building blocks that will need to produce the results. In practice like C and C++ claim to have the "write once, compile anywhere" (WOCA) is debatable [67-69].

C++ is a multi paradigm, statically type, and general purpose programming language which support procedural, generic, and object oriented programming (OOP) feature. C++ was developed by Bjarne Stroustrup preliminary in 1979 at Bell Labs, as an improvement to the C language and formerly named C with Classes but later it was renamed C++ in 1983. C++ is regarded as a middle level language as it is a combination of both high level and low level programming language features. Aim of C++ was to improve the quality of programs by assembly better design and programming technique simple to use and affordable. C++ completely supports OOP, including the four main object oriented development features such as: encapsulation, data hiding, inheritance, and polymorphism. C++ was designed to bring the flexibility and efficiency of C for system programming together. Major contribution of C++ is its support for defining abstract data types (ADTs) and generic programming. Number of systems supports ADT more than using object oriented (OO) features of language. For some systems use of C++ OO features is necessary to build extremely flexible and extensible software. C++ is also trendy because of its portability means program can be written on one computer and run on many other systems. Usually requires program is recompiled on each type of system but the program may require little or no change. C++ is being exceedingly used to write device drivers and other software that rely on direct management of hardware under real time constraints. Its key features build it stand out a strong, static type system, good performance, ability to utilize it in few programming styles, and expressiveness. So C++ is used by many developers in fundamentally every application area [70-72].

Java programming language was formerly developed by Sun Microsystems which was initiated by James Gosling and released in 1995. Java is a simple, OOP, robust, portable, and secure programming language which provide developers "write once, run anywhere" (WORA) means compiled Java code can run on all platforms that support Java without the recompilation. Java compiler generates an architecture neutral object file which makes the compiled code executable on many processors with the existence of Java runtime system. Java uses a usual garbage collector to manage memory in object lifecycle. Developer determines when objects are created and Java runtime is liable for recover memory once objects are no longer in use. Java is more dynamic language than C or C++ since it is designed to adjust to a developing environment. When we program for Java platform, we write source code in .java files and then compile them. Compiler checks code against the language syntax rules then writes out byte code in .class files. Byte code is a set of instructions used to run Java virtual machine (JVM). This level of abstraction is differs from other language compilers which write out instructions appropriate for the CPU chipset the program will run on. Using just in time compilers Java enables high performance. With Java multithreaded feature it is feasible to write programs that can perform lots of tasks concurrently. This design feature allows developers to assemble interactive applications that can run efficiently. So developers adore Java for its concurrency which is better than Python, great variety of library, and steadily good performance. Java programming language requires the existence of software platform to organize compile programs to be executed. Oracle supplies Java platform for use with Java. Android software development kit (SDK or devkit) is a different software platform used mainly for developing android applications [73-75].

Python was developed by Guido van Rossum in 1991 at the National Research Institute for Mathematics and Computer Science in the Netherlands. Python is a multi paradigm, functional, imperative, OO, structured, procedural styles, and general purpose dynamic programming language that emphasize code readability and enable developers to use fewer lines of code in contrast with C++ and Java. It supports multiple programming paradigms and has a large standard library. It uses simple keywords normally where other languages use punctuation. It has less syntactical construction than other languages. It can be used as a scripting language or can be compiled to byte code for developing large application. Python is usually implemented using a combination of a byte code compiler and interpreter. Compilation is completely performed as modules are loaded and numerous language primitives require the compiler to be accessible at run time. We can include low level modules to the Python interpreter. These modules enable developers to attach or customize their tools to be more efficient. Low level languages easily add to Python interpreter to enlarge the functionality of Python code. Python provide an enhanced structure and support large programs than shell scripting. It is an easy to use language that makes it easy to get program working. That makes Python perfect for prototype development and other ad hoc programming tasks without compromising maintainability. Python is a portable language. It can run on different platforms, operation systems, and provide similar interface. Python divide code in numerous modules which can be used later in other programs. It can be simply integrated with ActiveX, C, C++, Component Object Model (COM), Common Object Request Broker Architecture (CORBA), and Java. Developers use this language for good OOP support, clear syntax, and great shortcuts [76-78].

Mathematica is a mathematical software system to add a novel level of flexibility to incredible concept of programming. The Wolfram Language is a universal multi-paradigm programming language developed by Wolfram Research which is the programming language of the mathematical symbolic computation program Mathematica and the Wolfram Programming Cloud. Mathematica computer program is an all purpose system for doing mathematical computation which include a command language, programming language, and calculation environment that is oriented to symbolic as well as numeric mathematics. It was conceived by Stephen Wolfram and developed by Wolfram Research of Champaign. Mathematica stands out from traditional languages by simultaneously sustaining many programming paradigms such as: functional, procedural, pattern based, rule based, and more. Mathematica is more evolutionary than revolutionary. Mathematica provides library of mathematical elementary functions and special functions, tool for combinatoric problems and parallel programming, constrained and unconstrained local or global optimization, tool to connect to dynamic link library (DLL), Java, Structured Query Language (SQL), C++, .NET, Fortran, OpenCL, and Hypertext Transfer Protocol (HTTP) based systems, supporting procedural, functional, and object oriented constructs. Mathematica provides integrated tools that greatly enlarge the scope of cross domain projects [79-81].

10.6 OVERVIEW OF PROGRAMMING LANGUAGES

Table 10.3 provides an overview of these programming languages. Table 10.3 also characterize and compare all theses programming languages using vital context such as: when these programming are available, what are the purpose, programming paradigm, feature, area of application, stable release, disadvantage, file extension, and web framework of these languages, which platform are required to run particular language, and how language is executed. Table 10.4 shows the list of quality attributes of programming languages which is based on Table 10.3 [82-87].

Table 10.3 Overview of Programming Languages

Context	C	C++	Java	Python	Mathematica
About name & original purpose	Designed by Dennis Ritchie and used to re implement the Unix operating system	Designed by Bjarne Stroustrup, 'C with Classes' added features to the C compiler	Specifications developed by James Gosling and Sun Microsystems, which was later acquired by the Oracle corporation, named for Java coffee, created for interactive TV	Designed by Guido van Rossum, named for Monty Python, created as a scripting language to bridge the gap between the shell and C	Developed by Wolfram Research for mathematical symbolic computation programming
Appeared	1972	1983	1995	1991	1988
Paradigm	Imperative (proc edural),structure d	Multi paradigm (procedural, functional, object oriented, generic)	Imperative, structured, object oriented, functional style	Multi paradigm (object oriented, imperative, functional, procedural, reflective)	Knowledge based language, computer algebra system, symbolic programming environment
Language principles	Prone to defects, portable, flexible, modular, speed, case sensitive	Platform dependent language, case sensitive, OOP, speed, efficient use of pointers	Robust, secure, portable to run anywhere, high performance, multithreading for concurrency, dynamic, OOP, platform independence	Readability, explicit, OOP, namespaces, variable scoping, exception handling	Libraries of mathematical elementary functions and special functions
Used for creating	Scientific programming situations, computer applications, embedded software, operating systems, graphics, and game	Games, office applications, graphics, video editors, operating systems, and graphic user interface (GUI) based applications	Android apps, large websites, embedded devices, mobile phones, enterprise servers, and supercomputers	Math scripts, websites, game, scientific computing, web and Internet development	Many scientific, engineering, mathematical, and computing fields
Stable release	C18, June 2018	ISO/IEC 14882:2017, December 2017	Java 12, March 2019	Python 3.7.3 25 March 2019 2.7.16 4 March 2019	12 April 2019
Used especially by	Mobile (iOS, android and windows phone kernels),	High performance software, resource	Large companies, Banks, Ecommerce,	Academics, gaming, startups, Google	Medical, Complex number, matrix and data

	Databases (Oracle, MySQL), 3D movies	constrained programming, large scale software infrastructure, energy constrained environment (mobile and cloud)	Google		manipulation tools, tools fcr 2D and 3D image processing
Unique feature	Robust, undefined behavior, easy to extend	Well structured language, standard libraries, classes, portable	Well optimized Java Virtual Machine (JVM) for running code, simple, secure, compatibility focus, garbage collection	List comprehensions for creating lists based on other lists, interactive interpreter (REPL), indentation is significant, Generators and Coroutines	Automation, integrated all in one platform, hybrid symbolic-numeric methodology. multi paradigm language
Operating system (OS)	Cross platform	Cross platform	Windows, Solaris, Linux, OS X	Cross platform	Windows (7, 8, 10), macOS, Linux, Raspbian, online service
Cons	Not supported OOP feature, no runtime checking, no strict type checking	Complex, unsecure, difficult to debug, can't support garbage collection, can be heavy	Strict rules help catch errors but reduce flexibility and brevity, requires more boilerplate code than others, primitives	Somewhat slow, using whitespace looks nice, but may cause occasional issues, language split between Python 2 and 3	Not completely secure, No command history, error messages do not include a source line number
Filename extensions	.c, .h	.cc, .cpp, .cxx, .C, .c++, .h, .hh, .hpp, .hxx, .h++	.java, .jar	.py, .pyc, . pyd, .pyo	.nb, .cdf
How language is executed	Preprocessing, compilation, assembly, linking, loading	Preprocessing, compilation, assembly, linking, loading	Compiled to run on the JVM so you can write once, run anywhere	Commonly executed on CPython, the official Python implementation written in C	press Shift+Enter
Top web framework	Kore, Raphters	Tntnet, Poco	Spring MVC, Spark	Django, web2py	webMathematica

Tables show the important characteristics of programming languages. To develop a quality product it is vital to select most appropriate programming language to develop software product which is easy to read, easy to use, easy to understand, and easy to modify. In this chapter software coding and its factors is analyze to show how they are related to each other and software quality by using C, C++, Java, Python, and

Mathematica. This chapter analyzed the impact of coding on programming features, standardized code, and code clone which impact on readability, maintainability, stability, and reliability. Also analyzed scalability, security, robustness, portability, and flexibility as an important quality attribute for analysis as depicted in Table 10.4. Various examples taking into consideration to analyze programming languages where programming styles of developer is fix as per each programming language and central processing unit (CPU) statistics also consider to analyze programming languages because its impact on performance.

Table 10.4 List of Quality Attributes of Programming Languages

Quality Attribute	C	C++	Java	Python	Mathematica
Flexibility	✓	✓	✓	✓	✓
Maintainability	✓	✓	✓	✓	✓
Portability	✓	×	✓	✓	×
Readability	✓	×	✓	✓	✓
Reliability	✓	✓	✓	✓	✓
Robustness	×	×	✓	✓	×
Scalability	×	×	✓	✓	✓
Security	✓	×	✓	✓	×
Stability	✓	×	✓	✓	✓

10.7 CODE QUALITY ANALYSIS

Code quality analysis provides numerous benefits when implementing and maintaining applications within an infrastructure. Basically there are static and dynamic aspects of code quality analysis. Static code analysis is essential building block that facilitates to improve the readability and maintainability. Fixing code quality issues has an important effect on overall quality of code and ability to accurately calculate when the software can be delivered to users. Static and dynamic analysis is that the earlier runs in development environment and the later needs to be dynamic during the runtime of the function under analysis. Dynamic code analyzers outline system and monitor its strength. Dynamic analysis tools regularly instrument the code to add tracing of method calls, any other statistics they collect, and catch or notify about exceptions [88].

In static code analysis, code analyze without executing it which is used to find bugs, conformance to coding guidelines, and quality product. Basic example is a compiler which finds lexical, syntactic, and semantic mistakes. Static analysis tools used to maintain code quality. There are very few research papers which analyze code quality based on various programming language. Different parameters from source code are used to measure code quality within each quality attribute. There are various metric which is used for C, C++, and Java code quality analysis: number of statements (N_STMTS), cyclomatic complexity (VG), maximum levels (MAX_LVLS), number of paths (N_PATHS), unconditional jumps (UNCOND_J), comment frequency (COM_R), vocabulary frequency (VOC_F), program length (PR_LGTH), average size (AVG_S), number of inputs/outputs (N_IO), number of comments, and afferent connections per class (ACC), density of comments, coupling between objects (CBO), coupling factor (COF), depth of inheritance tree (DIT), lack of cohesion on methods/functions (LCOM), lines of code (LOC), lines per method/function (AMZ_Size), number of attributes/variables (NOV), number of children per class

(NOC), number of methods/functions (NOF), number of classes/module (NM), number of public attributes/variables (NPV), number of public methods/functions (NPF), and response for class (RFC) [89-91].

Since we could not control code quality or code size, so Code Quality Monitoring Method (CQMM) is used for systematically assess and improve the code quality of a software product [92]. A novel methodology used to control software development process, developer competency, and quantify how the selection of programming language impact on software quality and developer efficiency. Researchers in the earlier period have studied the impact of programming languages on software quality, though, there has been tiny or no study on the impact of multiple languages on quality of software product [93]. Flawed identifiers in Java classes were linked with source code originate to be of low quality. So, identifier naming to program comprehension is an important parameter to be consider [94]. It is likely to predict source code quality based on static analysis and machine learning [95]. Python used for broad scientific research and development. Python's numerous classes, data structures, iterators, nested functions, flexible function calling syntax, great scientific libraries, extensive kitchen-sink-included standard library, and outstanding documentation. While Mathematica is ideal for interactive work with pure math, it is not planned for common use coding [96].

There are various code quality analysis tools which are used to find out the code quality of different programming languages. Parasoft is the one of the best tool for static analysis. **Cppcheck** is an open source static analysis tool for C and C++ code developed by Daniel Marjamäki in May 2007. Cppcheck stable release on October 2017. It provides unique code analysis to identify bugs and focuses on detecting undefined behavior and unsafe coding constructs. Main goal is to detect real errors in code. There are command line interface and graphical user interface are available. Cppcheck is integrated with many popular development tools [97].

CppDepend is a static analysis tool for C and C++ code stable release in March 2017. It is a code rule through LINQ query (CQLinq) and SQL query engine. CppDepend embed by default Cppcheck, Clang analyzer, and Vera++. It could be simply extensible to sustain other static analysis tools using its API. Visual studio analyzer plug-in code source is available to show how to combine other tools [98].

IBM Rational Logiscope perform code coverage analysis, code checking against predefined programming rules, code measurement, and quality assessment which help developers to identify error prune module of software being developed. Logiscope can perform code analysis for software written in Ada, C, C++, and Java. Its tool performs source code measurement in order to perform quality analysis of software and develop quality report. Logiscope quality analysis divided into three levels: application level, class level, and function level. Application level provides information regarding entire application examined, class level for classes which are used in source code and function level about the functions found in the source code. All three levels are further separated into detailed metrics which in turn are further analyzed into other one. These are each distinct metrics or the result of a formula consisting of metrics measured from the source code directly. Other different products for analysis: IBM Rational

Logiscope RuleChecker for code rule checking, IBM Rational Logiscope QualityChecker for code quality metrics, IBM Rational Logiscope Code Reducer for finding code similarities, IBM Rational Logiscope TestChecker for dynamic test coverage analysis [99].

Coverity is a commercial static analysis tool founded in November 2002. This tool is logically fast and returns few false positives. Coverity scan is a cloud based tool. It works for C, C++, Java, Python, C#, and JavaScript. The latest version supports an option to select between three levels of aggressiveness with the number of reports growing (and the number of potential false positives going up) at the higher levels [100].

PMD is an extensible cross language static code analyzer. PMD does not officially stand for anything, there are numerous unofficial names most appropriate is programming mistake detector (PMD), stable release in July 2017. PMD is an open source code quality tool for C, C++, Java, C#, Groovy, PHP, Ruby, Fortran, JavaScript, PLSQL, Apache Velocity, Scala, Objective C, Matlab, Python, Go, and Swift which try to find potential bugs from exception handling statements and code problems such as: dead code, overcomplicated expressions, suboptimal code, and duplicated code. PMD focuses on preemptive defect detection. It is a rich and highly configurable set of rules that can simply configure and prefer which particular rules should be used for a given project. PMD can be used with Eclipse, Maven, IntelliJ IDEA, Gradle, and Jenkins [101].

FindBugs is an open source code quality tool created by Bill Pugh and David Hovemeyer on June 2006 to reveal bugs in Java programs. It is stable release in March 2015. It doesn't concern for formatting and coding standards but is simply marginally interested in best practices. It concentrate on detecting possible bugs, performance issues, and does a good job to detect a variety of several types of general hard-to-find coding mistakes, including thread synchronization problems, infinite recursive loops, null pointer dereferences, misuse of API methods etc. FindBugs operate on Java byte code rather than source code [102].

Checkstyle is an open source static code analysis tool used for checking whether Java code conforms to the coding conventions that established. It is stable release in February 2017. Checkstyle automates the vital but tedious task to check Java code. It is the one of the most accepted tool to automate the code review process which comes with predefined rules that help to maintain coding standards. These rules are an excellent starting point but they do not account for specific requirements. Main trick to increase a successful automated code review is to combine the built-in rules with custom ones. Checkstyle can be used as an Eclipse plugin to build systems such as: Ant, Gradle, and Maven to validate code and create report for coding standard violations [103-104].

SonarQube (formerly Sonar) unlike the above is a plug in metrics tool for Java. It integrates as plug-ins, a set of code measurement tool in a one application and presents overall results. SonarQube is an open source platform initially launched in 2007 and used to manage source code quality. It was designed to support global continuous improvement strategy on code quality within a company and consequently can be used as a shared central system for quality management. It is stable release in November

2017. SonarQube makes management of code quality possible to any developer. As an effect in recent years it has grow to be a world's leader in continuous inspection of code quality management systems. Sonar presently supports a wide variety of languages: C, C++, Java, C#, PHP, Groovy, Flex, JavaScript, Python, and PL/SQL and offers fully automated analyses tool and integrates well with Gradle, Maven, Ant, and other continuous integration tools. Sonar uses PMD, FindBugs, and Checkstyle to collect and analyze source code for bugs, bad code, and possible violation of code styles. It examine and evaluate different aspects of source code from potential bugs, minor styling details, lack of test coverage, code defects to the critical design errors, and excess complexity. Sonar produces metric values and statistics to reveal problematic areas in the source code that require inspection and improvement [105].

Codacy is free open source project and integrates with tools such as: Bandit, Brakeman, FindBugs, and others. It offers security patterns for languages: C, C++, Java, Python, Ruby, Scala, Javascript, and more. Codacy offers a selection of hundreds of Code Patterns. Codacy offers security and performance checks [106].

Pylint is static code analyzer for Python code developed by Sylvain Thénault in 2001. It is highly active, maintained, and quite stringent; includes many stylistic warnings. It is stable release in September 2017. Python source code analyzer looks for programming errors, helps enforcing a coding standard, and sniffs for some code smells. Pylint detects duplicated code. Main advantage with Pylint is that it is highly configurable, customizable, and easily writes a small plugin to add a feature. Pylint is integrated into various IDEs: Spyder, Editra, TextMate, and Eclipse with PyDev [107].

Pyflakes is a simple program which checks Python source files for errors. Pyflakes examines the syntax tree of each file individually and combined with a limited set of errors which makes it faster than Pylint. Pyflakes is limited in terms of the things it can check. While pyflakes doesn't do any stylistic checks, there is another tool that combines pyflakes by style checks against PEP8: Flake8. Flake8 aside from combining pyflakes and PEP8 also add per-project configuration ability [108].

MathCode C++ compile Mathematica function into highly efficient and readable C++ code. Mathematica with MathCode provide a platform for rapid development of quality code for heavy simulations and other expensive computations. Automatic code generation of MathCode also preserves against the typing errors and bugs encountered as using predictable methods for prototyping and development which speed up Mathematica code results in faster analysis and increased productivity. Using Mathematica environment and generating C++ code development time and costs reduce and export Mathematica function deliver to users with increase flexibility [109].

Wolfram MUnit Tester is a significant feature of modern software development. It is a way that confirms different parts of source code are working correctly. A streamlined unit testing framework makes tracking defects so effortless that the tests are accepted extremely often. Its helps to identify problems as they are introduce which helps to resolve them. Wolfram workbench contains a version of the MUnit Tester application which is used wolfram research for developing and verifying Mathematica code [110].

As various code quality analysis tool is discussed which have been used to analysis of various quality attribute. Table 10.5 summarized code quality analysis tools for various programming languages.

Table 10.5 Code Quality Analysis Tools for Programming Languages

Code Quality Analysis Tool	Quality Attribute	Programming Language
Cppcheck	Maintainability	C and C++
CppDepend	Maintainability	C and C++
IBM Rational Logiscope	Maintainability	C, C++, and Java
Coverity	Security	C, C++, Java, and Python
PMD	Readability	C, C++, Java, and Python
FindBugs	Performance	Java
Checkstyle	Maintainability	Java
SonarQube	Accuracy	Java
Codacy	Accuracy and Security	C, C++, Java, and Python
Pylint	Maintainability	Python
Wolfram MUnit Tester	Accuracy	Mathematica

Table 10.5 shows that code quality analysis have been used to improve the readability, security, accuracy, and maintainability of source code which is essential for software quality.

10.8 COMPARISON OF PROGRAMMING LANGUAGES

For comparison of these programming languages this chapter considers various programs assign as P1, P2, P3, P4, and P5.

P1: hello world, P2: print 1 to 10, P3: print square of 1 to 10, P4: factorial program, and P5: print Fibonacci number which is implemented using C, C++, Java, Python, and Mathematica as shown in Figure 10.7, 10.8, 10.9, 10.10, and 10.11 respectively. The conception of software quality is not as straightforwardly definable. There are various promising quality attributes of software [111]. Each and every programming has individual property which affects quality of software product.

P1: hello world

```
#include <stdio.h>
main ()
{
printf("Hello world");
}
```

P2: print 1 to 10

```
#include<stdio.h>
int main()
{
int i;
for(i=1;i<=10;i++)
printf("x=%d\n",i);
}
```

P3: print square of 1 to 10

```
#include <math.h>
int main()
{int i;
for(i=1;i<=10;i++)
{
printf("Square is %d\n",i*i);
}
}
```

P4: factorial program

```
#include<stdio.h>
#include<conio.h>
void main()
{
int n, f, i;
printf("Please enter the number:");
scanf("%d",&n);
if (n<=0)
{
printf("Factorial is not possible");
}
else
{
f=1;
for (i=1; i<=n; i++)
{
f=f*i;
}
printf("Factorial of given number
is:%d", f);
}
getch();
clrscr();
}
```

P5: print fibonacci number

```
#include<stdio.h>
void main()
{
 int n1=1,n2=1,n3,i;
 printf("\n%d %d",n1,n2);
 for(i=2;i<20;++i)
 {
 n3=n1+n2;
 printf(" %d",n3);
 n1=n2;
 n2=n3;
 }
 getch();
 }
```

Figure 10.7 P1, P2, P3, P4, and P5 C Program

P1: hello world

```
#include <iostream>
main()
{
std::cout << "Hello world\n";
}
```

P2: print 1 to 10

```
#include <iostream>
int main(){
using namespace std; int x;
for( x = 1; x <= 10; x++ ){
cout << "x=" << x << endl;
}
}
```

P3: print square of 1 to 10

```
#include<iostream>
int main()
{
 using namespace std;
 int x;
for(x=1;x<=10;x++)
 cout<< x <<"Square
is"<<x*x<<"\n";
}
```

P4: factorial program

```
#include<iostream.h>
#include<conio.h>
void main()
{
int i, n, f;
clrscr();
cout<<"\nPlease enter the number:";
cin>>n;
if (n<=0)
{
cout<<"Factorial is not possible";
}
else
{
f=1;
for (i=1; i<=n; i++)
{
f=f*i;
}
cout<<"Factorial of given number
is:"<<f;
}
getch();
}
```

P5: print fibonacci number

```
#include <iostream>
using namespace std;
int main() {
int n1=1,n2=1,n3,i;
cout<<n1<<" "<<n2<<" ";
for(i=2;i<20;++i)
{
 n3=n1+n2;
 cout<<n3<<" ";
 n1=n2;
 n2=n3;
}
 return 0;
}
```

Figure 10.8 P1, P2, P3, P4, and P5 C++ Program

P1: hello world

```
class Hello {
 public static void main(String[]
args) {
System.out.println("Hello world");}
}
```

P2: print 1 to 10

```
public class Main{
public static void main(String[] args)
{
for(int x = 1; x <= 10; x++){
System.out.println("x=" + x); }
}
}
```

P3: print square of 1 to 10

```
public class Main{
public static void main(String
args[]){
int i=1,n=10;
for(int x = 1; x <= 10; x++){
System.out.println(i + "square is" +
i*i);i++;}
}
}
```

P4: factorial program

```
import java.util.Scanner;
class Fact
{
public static void main(String args[])
{
int n, f, fact = 1;
System.out.print("Please enter the
number:");
Scanner in = new
Scanner(System.in);
n = in.nextInt();
 if ( n <= 0 )
System.out.println("Factorial is not
possible");
 else
{
for ( f = 1 ; f <= n ; f++ )
fact = fact*f;
System.out.println("Factorial of
given number is:" +fact);
}
}
}
```

P5: print fibonacci number

```
class Main{
public static void main(String args[])
{
int n1=1,n2=1,n3,i;
System.out.print(n1+" "+n2);
for(i=2;i<20;++i)
{
 n3=n1+n2;
System.out.print(" "+n3);
 n1=n2;
 n2=n3; }
 }
 }
```

Figure 10.9 P1, P2, P3, P4, and P5 Java Program

P1: hello world

```
print("Hello world")
```

P2: print 1 to 10

```
for x in range(1,11):
print ("x=", x)
```

P3: print square of 1 to 10

```
counter = 1
while counter * counter < 11:
for counter in range(1,11):
a = counter*counter
print ("square is", a)
```

P4: factorial program

```
num = int(input("Please enter the
number:"))
factorial = 1
if num <= 0:
print("Factorial is not possible")
else:
for i in range(1,num + 1):
factorial = factorial*i
print ("Factorial of given number
is:", factorial)
```

P5: print fibonacci number

```
first = 1
second = 1
print(first, ",", second, end=", ")
for i in range(2, 20):
next = first + second
print(next, end=", ")
first = second
second = next
```

Figure 10.10 P1, P2, P3, P4, and P5 Python Program

P1: hello world

```
Print ["Hello world"]
```

P2: print 1 to 10

```
Do[Print["x= ",  x],{x,10}]
```

P3: print square of 1 to 10

```
Do[Print[x, " square is " , x^2],
{x,10}]
```

P4: factorial program

```
n = Input ["Please enter the
number:"];
fact[n_] := If[n <= 0, " not
possible", f = 1;
For[i = 1, i <= n, i++, f = f*i]; f];
Print["Factorial is", fact[n]];
```

P5: print fibonacci number

```
Table[Fibonacci[n], {n,20}]
```

Figure 10.11 P1, P2, P3, P4, and P5 Mathematica Program

10.9 IMPACT ANALYSIS OF SOFTWARE CODING ON SOFTWARE QUALITY

Programming features, standardized code, and code clone are important coding factor which reflect software quality attribute. Programming features give overall characteristics of programming code which influence the readability and maintainability.

Code readability is generally related through comments and naming standards [112]. Standardized code implement required product and improve comparability of results, maintainability, and stability of software. Maintainability calculated through lines of code (LOC), Halstead complexity, and McCabe complexity. Stability can be measure by how easily software adopts the update and changes. Code clone provide reusability of code and save effort by using existing code. Code clone increases maintainability and reliability of software product. Using Mean Time between Failure (MTBF) reliability is calculated [113].

For analysis readability, maintainability, stability, and reliability are compared with C, C++, Java, Python, and Mathematica which is shown in Table 10.6. Now analyze how readability, maintainability, stability, and reliability capture by these programming languages which as shown in Table 10.7.

Readability refers to the simplicity by which a person can understand the purpose, operation, and control flow of source code. There are special elements which contribute to readability such as: comments, indent styles (whitespace), decomposition, naming conventions for objects (variables, classes, procedures). The easiest way to measure readability is the number of lines in a program [114].

Another method is Comments Ratio (CR), defined as in equation 1:

$$CR = LOC/LOM \qquad (1)$$

Where LOC refers to total lines of code (including comments lines) and LOM represents total lines of comments in the code. So readability can be estimated by finding percentage of comment lines in total code and lower ratio provide better is readability [115].

Readability calculated using number of lines in P1, P2, P3, P4, and P5 program. Number of line in C and C++ programs is: 5, 7, 8, 23, and 14 which is same for all programs. For Java 4, 6, 7, 19, and 13 which is less than C and C++ but more than Python as 1, 2, 5, 8, and 8. Mathematica has less than number of line from other languages as 1, 1, 1, 3, and 1 which is reported in Table 10.6. So Mathematica is easy to read, understand, and implement than others. While C easier to use and C++ added a number of features that would be more interesting to programmers. But sometime C++ is difficult to understand, due to polymorphism. Java is a simple, easy, and OO programming and its readability rate is good. Python is executable pseudo code and it is much easier to use, understand, and implement. Its code readability is more than others but less than Mathematica [116-117].

Maintainability is a measure of the ease with which a software, system, or component can be modified to correct faults, improve performance, and adapt changing environment. Maintainability index (MI) is calculated through certain formulae such as: LOC measures, Halstead complexity measures, and McCabe measures. A new method which is proposed by Najm to find MI using LOC only is very easy and useful method to calculate maintainability [118].

A four polynomial metric is used to calculate MI as in equation 2:

$$MI = 171 - 5.44 * In(aveVol) - 0.23 * aveV(g') - 16.2 * ln(aveLOC) + 50 * sin(sqrt(2.46 * aveCM)) \quad (2)$$

Where, aveVol is the average Halstead Volume per module, aveV(g') is the average extended cyclomatic complexity per module, aveLOC is the average lines of code per module, and aveCM is the average proportion of lines of comments per module [119].

The three metric is used to calculate MI as in equation 3:

$$MI = 171 - 5.2 * ln(HV) - 0.23 * V(G) - 16.2 * ln(LOC) \quad (3)$$

Where LOC is the number of lines of code, V(G) is the cyclomatic complexity and HV is the Halstead Volume, determined by the number of operators and operands, including identifiers [120].

Maintainability calculated using LOC of various programs. Number of line in C and C++ is equal; Java has less than LOC as per C and C++ code, but more than Python. Mathematica has less than LOC from other languages. It has been reported in Table 10.6. So Mathematica is easy to maintain than other languages. C and C++ are less maintainable than others because of its complex nature. Java maintainability rate is good because of its simple, interpreted, and scripting nature. Python is more maintainable than other languages but less than Mathematica [121-124].

Stability is a capability of the software to avoid unpredicted things from modification of the software. Stability is an extension of language features and functionality while preserving existing behavior which allows code and software to maintain the purpose as originally designed. Stability index (SI) is a summative measure of the rate at which code is change, how well it is coping with that rate of change and lower ratio provide better stability. The SI is a combination of three factors:

- The weighted defect increase and fix rates
- The ratio of the code base that is change
- The amount of the code base that is change repeatedly [125].

To measure and visualize changes in source code heat maps is used. Code is stable when changes are enclosed to interrelated part of code base (architectural components) and unstable when changes occur at unrelated part of code [126].

Stableness measurement model (SMM) is a mathematical model for measuring the stability of handheld application usage [127]. Historical version analysis used to calculate software library stability [128] and another method to calculate stability is coupling between object (CBO) metric [129].

Stability calculated using rate of code change. Rate of code change in C program P1, P2, P3, P4 and P5 is 2, 3, 3, 5, and 4 respectively, for C++ it is 3, 4, 4, 6, and 5, for Java it is 1, 1, 1, 2, and 2 which is good, for Python it is 1, 1, 2, 3, and 3, for

Mathematica 1, 2, 2, 4, and 3 which is reported in Table 4. So Java is more stable than others because of its robust, portable, and dynamic nature. Python is much good than remaining other languages because of it's easy to change and modify nature. Mathematica is good as compare of C and C++. C code stability is much better than C++ [130-131].

Reliability is ability of a system or component to perform its required functions under stated condition for a given time. Using the following formula (4-6), the probability of failure is calculated by testing a sample of all available input states. Reliability of a system is calculated as the failure rate, lower failure rate provide better reliability.

MTBF = Mean Time to Failure (MTTF) + Mean Time to Repair (MTTR) (4)

Reliability = MTBF / (1+MTBF) (5)

Probability = Number of failing cases / Total number of cases under consideration (6)

The set of all possible input states is called the input space. To find reliability of software, we need to find output space from given input space and software [71-72]. Reliability is calculated using failure rate. Failure rate in C program P1, P2, P3, P4, and P5 is 1, 1, 2, 3, and 3 respectively, in C++ it is 3, 4, 4, 5, and 5, for Java it is 2, 3, 3, 4, and 3, in Python it is 1, 1, 1, 2, and 2 which is good and 2, 3, 3, 4, and 3 for Mathematica as reported in Table 10.6. So Python is more reliable than others because of its low failure rate. C programming is good to develop a reliable program. C++ is less reliable than all. Java and Mathematica is equally reliable [132-136].

Table 10.6 Comparison of Programming Languages

Quality Attribute	**C**				
	P1	**P2**	**P3**	**P4**	**P5**
Readability	5	7	8	23	14
Maintainability	5	7	8	23	14
Stability	2	3	3	5	4
Reliability	1	1	2	3	3

Quality Attribute	**C++**				
	P1	**P2**	**P3**	**P4**	**P5**
Readability	5	7	8	23	14
Maintainability	5	7	8	23	14
Stability	3	4	4	6	5
Reliability	3	4	4	5	5

Quality Attribute	**Java**				
	P1	**P2**	**P3**	**P4**	**P5**
Readability	4	6	7	19	13
Maintainability	4	6	7	19	13
Stability	1	1	1	2	2
Reliability	2	3	3	4	3

Quality Attribute	Python				
	P1	P2	P3	P4	P5
Readability	1	2	5	8	8
Maintainability	1	2	5	8	8
Stability	1	1	2	3	3
Reliability	1	1	1	2	2

Quality Attribute	Mathematica				
	P1	P2	P3	P4	P5
Readability	1	1	1	3	1
Maintainability	1	1	1	3	1
Stability	1	2	2	4	3
Reliability	2	3	3	4	3

Table 10.7 Summary of Analysis

Quality Attributes	C	C++	Java	Python	Mathematica
Readability	Equal as C++	Equal as C	More than C and C++	More than others but less Mathematica	More than all
Maintainability	Equal as C++	Equal as C	More than C and C++	More than others but less Mathematica	More than all
Stability	More than C++	Less than all	More than all	More than others but less Java	More than C and C++
Reliability	More than others but less Python	Less than all	Equal as Mathematica	More than all	Equal as Java

Table 10.7 shows the summary of analysis. Each and every language has its own programming feature and characteristics which impact on software quality. Selection of programming language depends on requirement of software and quality. C program is extendable, flexible, secure, and reliable. C++ is object oriented programming language which is easy to maintain. Java is robust, stable, portable, and dynamic programming language. Python has fantastic ability to read like a description of the algorithm. It has indent which is important to speed reading and readability. Python codes are easily modified and improve, which enhance maintainability and scalability. Reliability of python code is much high because of its minimum failure rate. Mathematica is incredible language because of its low LOC. Its readability and maintainability is much good as compare than others.

There are another important quality attribute which is consider for analysis: scalability, security, robustness, portability, and flexibility. Software scalability is an ability to greatly increase in capacity with relative ease. In software development scalability generally means software should be functioning as it supposed to, still when it is working at an advanced scale of things. Scalability can be measured using different parameters such as: size of the program, number of processor operations per unit time, and number of current users. Scalability calculated using size of the program. Size of program of C and C++ programs is same for all programs. For Java, it is less than C and C++ but more than Python. Mathematica has less size of program from other languages [137-138].

In past years, the research community has ever more contributed to software quality, assurance, and software security. Security metric calculated by different measure such as:

- Percentage of defects revealed during coding that was injected in requirements specification, architecture, and design
- Percentage of software components subject to dynamic and static code analysis against identified vulnerabilities and weaknesses
- Percentage of software components subject to code reliability, integrity, and handling procedures such as: anti tampering, chain of custody verification, and code signing.

Security is calculated using percentage % of software components subject to code reliability. Python is more reliable language than others because of its low failure rate. C programming is good to develop a reliable program. Java and Mathematica is equally reliable language. C++ is less reliable than all [139-141].

Robustness is a degree in which a system or component can function correctly in the presence of invalid inputs or stressful environmental conditions. There are different techniques to measure a robust program. Robustness can be internal or external. Internal robust program include classes, functions, threads, and external robustness is depend on surrounding environment. Java is more robust than others because of its dynamic nature and performs function correctly. Python is good than remaining other languages because of it's easy to adopt changing environment. Mathematica is good as compare of C and C++. C code robustness is much better than C++ [142-145].

Portability of software can be explain as degree of ease or complexity to which a software component or application can be efficiently and effectively transferred from one hardware, software, and other operational or usage environment to another. There are many approaches achieving greater portability for given software. These techniques can be effectively guided by three key principles such as: control the interfaces, isolate dependencies, and think portable. Java is more portable than other languages because of its dynamic nature to easily usage environment. Python is good than remaining other languages because it's easy to adopt changing environment. C code portability is much better than C++ and Mathematica [146-148].

Software flexibility indicates software is easy to change and modify. Flexible software can effortlessly adjust to user requirement and environment changes during software development or after the software is deployed. Flexibility can be measured by different parameters such as: construction time, modification time, latency, dimensionality, program size by counting LOC, and cyclomatic complexity. Python easily change and modify software as per user requirement. Mathematica and Java perform much better for software development. C and C++ is a flexible language to adjust effortlessly [149-153].

Past researches shows that programming styles applied for creating new code or maintaining existing programs and have an impact on quality of the resulting software product. There is no developer who never made a mistake. The developer style is fixed as per programming language for more clean, readable, flexible, and maintainable code. Basically there are naming and formatting code, line length, whitespace, control structures, indentation, namespaces, methods functions, and comments etc which affect programming styles whereas programming styles are generally performance enhancements [154-155].

There are diverse programming styles of different developer so using fix programming styles for more accurate result. For example to print Fibonacci number using python programming there are different type names, comments, and format. Developer uses different statements, loops, and functions to print Fibonacci number. Here Fibonacci number print using different programming styles of developer as shown in Figure 10.12. Figure 10.12 shows developer style 1 and developer style 2 to print Fibonacci number. Same program shown in Figure 10.10: P5 program as developer style 3. All program print Fibonacci number but uses different developer style.

def fibonacci(n):
a = 1
b = 1
for i in range(0, n):
temp = a
a = b
b = temp + b
return a
for c in range(0, 20):
print(fibonacci(c))
 	
Developer Style 1
	def fibonacci(n):
	a = 1
	b = 1
	for i in range(0, n):
	print(a)
	temp = a
	a = b
	b = temp + b
	return a
	fibonacci(20)
	
	Developer Style 2

Figure 10.12 Different Programming Styles of Developer

Suppose if we want to convert decimal number to binary number and sum of first 100 natural numbers using C, C++, Java, and Python, and Mathematica programming, there are different programming styles of each programming. Code quality depends on programming and its styles. Basically selection of programming language depends on various aspect of software requirement and quality attributes.

C program: Convert decimal to binary number

```
#include <stdio.h>
int convert(int);
int main()
{
int dec, bin;
printf("Enter a decimal number: ");
scanf("%d", &dec);
bin = convert(dec);
printf("The binary equivalent of %d is %d.\n", dec, bin);
return 0;
}
int convert(int dec)
{
if (dec == 0)
{
return 0;
}
else
{
return (dec % 2 + 10 * convert(dec / 2));
}
}
```

C++ program: Convert decimal to binary number

```
#include<iostream>
using namespace std;
void binary(int num)
{
int rem;
if (num <= 1)
{
cout << num;
return;
}
rem = num % 2;
binary(num / 2);
cout << rem;
}
int main()
{
int dec, bin;
cout << "Enter the number : ";
cin >> dec;
if (dec < 0)
cout << dec << " is not a positive integer." << endl;
else
{
cout << "The binary form of " << dec << " is ";
binary(dec);
cout << endl;
}
return 0;
}
```

Java program: Convert decimal to binary number

```
import java.util.*;
class DecimalBinary
{
public static void main(String[] args)
{
Scanner in = new Scanner(System.in);
Stack<Integer> stack = new Stack<Integer>();
System.out.println("Enter decimal number: ");
int num = in.nextInt();
while (num != 0)
{
int d = num % 2;
stack.push(d);
num /= 2;
}
System.out.print("\nBinary representation is:");
while (!(stack.isEmpty() ))
{
System.out.print(stack.pop());
}
System.out.println();
}
}
```

Python program: Convert decimal to binary number

```
def binary(n):
if n > 1:
binary(n//2)
print(n % 2,end = '')
dec = int(input("Enter an integer: "))
binary(dec)
```

Mathematica program: Convert decimal to binary number

```
n = Input ["Please enter the number:"];
Print["Binary No:", BaseForm[n, 2]];
```

Every program give equivalent binary number of decimal number but each programming has different characteristics and LOC. In C program there are 22 LOC, C++ 28, Java 23, for Python 6, and in Mathematica it is 2. Mathematica work best in mathematical term, we can directly convert decimal to binary number using "BaseForm[40, 2]" of particular number.

C program: Sum of first 100 natural numbers

```
#include<stdio.h>
#include<conio.h>
void main()
{
int i,sum=0;
clrscr();
for(i=1;i<=100;i++)
sum+=i;
printf("Sum= %d\n", sum);
getch();
}
```

C++ program: Sum of first 100 natural numbers

```
#include <iostream>
using namespace std;
int main()
{
int n, sum = 0;
for (int i = 1; i <= 100; ++i) {
sum += i;
}
cout << "Sum = " << sum;
return 0;
}
```

Java program: Sum of first 100 natural numbers

```
public class Main {
public static void main(String[]
args) {
int sum = 0;
for(int i = 1; i <= 100; ++i)
{
sum += i;
}
System.out.println("Sum = " +
sum);
}
}
```

Python program: Sum of first 100 natural numbers

```
n = 101
sum = 0
for i in range(1,n):
    sum = sum + i
print (sum
```

Mathematica program: Sum of first 100 natural numbers

```
Sum[i,{i,1,100}]
```

There are different technical characteristics to compare programming languages: how many key words, type-checking facilities, maximum length of identifier, overriding, polymorphism, and so on. It is approximately infinitely perplexing. Important paramter are:

- Fit-for-purpose
- Ease of understanding
- Ease of learning
- Speed of development
- Performance of compiled code
- Help with enforcement of correct code
- Portability
- Supported platform environments

While Java is a excellent language and highly portable but it is unsuitable for some game programming and system programming. Game programming requires quick access to the screen-display hardware. This can be completed by using DirectX which is available only in C or C++. However majority of programmers are writing business applications in Java (or C#) is just fine. C still be hard to understand. OO languages allow to write more compact code because of code reuse. Java easier to learn than C++ and C is most likely even easier. With OO programming there are additional ways of tackling a problem than in conventional languages [156].

There are various way to solve a problem and there are various method, framework, methodology, programming, and tool to get the required output. In case of programming for enterprise services: C, C++, Java, Python, and PHP are best programming languages [157]. Programming language can be decide on on operating system. Software professional select programming language based on software development requirements and application deployment environment [158].

CPU time or statistics is the amount of time for which CPU was used for processing instructions of a computer program and operating system. CPU time is measured by clock seconds or ticks [159]. CPU performance depend on clock rate of CPU, number of available calculation units, size of on-die caches, design of CPU bus protocol, size and speed of external buses, and design of CPU sub-architecture. Large caches speed up memory access and lead to faster performance. Law of diminishing returns in cache size therefore bigger will not constantly give better performance. In calculation, certain workload is too big to fit into reasonable cache size so performance of CPU isn't affected by cache size for those workloads [160]. In this chapter P1, P2, P3, and P5 program take 1 to 2 second for completion, while P4 take 2 to 5 second to complete the task because of an input command.

10.10 SUMMARY

To see the impact of design pattern on quality attribute a method to assess design pattern is discussed. Design pattern assessment method is a combination of quantitative and qualitative analysis to evaluate quality attributes. Abstract factory pattern is use to illustrate the applicability of method to evaluate reusability, composability, completeness, maintainability, stability, understandability, flexibility, simplicity, and expandability. Design patterns do not always improve the quality attribute. Each and every attribute affected by design pattern as positively and negatively. After assessment maintainability is acceptable. Reusability is good, understandability is negative, and flexibility is ok. Composability is suitable, completeness as per task, stability is appropriate, simplicity is good, and expandability as per user requirement.

This chapter provides an overview of programming languages and its impact on software quality. It is hard to determine which programming languages are most widely used, better, and what usage resources vary by perspective. One language may inhabit the greater amount of programmer hours, different one have more lines of code, and third might consume the most CPU time. For comparison and analysis purpose various programs in C, C++, Java, Python, and Mathematica have been developed and compared. Software coding affects quality of software products. Some languages are incredibly accepted for particular kind of applications. Quality attributes have been calculated using mathematical formation. Code quality analysis shows that every programming language has its own characteristics and importance which impact on various attributes of software quality and final product. Selection of programming language affects software product and its quality. For readability, maintainability, and scalability Mathematica is good because of its minimum number of line in a program and LOC. Java is a stable, portable, and robust language because it has the ability to avoid unpredicted things after update and modification. Python is a good for scalability, security, and develops a reliable product because of its minimum failure rate and C is secure, flexible, and reliable programming language due to its structure and high level abstraction.

10.11 REFERENCES

1. Gamma, E.; Helms, R.; Johnson, R.; Vlissides, J. (1995). *Design patterns: elements of reusable object-oriented software.* Addison-Wesley Professional.
2. McDonald, J. (2008). Design patterns. DZone Incorporated.
3. Singh, B.; Gautam, S. (2016). The impact of software development process on software quality: a review. IEEE 8th International Conference on Computational Intelligence and Communication Networks (CICN 2016), pp. 666-672.
4. McNatt, W. B.; Bieman, J. M. (2001). Coupling of design patterns: common practices and their benefits. IEEE 25th Annual International Computer Software and Applications Conference (COMPSAC 2001), pp. 574-579.

5. Wendorff, P. (2001). Assessment of design patterns during software reengineering: lessons learned from a large commercial project. IEEE Fifth European Conference on Software Maintenance and Reengineering, pp. 77-84.
6. Halkidis, S. T.; Chatzigeorgiou, A.; Stephanides, G. (2006). A qualitative analysis of software security patterns. Computers & Security, Vol. 25, Issue 5, pp. 379-392.
7. Khomh, F.; Guéhéneuc, Y. G.; Antoniol, G. (2009). Playing roles in design patterns: an empirical descriptive and analytic study. IEEE International Conference on Software Maintenance (ICSM 2009), pp. 83-92.
8. Huston, B. (2001). The effects of design pattern application on metric scores. Journal of Systems and Software, Vol. 58, Issue 3, pp. 261-269.
9. Bansiya, J.; Davis, C. G. (2002). A hierarchical model for object-oriented design quality assessment. IEEE Transactions on Software Engineering, Vol. 28, Issue 1, pp. 4-17.
10. Ampatzoglou, A.; Chatzigeorgiou, A. (2007). Evaluation of object-oriented design patterns in game development. Information and Software Technology, Vol. 49, Issue 5, pp. 445-454.
11. Sant'Anna, C.; Garcia, A.; Kulesza, U.; Lucena, C.; Staa, A. V. (2004). Design patterns as aspects: a quantitative assessment. Journal of the Brazilian Computer Society, Vol. 10, No. 2, pp. 42-55.
12. Garcia, A.; Sant'Anna, C.; Figueiredo, E.; Kulesza, U.; Lucena, C.; Staa, A. von (2006). Modularizing design patterns with aspects: a quantitative study. Transactions on Aspect-Oriented Software Development I, Springer Berlin Heidelberg, pp. 36-74.
13. Hsueh, N. L.; Chu, P. H.; Chu, W. (2008). A quantitative approach for evaluating the quality of design patterns. Journal of Systems and Software, Vol. 81, Issue 8, pp. 1430-1439.
14. Khomh, F.; Gueheneuce, Y. G. (2008). Do design patterns impact software quality positively? IEEE 12th European Conference on Software Maintenance and Reengineering (CSMR 2008), pp. 274-278.
15. Cacho, N.; Sant'anna, C.; Figueiredo, E.; Dantas, F.; Garcia, A.; Batista, T. (2014). Blending design patterns with aspects: a quantitative study. Journal of Systems and Software, Vol. 98, pp. 117-139.
16. Rasool, G.; Maeder, P.; Philippow, I. (2011). Evaluation of design pattern recovery tools. Procedia Computer Science, Vol. 3, pp. 813-819.
17. Misra, K. B. (2013). Sustainable designs of products and systems: a possibility. International Journal of Performability Engineering, Vol. 9, No. 2, pp.175-190.
18. Singh, B.; Gautam, S. (2016). Hybrid spiral model to improve software quality using knowledge management. International Journal of Performability Engineering, Vol. 12, No. 4, pp. 341-352.
19. Hannemann, J.; Kiczales, G. (2002). Design pattern implementation in Java and AspectJ. ACM Sigplan Notices, Vol. 37, No. 11, pp. 161-173.
20. Blewitt, A.; Bundy, A.; Stark, I. (2005). Automatic verification of design patterns in Java. IEEE/ACM 20th International Conference on Automated Software Engineering, pp. 224-232.
21. Niere, J.; Schäfer, W.; Wadsack, J. P.; Wendehals, L.; Welsh, J. (2002). Towards pattern-based design recovery. In Proceedings of 24th International Conference on Software Engineering, pp. 338-348.
22. Burmester, S.; Giese, H.; Niere, J.; Tichy, M.; Wadsack, J. P.; Wagner, R.; Wendehals, L.; Zündorf, A. (2004). Tool integration at the meta-model level: the Fujaba approach. International Journal on Software Tools for Technology Transfer, Vol. 6, No. 3, pp. 203-218.
23. Shi, N.; Olsson, R. A. (2006). Reverse engineering of design patterns from java source code. IEEE/ACM 21st International Conference on Automated Software Engineering (ASE'06.), 123-134.
24. Vandyke, C. (2009). A design pattern generation tool. Doctoral dissertation Department of CS, Worcester Polytechnic Institute.

25. Arcelli, F.; Perin, F.; Raibulet, C.; Ravani, S. (2008). Design pattern detection in Java systems: a dynamic analysis based approach. Springer International Conference on Evaluation of Novel Approaches to Software Engineering, pp. 163-179.
26. Fontana, F. A.; Zanoni, M. (2011). A tool for design pattern detection and software architecture reconstruction. Information Sciences, Vol. 181, Issue 7, pp. 1306-1324.
27. Binun, A.; Kniesel, G. (2012). Joining forces for higher precision and recall of design pattern detection. Uni. Bonn, Germany, CS Department III, Technical report IAI-TR-2012-01.
28. Dabain, H.; Manzer, A.; Tzerpos, V. (2015). Design pattern detection using FINDER. ACM 30th Annual Symposium on Applied Computing, pp. 1586-1593.
29. https://www.percerons.com/designPAD/index.php
30. Ampatzoglou, A.; Frantzeskou, G.; Stamelos, I. (2012). A methodology to assess the impact of design patterns on software quality. Information and Software Technology, Vol. 54, Issue 4, pp. 331-346.
31. Freeman, E.; Robson, E.; Bates, B.; Sierra, K. (2004). *Head first design patterns*. O'Reilly Media Inc.
32. Bieman, J. (2000). Do design patterns create more maintainable systems? Citeseer, pp. 1-2.
33. Hegedűs, P.; Bán, D.; Ferenc, R.; Gyimóthy, T. (2012). Myth or reality? analyzing the effect of design patterns on software maintainability. Computer Applications for Software Engineering, Disaster Recovery, and Business Continuity, Springer Berlin Heidelberg, pp. 138-145.
34. Alghamdi, F. M.; Qureshi, M. R. J. (2014). Impact of design patterns on software maintainability. International Journal of Intelligent Systems and Applications, Vol. 6, No. 10, p. 41.
35. Najm, N. (2014). Measuring maintainability index of software depending on line of code only. IOSR Journal of Computer Engineering (IOSR-JCE), Vol. 16, No. 2, pp. 64-69.
36. Gamma, E.; Helm, R.; Johnson, R.; Vlissides, J. (1993). Design patterns: abstraction and reuse of object-oriented design. In Springer European Conference on Object-Oriented Programming, pp. 406-431.
37. Schmidt, D. C. (1995). Using design patterns to develop reusable object-oriented communication software. Communications of the ACM, Vol. 38, No. 10, pp. 65-74.
38. Ampatzoglou, A.; Kritikos, A.; Kakarontzas, G.; Stamelos, I. (2011). An empirical investigation on the reusability of design patterns and software packages. Journal of Systems and Software, Vol. 84, Issue 12, pp. 2265-2283.
39. Yacoub, S.; Ammar, H. H.; Mili, A. (2000). Constructional design patterns as reusable components. In Springer International Conference on Software Reuse, pp. 369-387.
40. Singh, A.; Tomar, P. (2014). Estimation of component reusability through reusability metrics. International Journal of Computer, Information, Systems and Control Engineering, Vol. 8, No. 11, pp. 1965-1972.
41. Garzás, J.; García, F.; Piattini, M. (2007). Do patterns affect design understandability? kybeleconsulting, pp. 1-11.
42. Muller, G. (2016). Design objectives and design understandability. Buskerud University College.
43. Lin, J. C.; Wu, K. C. (2006). A model for measuring software understandability. IEEE Sixth International Conference on Computer and Information Technology (CIT'06), pp. 192-192.
44. Larsson, T.; Sandberg, M. (2000). Building flexible components based on design patterns. Department of Computer Science, Malardalen University, Sweden, pp. 1-9.
45. Ballesteros, F. J.; Kon, F.; Patino, M.; Jiménez, R.; Arévalo, S.; Campbell, R. H. (2009). Batching: a design pattern for efficient and flexible client/server interaction, In Transactions on Pattern Languages of Programming I, Springer Berlin Heidelberg, pp. 48-66.

46. Rasool, G.; Mäder, P. (2011). Flexible design pattern detection based on feature types. IEEE/ACM 26th International Conference on Automated Software Engineering (ASE 2011), pp. 243-252.
47. Eden, A. H.; Mens, T. (2006). Measuring software flexibility. IEE Proceedings-Software, Vol. 153, No. 3, pp. 113-125.
48. Cacho, N. A. A.; Figueiredo, E. M. L.; SantþAnna, C. N.; Garcia, A. F.; Batista, T. V.; de Lucena, C. J. P. (2005). Aspect-oriented composition of design patterns: a quantitative assessment. PUC.
49. Zimmer, W. (1995). Relationships between design patterns. Pattern Languages of Program Design, Vol. 57, pp. 345- 364.
50. Bayley, I.; Zhu, H. (2008). On the composition of design patterns. IEEE Eighth International Conference on Quality Software (QSIC'08), pp. 27-36.
51. Zhu, H.; Bayley, I. (2015). On the composability of design patterns. IEEE Transactions on Software Engineering, Vol. 41, Issue 11, pp. 1138-1152.
52. DeHon, A.; Adams, J.; DeLorimier, M.; Kapre, N.; Matsuda, Y.; Naeimi, H.; Vanier, M.; Wrighton, M. (2004). Design patterns for reconfigurable computing. IEEE 12th Annual Symposium on Field-Programmable Custom Computing Machines (FCCM 2004), pp. 13-23.
53. Pousman, Z.; Stasko, J. (2006). A taxonomy of ambient information systems: four patterns of design. ACM Conference on Advanced Visual Interfaces, pp. 67-74.
54. Fayad, M. E.; Rajagopalan, J.; Ranganath, A. (2003). AnyLog stable design pattern. IEEE International Conference on Information Reuse and Integration (IRI 2003), pp. 566-571.
55. S. K. Singh and M. E. Fayad, The AnyCorrectiveAction stable design pattern. ACM 17th Conference on Pattern Languages of Programs, p. 24, Oct 2010.
56. Fayad, M. E.; Singh, S. K.; Basia, A.; Ajesh, A.; Kasi, P. (2011). AnyEvidence stable design pattern. pp. 1-20.
57. Elish, M. (2006). Do structural design patterns promote design stability? IEEE 30th Annual International on Computer Software and Applications Conference (COMPSAC'06), Vol. 1, pp. 215-220.
58. Gore, D. (2009). A stable design for the state design pattern. International Journal of Software Engineering and Its Applications, Vol. 3, No. 4, pp. 33-50.
59. Ampatzoglou, A.; Chatzigeorgiou, A.; Charalampidou, S.; Avgeriou, P. (2015). The effect of GoF design patterns on stability: a case study. IEEE Transactions on Software Engineering, Vol. 41, Issue 8, pp. 781-802.
60. Nesbitt, K. V. (2013). Simplicity: a design pattern for ideas. NOVA - The University of Newcastle's Digital Repository.
61. Khomh, F.; Guéhéneuc, Y. G. (2007). Perception and reality: what are design patterns good for? In Proceedings of Springer 11th ECOOP Workshop on Quantitative Approaches in Object Oriented Software Engineering (QAOOSE), p. 7.
62. Dascalu, S.; Hao, N.; Debnath, N. (2005). Design patterns automation with template library. IEEE Fifth International Symposium on Signal Processing and Information Technology, pp. 699-705.
63. Cass, S. (2016). The 2016 Top Programming Languages, IEEE Spectrum, 26 Jul 2016.
64. https://spectrum.ieee.org/computing/software/the-2016-top-programming-languages
65. Cass, S. (2017). The 2017 Top Programming Languages, IEEE Spectrum, 18 Jul 2017.
66. https://spectrum.ieee.org/computing/software/the-2017-top-programming-languages
67. Cass, S.; Bulusu, P. (2018). Interactive: The Top Programming Languages 2018, IEEE Spectrum, 31 Jul 2018.
68. https://spectrum.ieee.org/static/interactive-the-top-programming-languages-2018
69.
70. McLoone, J. (2012). Code Length Measured in 14 Languages, Wolfram Blog, 14 Nov 2012.
71. https://blog.wolfram.com/2012/11/14/code-length-measured-in-14-languages/
72. Ritchie, D. M. (1996). The development of the C programming language. In History of Programming languages II, pp. 671-698, ACM.

73. University of Leicester, Why use C? https://www.le.ac.uk/users/rjm1/cotter/page_04.htm
74. Andahi, A. (2017). C Programming For Beginners. ACM.
75. Stroustrup, B. (1999). An overview of the C++ programming language. Handbook of object technology.
76. Infocodify, Overview of C++ programming.
77. http://www.infocodify.com/cpp/cpp_overview
78. Stroustrup, B. The C++ Programming Language. http://www.stroustrup.com/C++.html
79. Arnold, K.; Gosling, J.; Holmes, D. (2000). The Java programming language (Vol. 2), Addison wesley.
80. Jecan, A. (2017). Java 9 Modularity Revealed. Springer.
81. Perry, J. S. (2010). Introduction to Java programming. Part 1, IBM.
82. https://www.ibm.com/developerworks/java/tutorials/j-introtojava1/index.html
83. Rossum, G. Van (2007). Python Programming Language. In USENIX Annual Technical Conference, Vol. 41, p. 36.
84. Zelle, J. M. (2004). Python programming: an introduction to computer science. Franklin, Beedle & Associates, Inc.
85. Python, Python programming language. Python Software Foundation, 2009.
86. Fateman, R. J. (1992). A review of Mathematica. Journal of Symbolic Computation, Vol. 13, Issue 5, pp. 545-579.
87. Trott, M. (2013). The Mathematica GuideBook for Programming. Springer.
88. Zotos, K. (2007). Performance comparison of Maple and Mathematica. Applied mathematics and computation, Vol. 188, Issue 2, pp. 1426-1429.
89. Stroustrup, B. (2002). C and C++: Siblings. CC Plus Plus Users Journal, Vol. 20, No. 7, pp. 28-36.
90. Oliphant, T. E. (2007). Python for scientific computing. Computing in Science & Engineering, Vol. 9, No. 3.
91. Learneroo, (2016). What Programming Language Should You Learn? Learneroo Blog, 2016.
92. https://www.learneroo.com/modules/9/nodes/620
93. Wolfram, S. (1990). Who uses mathematica? The Mathematical Journal, Addison Wesley Publishing Co, Vol. 1, Issue 1.
94. https://stephenwolfram.com/publications/academic/who-uses-mathematica.pdf
95. Singh, B.; Kannojia, S. P. (2012). Languages and Their Importance in Quality Software. In IEEE International Conference on Communication Systems and Network Technologies (CSNT), pp. 956-959.
96. Sujatha, P.; Sankar, G. V.; Rao, A. S.; Satyanarayana, T. (2001). The Role of Software Verification and Validation in Software Development Process. IETE Technical Review, Vol. 18, Issue 1, pp. 23-26.
97. The Wise Developers' Guide to Static Code Analysis featuring FindBugs, Checkstyle, PMD, Coverity and SonarQube.
98. https://zeroturnaround.com/rebellabs/developers-guide-static-code-analysis-findbugs-checkstyle-pmd-coverity-sonarqube/
99. Stamelos, I.; Angelis, L.; Oikonomou, A.; Bleris, G. L. (2002). Code quality analysis in open source software development. Info Systems J., Vol. 12, pp. 43-60.
100. Meirelles, P.; Jr., C. S.; Miranda, J.; Kon, F.; Terceiro, A.; Chavez, C. (2010). A Study of the Relationships between Source Code Metrics and Attractiveness in Free Software Projects. In IEEE Brazilian Symposium on Software Engineering (SBES), pp. 11-20.
101. Iqbal, T.; Iqbal, M.; Asad, M.; Khan, A. (2016). A Source Code Quality Analysis Approach. IEEE 10th International Conference on Software, Knowledge, Information Management & Applications (SKIMA), pp. 142-145.
102. Plosch, R.; Gruber, H.; Korner, C.; Saft, M. (2010). A method for continuous code quality management using static analysis. In IEEE Seventh International Conference on the Quality of Information and Communications Technology (QUATIC), pp. 370-375.

103. Kochhar, P. S.; Wijedasa, D.; Lo, D. (2016). A large scale study of multiple programming languages and code quality. In IEEE 23rd International Conference on Software Analysis, Evolution, and Reengineering (SANER), Vol. 1, pp. 563-573.
104. Butler, S.; Wermelinger, M.; Yu, Y.; Sharp, H. (2010). Exploring the influence of identifier names on code quality: An empirical study. In IEEE 14th European Conference on Software Maintenance and Reengineering (CSMR), pp. 156-165.
105. Barstad, V.; Goodwin, M.; Gjøsæter, T. (2014). Predicting Source Code Quality with Static Analysis and Machine Learning. In NIK.
106. Koepke, H. (2011). Why Python rocks for research. Hacker Monthly, No. 8.
107. http://cppcheck.sourceforge.net
108. https://www.cppdepend.com
109. https://www-01.ibm.com/common/ssi/cgi-bin/ssialias?infotype=OC&subtype=NA&html fid=897/ENUS5724-X69&appname=totalstorage
110. http://scan.coverity.com
111. http://pmd.sourceforge.net
112. http://findbugs.sourceforge.net
113. http://checkstyle.sourceforge.net
114. 11 Best Practices and Tools to Improve the Java Code Quality. https://www.romexsoft.com/blog/improve-java-code-quality
115. https://www.sonarqube.org
116. https://www.codacy.com
117. https://github.com/PyCQA/pylint
118. https://github.com/PyCQA/pyflakes
119. https://www.wolfram.com/products/applications/mathcode
120. https://reference.wolfram.com/workbench/index.jsp?topic=/com.wolfram.eclipse.help/html/tasks/tester/tester.html
121. Singh, B.; Kannojia, S. P. (2012). A model for software product quality prediction. Journal of Software Engineering and Applications, Vol. 5, pp. 395-401.
122. Tashtoush, Y.; Odat, Z.; Alsmadi, I.; Yatim, M. (2013). Impact of programming features on code readability. International Journal of Software Engineering and Its Applications, Vol. 7, Issue 6, pp. 441-458.
123. Roy, C. K.; Cordy, J. R.; Koschke, R. (2009). Comparison and evaluation of code clone detection techniques and tools: A qualitative approach. Science of Computer Programming, Vol. 74, Issue 7, pp. 470-495.
124. DeYoung, G. E.; Kampen, G. R. (1979). Program Factors as Predictors of Program Readability. In Proceedings of IEEE Third International Conference on Computer Software and Applications (COMPSAC 79), pp. 668-673.
125. Aggarwal, K. K.; Singh, Y.; Chhabra, J. K. (2002). An Integrated Measure of Software Maintainability. In IEEE Annual Proceedings of Reliability and maintainability symposium, pp. 235-241.
126. Buse, R. P.; Weimer, W. R. (2008). A metric for software readability. In Proceedings of ACM international symposium on Software testing and analysis, pp. 121-130.
127. Lee, T.; Lee, J. B.; In, H. P. (2013). A study of different coding styles affecting code readability. International Journal of Software Engineering and Its Applications, Vol. 7, No. 5, pp. 413-422.
128. Beazley, D. M. (2000). Scientific computing with Python. In Astronomical Data Analysis Software and Systems IX, Vol. 216, p. 49.
129. M.A.M. Najm, N. (2014). Measuring Maintainability Index of a Software Depending on Line of Code Only. IOSR Journal of Computer Engineering (IOSR-JCE), Vol. 16, Issue 1, pp. 64-69.
130. Pearse, T.; Oman, P. (1995). Maintainability Measurements on Industrial Source Code Maintenance Activities. In Proceedings of IEEE International Conference on Software Maintenance, pp. 295-303.

131. Butler, S.; Wermelinger, M.; Yu, Y.; Sharp, H. (2010). Exploring the Influence of Identifier Names on Code Quality: An empirical study. In 14th European Conference on Software Maintenance and Reengineering, Madrid, Spain, pp. 156-165.
132. Linos, P. K.; Courtois, V. (1996). A toolset for maintaining hybrid C++ programs. Journal of Software: Evolution and Process, Vol. 8, No. 6, pp. 389-419.
133. Heitlager, I.; Kuipers, T.; Visser, J. (2007). A practical model for measuring maintainability. IEEE 6th International Conference on Quality of Information and Communications Technology (QUATIC 2007), pp. 30-39.
134. Seref, B.; Tanriover, O. (2016). Software Code Maintainability: A Literature Review. International Journal of Software Engineering & Applications (IJSEA), Vol.7, No. 3.
135. Cheaito, R.; Frappier, M.; Matwin, S.; Mili, A.; Crabtree, D. (1995). Defining and Measuring Maintainability. Citeseer.
136. Smith, T. (2013). Measuring Code Stability- Part I. Perforce.
137. https://www.perforce.com/blog/130919/measuring-code-stability-part-i
138. Staron, M.; Hansson, J.; Feldt, R.; Meding, W.; Henriksson, A.; Nilsson, S.; Höglund, C. (2013). Measuring and visualizing code stability- a case study at three companies. In IEEE Joint Conference of the 23rd International Workshop on Software Measurement and Eighth International Conference on Software Process and Product Measurement (IWSM-MENSURA), pp. 191-200.
139. Fadzlah, A. F. A. (2013). Stableness Measurement Model: A Mathematical Approach for Measuring the Stability of Handheld Application Usage. International Journal of Software Engineering & Applications, Vol. 4, No. 5, pp. 49-61.
140. Raemaekers, S.; van Deursen, A.; Visser, J. (2012). Measuring software library stability through historical version analysis. In IEEE 28th International Conference on Software Maintenance (ICSM), pp. 378-387.
141. Asadi, M.; Rashidi, H. (2016). A Model for Object-Oriented Software Maintainability Measurement. International Journal of Intelligent Systems and Applications, Vol. 8, No. 1, pp. 60-66.
142. Grosser, D.; Sahraoui, H. A.; Valtchev, P. (2003). An analogy-based approach for predicting design stability of Java classes. In Proceedings of IEEE Ninth International Software Metrics Symposium, pp. 252-262.
143. Baggen, R.; Correia, J. P.; Schill, K.; Visser, J. (2012). Standardized code quality benchmarking for improving software maintainability. Software Quality Journal, Vol. 20, Issue 2, pp. 287-307.
144. Jalote, P.; Murphy, B.; Garzia, M.; Errez, B.; Way, O. R. (2004). Measuring reliability of software products. In ISSRE 2004 Conference, France.
145. Wohlin, C.; Höst, M.; Runeson, P.; Wesslén, A. (2001). Software reliability. Encyclopedia of Physical Sciences and Technology, Vol. 15, pp. 25-39.
146. Rajaraman, C.; Lyu, M. R. (1992). Reliability and maintainability related software coupling metrics in C++ programs. In IEEE Proceedings on Third International Symposium Software Reliability Engineering, pp. 303-311.
147. Hunt, J.; Long, F. (1998). Java's reliability: An analysis of software defects in Java. IEE Proceedings Software, Vol. 145, No. 2, pp. 41-50.
148. Measuring the System Scalability http://www.omsar.gov.lb/ICTSG/105OS/5.10_Measuring_the_System_Scalability.htm
149. Weyuker, E. J.; Avritzer, A. (2002). A metric to predict software scalability. In IEEE eighth symposium on Software metrics, pp. 152-158.
150. Allen, J. (2009). Measuring software security. CERT Research Annual Report.
151. McGraw, G. (2004). Software security. IEEE Security & Privacy, Vol. 2, No. 2, pp. 80-83.
152. Potter, B.; McGraw, G. (2004). Software security testing. IEEE Security & Privacy, Vol. 2, No. 5, pp. 81-85.
153. Abdalla, A. M.; Abdallah, M. M.; Salah, M. I. (2017). A brief program robustness survey. International Journal of Software Engineering & Applications (IJSEA), Vol. 8, No. 1.

154. Aggarwal, K. K.; Singh, Y.; Kaur, A.; Malhotra, R. (2007). Software Design Metrics for Object Oriented Software. Journal of Object Technology, Vol. 6, No. 1, pp. 121-138.
155. Winter, S.; Sârbu, C.; Suri, N.; Murphy, B. (2011). The impact of fault models on software robustness evaluations. In Proceedings of ACM 33rd International Conference on Software Engineering, pp. 51-60.
156. Why java is called robust?
157. http://interactivejava.blogspot.in/2013/01/why-java-is-called-robust.html
158. Mooney, J. D. (2004). Developing Portable Software. In IFIP Congress Tutorials, pp. 55-84.
159. Mooney, J. D. (1993). Issues in the specification and measurement of software portability. In 15th International Conference on Software Engineering.
160. Programming for Portability
161. http://www.catb.org/esr/writings/taoup/html/ch17s05.html
162. Shen, L.; Ren, S. (2006). Analysis and measurement of software flexibility based on flexible point. In 3th Software Measurement European Forum, Italy, pp. 331-341.
163. Wang, E. T.; Ju, P. H.; Jiang, J. J.; Klein, G. (2008). The effects of change control and management review on software flexibility and project performance. Information & Management. Vol. 45, Issue 7, pp. 438-443.
164. Davis, D. (2013). Modelled on Software Engineering: Flexible Parametric Models in the Practice of Architecture. PhD dissertation, RMIT University.
165. Liu, J. Y.; Chen, V. J.; Chan, C. L.; Lie, T. (2008). The impact of software process standardization on software flexibility and project management performance: Control theory perspective. Information and Software Technology, Vol. 50, No. 9, pp. 889-896.
166. On the Flexibility of Programming Languages http://gerbil.org/tom/papers/flex.shtml
167. How to deal with different programming styles in a team? https://softwareengineering.stackexchange.com/questions/119455/how-to-deal-with-different programming-styles-in-a-team
168. Capiluppi, A.; Boldyreff, C.; Beecher, K.; Adams, P. J. (2009). Quality Factors and Coding Standards - a Comparison Between Open Source Forges. Electronic Notes in Theoretical Computer Science Vol. 233, pp. 89-103.
169. https://msdn.microsoft.com/en-us/library/cc168615.aspx
170. http://www.theserverside.com/feature/Best-programming-languages-for-enterprise-development
171. http://careerdrill.com/blog/career-guide/how-to-select-programming-language-in-software-industry/
172. How to determine CPU and memory consumption from inside a process? https://stackoverflow.com/questions/63166/how-to-determine-cpu-and-memory-consumption-from-inside-a-process
173. What does CPU performance depend on?
174. https://www.quora.com/What-does-CPU-performance-depend-on

Chapter - 11

INTEGRATING USABILITY THROUGH SOFTWARE PROCESS

Software quality is most imperative aspect of software product which can be accomplish using software process. Success of software product directly related to users who use the software. Usability of software product is the one of the most important quality attribute for user. Usability of software product can be achieved using software process. Usability requirement analysis is focused to show the important aspects of usability and its attributes. Usability requirement analyzed based on questionnaire where response from user perspective. After analysis user satisfaction, performance, accuracy, usefulness, and efficiency are most vital usability attribute from user point of view which is illustrated as an example to show the integration of usability through software process. Section I describes the introduction of this chapter. Section II describes the related work. Section III describes the usability framework to achieve usability. Section IV describes the software usability requirement analysis. Section V describes the usability evaluation. Section VI describes the summary of this chapter.

11.1 INTRODUCTION

Software quality is most important aspect of software product which can be accomplish using software process. Usability of software product is the one of the most important quality attribute for user. Usability of software product can be achieved using software process. It can be brought at the development process and considered at requirement analysis. Addressing usability at requirements analysis has more benefit on early development process. The earlier usability attribute are identified and built into the software [1-3]. The quality of final product is highly dependent on effective requirement identification process.

Accurately identified requirements result from effective communication and collaboration among all member of the project team and provide a best chance of creating a system that fully satisfies the needs and requirement of the users [4]. After requirement software design accordingly then software coding is start for executable version of software next testing of software is started to check the software errors and bugs. Usability impact on the development process because usability requirements turn into quantified usability specification which will have impact on design outcome [5].

Software developer will have to conduct usability evaluations to ensure that the final product will meet these usability goals [6]. So usability questionnaire is used to find out the important aspects of usability, its attribute, and requirement. Main aim of this chapter is to show the importance of software process to achieve usability in final product. Each and every phase is important for software development. At a time of development if we analyze the user task completely then design and implement accordingly, after that the output will be a usable product.

11.2 RELATED WORK

The software process consists of the activities and associated information that are required to develop software [7]. The purpose of the outline of software process and usability is to show that software process play an important role to develop a usable product within given time and cost.

11.2.1 OVERVIEW OF SOFTWARE PROCESS

Software process is sequence of activities to develop a software product. Software process provides a framework for managing activities that can very easily get out of control in software development. Different software projects require different software process. One of the best known software development process models is waterfall model formally described by Winston Royce in 1970 suggests the implementation steps to deliver a small computer program for internal operations and large computer program for delivery to a customer. The waterfall model is used when requirements are very well known and fixed [8]. Basili and Turner proposed an iterative approach for developing software in 1975. Iterative model helps when requirements are not completely clear. Iterative approach maintains efficient learning [9].

The spiral model proposed by Barry Boehm represents a risk driven approach for assessment of risks [10]. The spiral model combines waterfall model, iterative model, and prototyping. The V model represents the sequence of steps in a software development. The left side of the V represents the requirements and design. The right side of the V represents the validation and verification. The V model provides assistance how to implement requirement consequently. All requirements are known up front which can be modified to handle changing requirements [11].

Rapid application development (RAD) is a software development model which favors iterative development and rapid implementation of prototypes in place of large amounts of straight planning and reasonably well known requirements [12]. In agile software development requirements and solutions change during development using self organizing and cross functional teamwork. It promotes early delivery using adaptive planning and evolutionary development. The agile model encourages quick and flexible response to change [13].

Hybrid spiral model is an integration of spiral model and knowledge management which is based on knowledge flow during development process. Hybrid spiral model is a modified version of spiral model if we use tacit knowledge in requirement and design than quality based product is develop [14]. Every development process follows requirement, design, coding, and testing. In this chapter requirement (usability) is fix. There are few development models which work on fixed and well known requirements such as: waterfall model, V model, and rapid application development (RAD). Using waterfall model we integrate the usability through software process because it is most appropriate for fixed requirement and small scale software. Software subsystem may be developed in parallel by different development team that requires software specification to be defined early in the development process from user point of view.

11.2.2 SOFTWARE USABILITY

Software usability is one of the most important quality attribute. Usability definition is different from one user to another. For many, user satisfaction, ease of use, performance, user friendly, and learnability are important attributes of usability. Usability is the measure of how usable something is? Usability of software should measure both user preference and performance [15].

IEEE defined usability as, "The ease with which a user can learn to operate, prepare inputs for, and interpret outputs of a system or component" [16]. Shackel defined usability as, "the usability of a system is the capability in human functional terms to be used easily and effectively by the specified range of users, given specified training and user support, to fulfill the specified range of tasks, within the specified range of scenarios" [17].

ISO 9126 defined usability as, "a set of attributes of software which bear on the effort needed for use and on the individual assessment of such use by a stated or implied set of users" [18]. Nielsen defined usability to consist of five kinds of attributes: Learnability, Efficiency, Memorability, Errors, and Satisfaction [19]. ISO 9241 defined usability as, "the extent to which a product can be used by specified users to achieve specified goals with effectiveness, efficiency and satisfaction in a specified context of use" [20].

Several software development models have been proposed to develop software. In system centered method usability is not addressed. For integrating usability through software process, user centred design (UCD) used as a key factor for leading towards the development of successful interface [21]. Using UCD final users are involved from the early planning stage and identify user requirement. Early involvement of users is important for usability. UCD are related to completeness of system functionality, repair and effort saving, and user satisfaction. Involving users from early stages allows the system on what is needed. The basic steps of UCD are: (1) analyze users and task; (2) design and implement the system iteratively using prototypes; (3) evaluate design and prototypes with users. UCD requires understanding of who will use the system; how, where, and to do what.

To create usable interactive systems, it is necessary to software development process to address usability. For that purpose we need to shift from the system centered (classical waterfall model) to user centered design (revised waterfall model) that may lead to usable systems. So revised waterfall model is useful to achieve usability in finished product which is same as classical waterfall model but some modification [22].

The user interface is the most important part of the software from the user's point of view [23]. For illustration purpose Library Management System (LMS) is developed using Java language to show how to integrate usability through software process. Main objective of this chapter to integrate usability through software process, analyze the important aspects of usability, and its attributes.

11.3 USABILITY FRAMEWORK TO ACHIEVE USABILITY

Various methods, models, and standard are used to develop software. To achieve quality in finished product well defined process is required because quality of product completely related to quality of process. So effective software development process is used to accomplish the desired goal and develop a quality product. ISO 13407 provides guideline for achieving usability through UCD to develop interactive software [24].

The standard describes four UCD activities that required at the earliest planning of project. These are:

- Understand and specify the context of use
- Specify the user requirements
- Develop design solutions
- Evaluate designs against requirements.

The iterative nature of these activities is shown in Figure 11.1 [25].

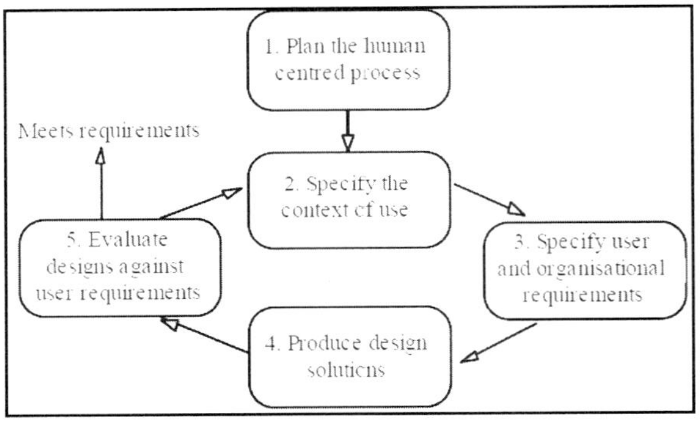

Figure 11.1 ISO 13407 activities

Based on these steps usability framework is used to achieve usability through software process. This framework applicable for small scale software such as: library management system, online reservation system, and online examination system where requirements are fix. Usability framework follows simple steps to incorporate usability in software product as shown in Figure 11.2, which has been described as:

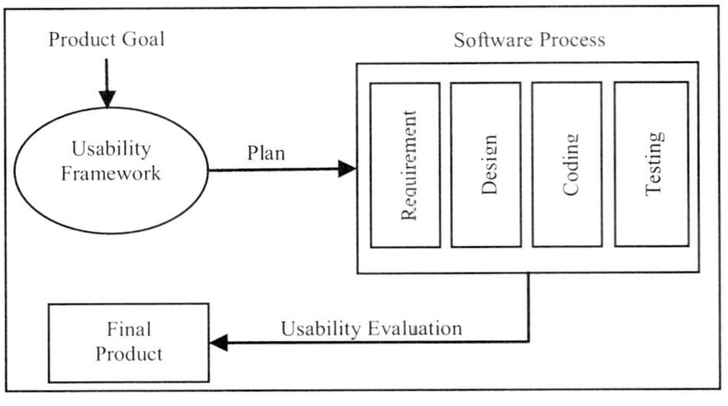

Figure 11.2 Usability Framework

Usability framework accomplishes the product goal to achieve usability in final product. **Planning** is the one of the most important step to successfully accomplish software development process [26]. Planning include assessing the usability capability maturity and identify where improvements are required [27]. This chapter uses planning as a part of usability framework then **requirement** is consider. Various techniques are proposed by researchers to collect user requirement [28-30]. User requirement are developed using questionnaire [31]. After that software **design** is started to capture the functional and non functional requirements [32]. Basically there is function oriented design and object oriented design [33-34]. Design software (LMS) using object oriented design [35].

Software **coding** uses programming to create an executable version of software. There are structured programming and functional programming [36-37]. The selection affects result and quality of software [38]. Java programming has been used to develop software for illustration to integrate usability [39]. After that software **testing** is started. There are two types of software testing: reviews, walkthroughs, and inspections are static testing; whereas executing programmed code with a given set of test cases is dynamic testing [40]. Software testing methods are divided into white-box and black-box testing [41]. Rubin and Chisnell proposed a usability testing which has been used to test software in specified environment [42]. Finally **usability evaluation** is considered to evaluate software using user satisfaction, performance, and feedback [43-44]. After that final product is usable software.

Usability framework is used to drive usability in software through software process and usability evaluation as shown in Figure 11.2. Using usability framework any small scale software is developed to achieve usability in final product. This framework is appropriate for academic and industrial objective where requirements are fix.

11.4 SOFTWARE USABILITY REQUIREMENT ANALYSIS

Software quality plays a major role for project success. Success of any project directly related to user satisfaction. There are various quality attributes which affect the project success and performance [45]. Important quality attributes are usability, reliability, functionality, maintainability, portability, and security. Software usability is the one of the most important quality attribute from user point of view. So main focus on important aspects of usability because usability of software increases user satisfaction and project success [46].

To find out usability requirement for software questionnaire is prepared because requirement is first and most important step to integrate usability [47]. A questionnaire was designed with 50 questions. This chapter considered various objectives to prepare the questionnaire, which is as: **Part 1:** To find out important aspect of usability as shown in Table 1, **Part 2:** To find out important attribute of usability as shown in Table 2, and **Part 3:** To find out basic requirement for usability as shown in Table 5. This questionnaire fills out in two international conferences: International Conference on Advanced Computing and Software Engineering (ICACSE-16) and IEEE 8th International Conference on Computational Intelligence and Communication Networks (CICN 2016). The participants from academic and industry area such as: researcher, scientists, and developer. After two months the questionnaire was closed with 70 responses.

After completing the questionnaire, usability can be explained in single word "user satisfaction". Usability can be defined as, the ability of user to perform operation/function easily and effectively by specified range of input, environment, and task. Usability refers to a set of multiple concepts such as: user satisfaction, performance, accuracy, usefulness, efficiency, consistency, understandability, effectiveness, and learnability. The modified relationship between the usability factors is shown in Figure 11.3 [48].

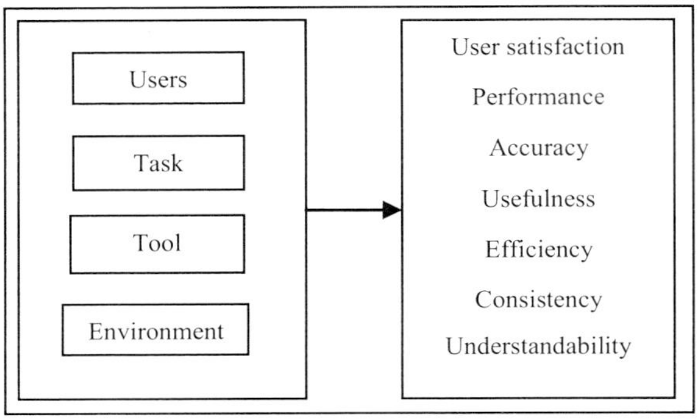

Figure 11.3 Usability Factors

User, task, tool, and environment affect usability factors. There are various usability attributes which reflect software performance.

Standards provide a guideline to achieve usability. Standards related to usability can be categorized as:

1. The process used to develop the product.
2. The capability to apply UCD.
3. The user interface and interaction.
4. The use of the product (effectiveness, efficiency, and user satisfaction)

This modified version of category of usability standard is shown in Figure 11.4 which shows the importance of process quality to product quality [49].

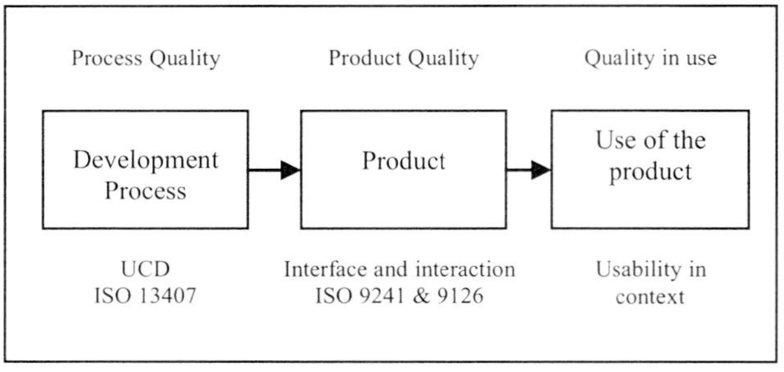

Figure 11.4 Category of usability standard

ISO 13407:1999 (Human-centred design processes for interactive systems) has been revised by ISO 9241-210:2010 (Ergonomics of human-system interaction- Part 210:

Human-centred design for interactive systems). ISO/IEC 9126-1:2001 (Software engineering- Product quality- Part 1: Quality model) has been revised by ISO/IEC 25010:2011 (Systems and software engineering- Systems and software Quality Requirements and Evaluation (SQuaRE)- System and software quality models) [50].

User has different software requirements. For many it is most important that the users perform efficiently and for others reliability. Requirements analysis identifies and prioritizes these goals [51]. To optimize the interaction between software and users, it is necessary to be aware of user requirements, user characteristics, usage, and systems aspects [52]. Various people are involved for collecting requirement such as: systems analysts, project manager, business analysts, users, and designers [53]. The analysis of usability framework is based on usability questionnaire so we prepare a questionnaire to find out the user perception on usability.

Part 1: To find out important aspect of usability

The objective of part 1 (Q1-10) used to find out important aspect of usability and there are total 70 response for each question. The important aspect of usability is shown in Table 11.1.

Table 11.1 Important aspect of usability

Q. No.	Usability Aspects	No. of response out of 70
1.	Usability defines as: a) Learnability b) Accuracy	 42 28
2.	What is the development process do you usually follow for usability? a) Spiral Development b) Prototyping	 26 44
3.	Usability satisfaction metric measured by: a) Task Level Satisfaction b) Test Level Satisfaction	 44 26
4.	Learnability can be increase by: a) User documentation and help b) System should be easy to learn	 41 29
5.	Which kind of interface is more useful? a) Consistent b) Attractive	 44 26
6.	Which is more important factor for usability? a) User satisfaction b) Ease of use	 43 27
7.	Which method is best to check usability of software? a) Paper based methods such as: Prototyping, Survey, and Walk-Through b) Usability Testing	 36 34
8.	To increase operability which is more important? a) Error messages b) User guide	 28 42

9.	How error message is helpful to use software easily? a) Provides information about what should be done. b) Suggestion to input a correct data.	40 30
10.	Understandability means: a) Interface elements (menu) should be easy to understand. b) The operation of the system should be easily understandable.	40 30

Figure 11.5 shows the important aspect of usability in graph; based on data available in Table 11.1. There are total 10 questions and 70 responses for each question which is shown in horizontal and vertical axis respectively.

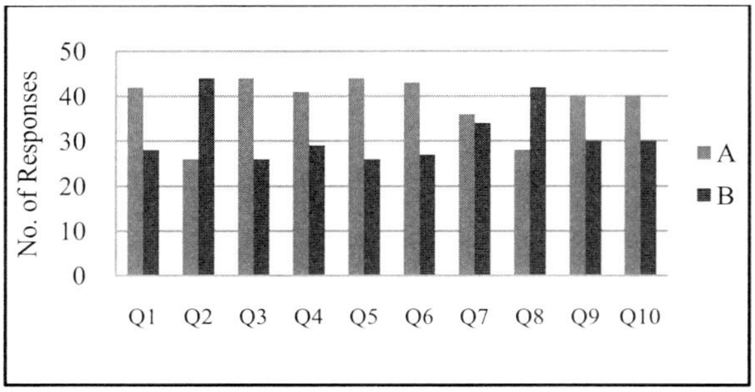

Figure 11.5 Important aspect of usability

Explanation of important aspect of usability as: usability defined as learnability (60%) because there are total 42 responses out of 70. Prototyping (62.85%) is an important development process to incorporate usability in final product. There are total 44 responses out of 70. Usability satisfaction metric measured by, task level satisfaction (62.85%) where total 44 responses out of 70. Learnability can be increase by user documentation and help (58.57%) because there are total 41 responses out of 70. Consistent (62.85%) interface is more useful than attractive because there are total 44 responses out of 70. User satisfaction (61.42%) is more important factor than ease of use because there are total 43 responses out of 70. Paper based methods (51.42%) and usability testing (48.57%) equally important for usability testing because there are total 36 and 34 responses out of 70 respectively. To increase operability user guide (60%) is more important than error messages because there are total 42 responses out of 70. To use software easily information about what should be done (57.14%) is more helpful than suggestion to input a correct data because there are total 40 responses out of 70. Understandability means interface elements (menu) should be easy to understand (57.14%) there are total 40 responses out of 70.

Part 2: To find out important attribute of usability

In previous research various researcher describe the usability and its attributes in user performance (objective) and user view (subjective). Shackel's [17] definition of usability: effectiveness, learnability, flexibility, and attitude as shown in Figure 11.6.

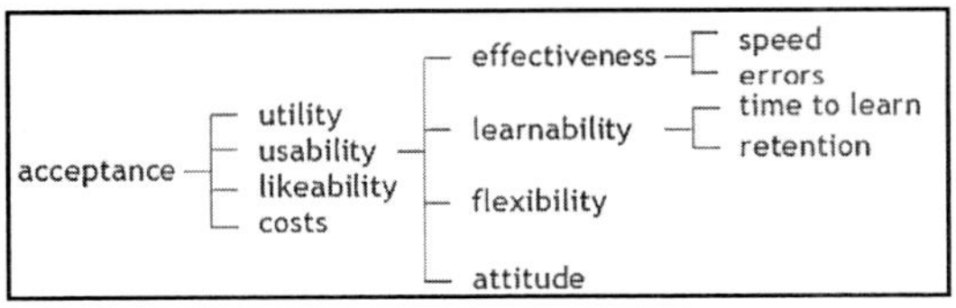

Figure 11.6 Shackel's definition of usability

ISO 9126 [18] define usability attributes as: understandability, learnability, operability, and attractiveness. ISO 9126 quality model usability measurable attributes are shown in Figure 11.7.

Usability

Understandability
Learnability
Operability
Attractiveness

Figure 11.7 ISO 9126 usability attributes

Nielsen [19] defines usability using five attributes as shown in Figure 11.8.

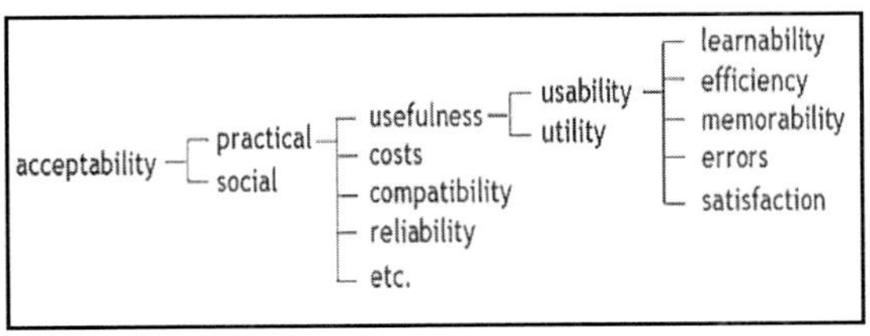

Figure 11.8 Nielsen's definition of usability

ISO 9241-11 [20] define three attributes of usability which is shown in Figure 11.9.

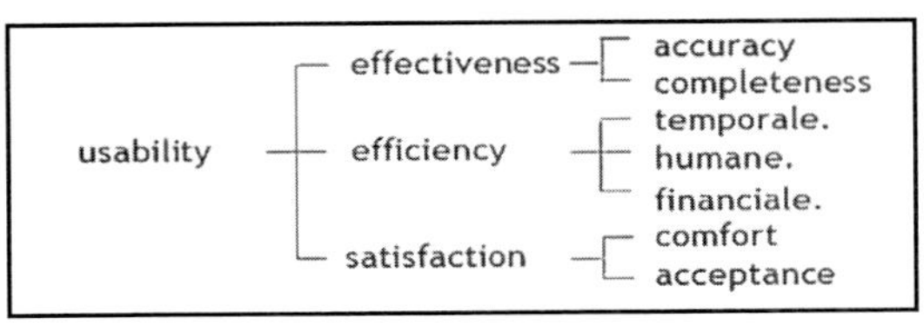

Figure 11.9 ISO 9241-11 definition of usability

To find out important attributes of usability part 2 (Q1-20) is listed and 1 to 5 rating is used to identify which attribute is more important or less important from 1 (High), 2 (Less High), 3 (Medium), 4 (Less Low), and 5 (Low) there are total 70 response for each attribute. List of usability attributes are shown in Table 11.2.

Based on Table 11.2 data, top 10 usability attributes is extracted from user point of view which is shown in Table 11.3.

Table 11.2 List of usability attributes

Q. No.	Usability Attribute	1 (H)	2 (LH)	3 (M)	4 (LL)	5 (L)
1.	Ease of Use	37	25	7	1	0
2.	User Satisfaction	47	19	1	1	2
3.	Usefulness	42	20	3	3	2
4.	Ease of Learning	33	24	11	1	1
5.	Complexity	14	16	13	16	11
6.	Efficiency	42	18	5	3	2
7.	Understandability	39	26	2	1	2
8.	Accuracy	45	18	4	2	1
9.	Learnability	33	20	8	8	1
10.	Consistency	40	21	4	4	1
11.	Effectiveness	37	22	3	4	4
12.	Execution Time	31	12	11	9	7
13.	Reliability	40	20	5	4	1
14.	Response Time	38	15	10	2	5
15.	Performance	47	6	5	6	6
16.	User Interface	34	30	5	1	0
17.	Attractiveness	21	23	19	5	2
18.	Flexibility	33	26	4	4	3
19.	Memorability	23	27	16	3	1
20.	Compatibility	24	30	13	2	1

Table 11.3 Top 10 usability attributes

Rank	Usability Attributes	No. of Response
1.	User Satisfaction	47 (67.14%)
2.	Performance	47 (67.14%)
3.	Accuracy	45 (64.28%)
4.	Usefulness	42 (60%)
5.	Efficiency	42 (60%)
6.	Consistency	40 (57.14%)
7.	Reliability	40 (57.14%)
8.	Understandability	39 (55.71%)
9.	Response Time	38 (54.28%)
10.	Effectiveness	37 (52.85 %)

Figure 11.10 show the important usability attributes using Table 11.3 data where usability attributes and number of response shown in horizontal and vertical axis respectively. Top 5 usability attributes are used to integrate usability through software process.

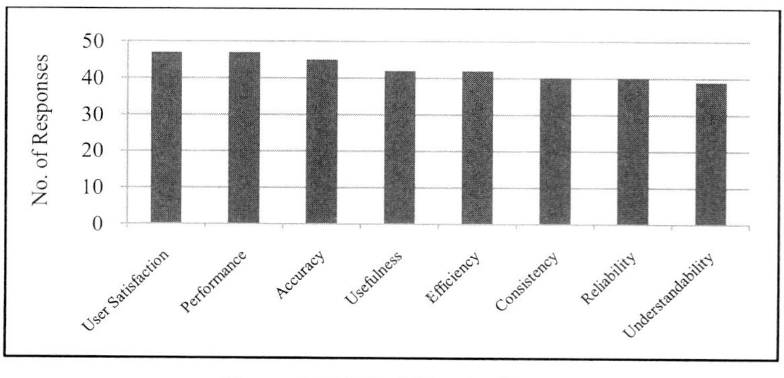

Figure 11.10 Usability Attributes

Figure 11.10 shows that user satisfaction and performance (67.14%) is more important usability attribute for user because there are total 47 responses out of 70. There is maximum response by users that show user satisfaction and performance is most important attribute than other.

Shackel is one of the first authors in the field of usability; define a model where product acceptance is the highest concept on four dimensions. ISO 9126 define product oriented role of usability using four attributes. ISO 9241 define process oriented role of usability using three attributes. Nielsen is another initiator of usability defines usability as product acceptance same as shackle consisting five attributes [54]. For usability framework, top 6 usability attributes from Table 11.3 are used for modified usability attributes from user point of view which has been compared in Table 11.4.

Table 11.4 shows the overview of usability attributes from different perspective. In usability framework user satisfaction is most important usability attribute from user perspective.

Table 11.4 Modified usability attributes

Shackel	**ISO 9126**	**ISO 9241-11**	**Nielsen**	**Usability Framework**
Effectiveness	Understandability	Effectiveness	Learnability	User Satisfaction
Learnability	Learnability	Efficiency	Memorability	Performance
Flexibility	Operability	Satisfaction	Efficiency	Accuracy
Attitude	Attractiveness		Satisfaction	Usefulness
			Errors	Efficiency
				Consistency

Part 3: To find out basic requirement for usability

There is various usability requirements that should be consider at the time of development. Part 3 (Q1-20) used to find out the basic requirement for usability. There are total 70 responses for each requirement. List of basic requirement for usability are shown in Table 11.5.

Table 11.5 Basic requirement for usability

Q. No.	**Usability Requirement**	**Yes**	**No**
1.	Help and documentation is useful to use software.	70	0
2.	Simple and easy dialogue saves user time.	69	1
3.	Feedback can improve software performance.	62	8
4.	Minimum memory load encourage user to use software.	58	12
5.	Error messages help user to give correct and required input.	63	7
6.	Consistency of screen and display play a major factor for usability.	55	15
7.	Easy navigation in menu is important for usable software.	66	4
8.	System/software speed affects user satisfaction.	60	10

9.	Using learnability we perform common tasks or operations.	47	23
10.	Understandability increases the satisfaction rate of user.	58	12
11.	Efficiency of software can be measured in terms of task time.	54	16
12.	Frequent execution time is important parameter for consideration.	52	18
13.	Response Time can be measured by in term of turnaround time.	58	12
14.	Traceability defines the relationship between the requirements and design.	57	13
15.	Accuracy of software can be measured by in term of correctness.	55	15
16.	For attractive user interface screen layout and color should be appealing.	53	17
17.	For operability interface actions and elements should be consistent.	60	10
18.	Flexibility of data entry is useful for direct manipulation.	57	13
19.	Task completion rate is important factor for user to perform or work on.	65	5
20.	Help messages useful for using and operating software.	59	11

Figure 11.11 and 11.12 show the basic requirement for usability in graph respectively based on data available in Table 11.5 where usability requirement question and number of response shown in horizontal and vertical axis respectively.

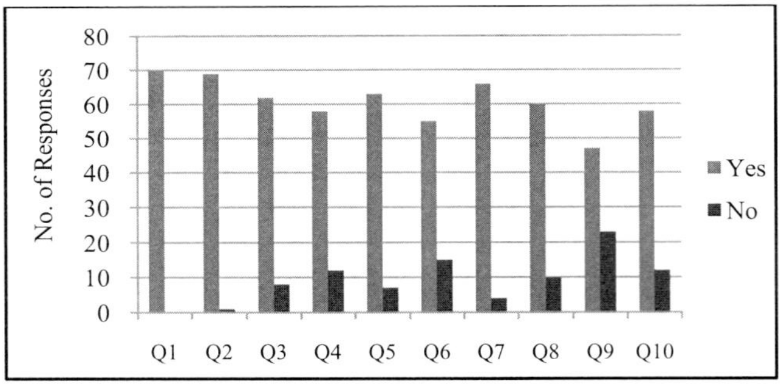

Figure 11.11 Basic requirements for usability

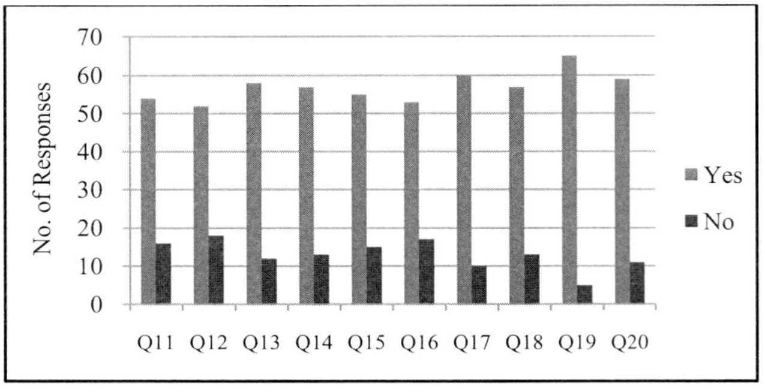

Figure 11.12 Basic requirements for usability

Figure 11.11 and 11.12 shows basic requirement for usability where help and documentation is most important and useful to use software because there are total 70 responses out of 70. Simple and easy dialogue saves user time there are total 69 responses out of 70. Easy navigation in menu is important for usable software and task completion rate is important factor for user to perform because there are total 66 and 65 responses out of 70 respectively.

Outcome of this analysis is summaries as for part 1: usability defined as learnability means system easy to learn. Usability satisfaction metric measured by task level satisfaction, after task completion we measure user satisfaction easily. We conclude that user satisfaction is more important factor than ease of use because satisfy user increases success rate.

For part 2: after analysis we extract main 10 usability attributes such as: user satisfaction, performance, accuracy, usefulness, efficiency, consistency, reliability, understandability, response time, and effectiveness from user point of view.

For part 3: help and documentation is one of most important basic requirement for usability because it provides the useful information to perform task easily. Simple and easy dialogue saves user time because user knows how to perform task. Easy navigation in menu is important for usable software because it improve user satisfaction rate. Task completion rate is important factor to consider because it reflects user performance.

11.5 USABILITY EVALUATION

Usability evaluation is the extent to which an interactive system is pleasant and easy to use. Basically there are three types of usability evaluation methods [55-56]: testing, inspection, and inquiry [57-58]. Assessing the usability of software used to identify usability problem and measures usability. The purpose of usability evaluation to improve the usability of product as part of development (formative evaluation), and to evaluate the extent to which usability objectives have been achieved (summative evaluation) [59].

When usability inspection and testing is carried out at the end of the design, modification in interface can be complex and expensive to implement so it is necessary to involve usability at early development process. Table 11.6 depicts the summary of inspection methods (without users) and testing methods (with users) [60]. Both inspection and testing methods are used for usability evaluation [61].

All these methods used to evaluate and measure usability. For usability evaluation 10 research scholars and software developer are participated for measurement which has minimum five year experience.

User satisfaction is the one of the most important usability attribute and key measure of project success [62]. User satisfaction is a measure of how well a product meets user expectations which is calculated using subjective metrics. ISO 9241 seem user satisfaction is extremely tough to compare to performance since user are not conscious of their actual performances [63].

Table 11.6 Summary of inspection methods and testing methods

	Inspection Methods			**Testing Methods**		
	Action Analysis	Cognitive Walkthrough	Heuristic Evaluation	Field Observation	Questionnaires	Thinking Aloud
Applicably in Phase	Design	All	All	Final Testing	All	Design
Required Equipment	Low	Low	Low	Medium	Low	High
Required Time	High	Medium	Low	Medium	Low	High
Needed Users	None	None	None	20+	30+	3+
Required Evaluators	1-2	3+	3+	1+	1	1
Required Expertise	High	High	Medium	High	Low	Medium
Intrusive	No	No	No	Yes	No	Yes

In most user satisfaction surveys user attitude and knowledge prior to measuring satisfaction. User satisfaction is measured throughout standardized satisfaction questionnaire which can be administered after each task (task level satisfaction) and test (test level satisfaction) session. Another method to calculate user satisfaction, task performance without error rate metric which is analyze using satisfaction ratio with questionnaire and survey [64-65].

Task level satisfaction measure after users attempt a task and how much difficulty and problem face to perform a particular task which is calculated using up to 5 questions with Likert scale questionnaire. The most popular post-task questionnaires are:

1. NASA-TLX: NASA's task load index is a measure of mental effort (5 questions)
2. ASQ: After Scenario Questionnaire (3 questions)
3. SMEQ: Subjective Mental Effort Questionnaire (1 question)
4. UME: Usability Magnitude Estimation (1 question)
5. SEQ: Single Ease Question (1 question)

Test Level Satisfaction is measured by formalized questionnaire to each user at the end of the test session. The most useful questionnaires are:

1. SUMI: Software Usability Measurement Inventory (50 questions)
2. QUIS: Questionnaire For User Interaction Satisfaction (24 questions)
3. CSUQ: Computer System Usability Questionnaire (19 questions)
4. SUPR-Q: Standardized User Experience Percentile Rank Questionnaire (13 questions)
5. SUS: System Usability Scale (10 questions)

These questionnaires are used for usability measurement [66-67]. We calculate user satisfaction after task completion (task level satisfaction) using SEQ as shown in Figure 11.13.

Very Difficult 1	2	3	4	5	6	Very Easy 7
○	○	○	○	○	○	○

Figure 11.13 SEQ with 7 point rating scale

SEQ is a 7 point rating scale used to record how difficulty users face to perform a particular task [68]. Across the over 20 tasks with 10 users we find the average score between 5.6.

Performance is another important factor from user perspective. User satisfaction and performance are directly related to usability [69-70]. Performance (throughput or quality of service) calculated using objective metrics. To calculate objective performance metrics, task completion time, number of clicks, and error rate used [71].

For performance computation, when user performs particular tasks then how long it takes them and success rate (whether users can perform the task at all) are recorded [72]. To calculate software performance using task completion time. The user completed their desire task in just seconds (eg: user login in 5-10 seconds).

Accuracy of software product is another important attribute from user point of view. Accuracy is a qualitative evaluation of correctness and freedom from error [73]. Accuracy of software is calculated in terms of correctness. The software correctly developed as per user requirement and user pleased with software product.

Next important usability attribute is usefulness. Usefulness is more important than ease of use. Usefulness (value) is the assessment of solution approved by users. "How much useful" are component, features, and function of software product. For assessing the usefulness of a presented product help predicting the usefulness of a novel product. The participants/user are asked to provide their level of agreement on a 5 point scale (1 for strongly disagree and 5 for strongly agree) [74-75]. To calculate usefulness 5 point scale method is used. After completing 20 tasks with 10 users the average score is 3.8.

Efficiency is an ability to complete task well, successfully, and utilize resource. According to ISO 9241, product efficiency is defined as "resources spent by user in order to ensure accurate and complete achievement of the goals". In software products resource normally measured by time spent to accomplish the task. Thus, efficiency can be calculated as user effectiveness divided by the time spent by user [48].

Efficiency is calculated in terms of task time. The time (seconds and minutes) the participant takes to successfully complete a task. The time taken to complete a task can be measured by simply subtracting the start time from the end time as shown in the equation (1):

$$\text{Task Time} = \text{End Time} - \text{Start Time} \qquad (1)$$

Efficiency can then be calculated in the form Time Based Efficiency and Overall Relative Efficiency [76]. To calculate efficiency using task time. Users complete their task easily in 1-2 seconds and more.

Illustration of usability framework as describe in section 11.3 is as: usability framework used to integrate usability through software process. To show the integration Library Management System (LMS) is developed using usability framework. The objective of LMS is to perform library operation easily and effectively. There are user login using user name and password. We can save the record of new books and retrieve the details of books available and issue in the library.

We can add, save, and modify the member details. We can issue and return the books. Transaction report display the detail of member, book, issue date, and receive date.

Using usability framework LMS is planned, for usable development it is important to consider usability at early development. Usability requirement collected using questionnaire because this paper based method consist of open or closed questions that makes it an efficient way to collect requirement from user and make design document quickly and easily as describe in section 11.4. To design LMS object oriented design method is used because it provide a complete representation of design and group functions together for direct access and modification. Then LMS is implemented using Java language because it is a high-level language which is designed to run any operating system (OS) because Java Virtual Machine (JVM). It is a powerful and flexible language includes interaction with the user, creating graphical programs, reading from files, cross-platform compatibility, and more. After coding usability testing is started to test LMS in specified environment. Usability testing performed in windows OS, Intel i3 processor, 2 GB RAM, and 100 GB Hard Disk. Next usability evaluation is started to evaluate LMS.

There are various tasks which are used to perform library operation easily and effectively. We consider 20 tasks which are useful to perform library operation. All these tasks are essential to perform usable library operation such as:

Task 1: To login in LMS simple login form is required. Using user name and password we enter LMS.

Task 2: Using user master, we add and update the detail of users and manage the user information using first, previous, next, last, new, edit, save, and close command.

Task 3: Change password used to modify user password.

Task 4: Book master used to add and update the detail of books and manage the book information using first, previous, next, last, new, edit, save, and close command.

Task 5: Search book used to search book using book ID or title or keyword.

Task 6: Book status depicts the status of book such as: issue and available.

Task 7: Member master used to add and update the detail of members and manage the member information using first, previous, next, last, new, edit, save, and close command.

Task 8: Search member used to search member using member ID or name or mobile.

Task 9: Issue book used to issue book to member.

Task 10: Receive book used to receive book from member.

Task 11: Transaction report summarizes the whole transaction on library with the detail of member, book, issue date, and receive date.

Task 12: First command used to move to first position.

Task 13: Previous command used to move to previous position.

Task 14: Next command used to move to next position.

Task 15: Last command used to move to last position.

Task 16: New command used to enter a new value such as: user, book, and member.

Task 17: Edit command used to update the information of user, book, and member.

Task 18: Save command used to save the information of user, book, and member.

Task 19: Close command used to close the running window.

Task 20: Exit command used to exit from LMS

Usability of LMS is evaluated by user satisfaction, performance, accuracy, usefulness, and efficiency as shown in Table 11.7. Using summary of usability evaluation we show the 20 tasks which is perform by 10 users.

Table 11.7 Summary of usability evaluation

S. No.	No. of Tasks	User Satisfaction (7 Point Scale)	Performance (Task Completion Time)	Accuracy (Correctness)	Usefulness (5 Point Scale)	Efficiency (Task Time)
1.	Task 1	5	5-10 Sec.	correct	4	5-10 Sec.
2.	Task 2	5	15-30 Sec.	ok	3	15-20 Sec.
3.	Task 3	6	15-20 Sec.	correct	4	15-20 Sec.
4.	Task 4	5	15-30 Sec.	correct	3	15-20 Sec.
5.	Task 5	5	10-15 Sec.	correct	4	5-10 Sec.
6.	Task 6	5	1-5 Sec.	correct	4	1-5 Sec.
7.	Task 7	5	15-30 Sec.	ok	3	15-20 Sec.
8.	Task 8	5	10-15 Sec.	correct	4	5-10 Sec.
9.	Task 9	5	15-30 Sec.	correct	4	15-20 Sec.
10.	Task 10	5	15-30 Sec.	correct	4	15-20 Sec.
11.	Task 11	6	2-5 Sec.	correct	4	1-5 Sec.
12.	Task 12	6	1-5 Sec.	correct	4	1-5 Sec.
13.	Task 13	6	1-5 Sec.	correct	4	1-5 Sec.
14.	Task 14	6	1-5 Sec.	correct	4	1-5 Sec.
15.	Task 15	6	1-5 Sec.	correct	4	1-5 Sec.
16.	Task 16	5	30-50 Sec.	correct	3	20-30 Sec.
17.	Task 17	6	20-40 Sec.	correct	4	10-30 Sec.
18.	Task 18	6	15-30 Sec.	correct	4	15-20 Sec.
19.	Task 19	7	1-2 Sec.	correct	4	1-2 Sec.
20.	Task 20	7	1-2 Sec.	correct	4	1-2 Sec.

Usability of LMS evaluated by 20 tasks, 10 user, and 5 attributes as shown in Table 11.7. User satisfaction measured by 7 point scale, performance in task completion time, accuracy in terms of correctness, usefulness using 5 point scale, and efficiency by task time.

11.6 SUMMARY

To integrate usability through software process usability framework is used. Usability framework applicable for small scale software where requirements are fix. For usability requirement analysis, usability questionnaire is used to find out the important aspects of usability, its attribute, and requirement where responses have been taken from researcher, scientists, and developer. After analysis user satisfaction, performance, accuracy, usefulness, and efficiency are most important usability attribute from user point of view. It has been illustrated in Library Management System (LMS) to show the integration of usability through software process.

11.7 REFERENCES

1. Juristo, N. (2009). Impact of Usability on Software Requirements and Design. Springer LNCS, pp. 55-77.
2. Lewis, J. R. (2014). Usability: lessons learned... and yet to be learned. International Journal of Human-Computer Interaction, Vol. 30, No. 9, pp. 663-684.
3. Tractinsky, N. (2017). The Usability Construct: A Dead End? Human-Computer Interaction, pp. 1-47.
4. Coughlan, J.; Macredie, R. D. (2002). Effective communication in requirements elicitation: a comparison of methodologies. Requirements Engineering, Vol. 7, Issue 2, pp. 47-60.
5. Vredenburg, K.; Mao, J. Y.; Smith, P. W.' Carey, T. (2002). A survey of user-centered design practice. In Proceedings of the SIGCHI conference on Human factors in computing systems, ACM, pp. 471-478.
6. Gulliksen, J.; Boivie, I.; Persson, J.; Hektor, A.; Herulf, L.; (2004). Making a difference: a survey of the usability profession in Sweden. In Proceedings of the third Nordic conference on Human-computer interaction, ACM, pp. 207-215.
7. Sommerville, I. (1996). Software Process Models. ACM Computing Surveys, Vol. 28, No. 1, pp. 269-271.
8. Royce, W. (1970). Managing the development of large software systems. In Proceedings of IEEE WESCON, Vol. 26, Issue 8, pp. 328-338.
9. Basili, V. R.; Turner, A. J. (1975). Iterative enhancement: A practical technique for software development. IEEE Transactions on Software Engineering, Vol. SE-1, Issue 4, pp. 390-396.
10. Boehm, B. W. (1988). A spiral model of software development and enhancement. IEEE Computer, Vol. 21, Issue 5, pp.61-72.
11. Bucanac, C. (1999). The V-model. University of Karlskrona/Ronneby, enero.
12. Martin, J. (1991). Rapid application development. Vol. 8, New York, Macmillan.
13. Highsmith, J.; Cockburn, A. (2001). Agile software development: The business of innovation. IEEE Computer, Vol. 34, Issue 9, pp.120-127.
14. Singh, B.; Gautam, S. (2016). Hybrid Spiral Model to Improve Software Quality Using Knowledge Management. International Journal of Performability Engineering (IJPE), Vol. 12, No. 4, pp. 341-352.
15. Nielsen, J.; Levy, J. (1994). Measuring Usability Preference vs. Performance. Communication of ACM, Vol. 37, Issue 4, pp. 66-75.
16. Radatz, J.; Geraci, A.; Katki, F. (1990). IEEE standard glossary of software engineering terminology. IEEE Std. 610.12-1990.
17. Shackel, B. (1991). Usability- context, framework, design and evaluation. In: B. Shackel and S. Richardson, (Eds.), Human Factors for Informatics Usability. Cambridge University Press, Cambridge, pp. 21-38.
18. ISO 9126, 1991. Software Product Evaluations- Quality Characteristics and Guidelines for their Use, ISO DIS 9126.
19. Nielsen, J. (1993). *Usability Engineering*. Academic press, San Diego, CA.
20. ISO/IEC 9241-11, 1998. Ergonomic requirements for office work with visual display terminals (VDT)s - Part 11 Guidance on usability, ISO/IEC 9241-11:1998 (E).
21. Rubin, J.; Chisnell, D. (2008). *Handbook of Usability Testing*. Second Edition: How to Plan, Design, and Conduct Effective Tests, Wiley Publishing, Inc.
22. Costabile, M. (2001). *Usability in the Software Life Cycle*. Handbook of Software Engineering and Knowledge Engineering, in Chang, S. (ed.), Vol. 1, World Scientific Publishing Co. Pte. Ltd.
23. Chang, S. K. (2001). *Handbook of Software Engineering & Knowledge Engineering*. Vol. 1 Fundamentals, World Scientific Publishing Co. Pte. Ltd.
24. ISO DIS 13407, 1997. User centred design process for interactive systems.

25. Bevan, N. (2003). UsabilityNet methods for user centred design. Human-Computer Interaction: Theory and Practice, (Part 1), Vol. 1, pp. 434-438.
26. Dvir, D.; Raz, T.; Shenhar, A. J. (2003). An empirical analysis of the relationship between project planning and project success. International journal of project management, Vol. 21, Issue 2, pp. 89-95.
27. Bevan, N.; Curson, I. (1999). Planning and Implementing User-Centred Design. In CHI'99 Extended Abstracts on Human Factors in Computing Systems, ACM, pp. 137-138.
28. Davis, A.; Overmyer, S.; Jordan, K.; Caruso, J.; Dandashi, F.; Dinh, A.; Kincaid, G.; Ledeboer, G.; Reynolds, P.; Sitaram, P.; Ta, A. (1993). Identifying and measuring quality in a software requirements specification. In Proceedings of IEEE First International Software Metrics Symposium, pp. 141-152.
29. Davis, A.; Dieste, O.; Hickey, A.; Juristo, N.; Moreno, A. M. (2006). Effectiveness of requirements elicitation techniques: Empirical results derived from a systematic review. In 14th IEEE International Requirements Engineering Conference (RE'06), pp. 179-188.
30. Achimugu, P.; Selamat, A.; Ibrahim, R.; Mahrin, M. N. R. (2014). A systematic literature review of software requirements prioritization research. Information and Software Technology, Vol. 56, Issue 6, pp. 568-585.
31. Wiegers, K.; Beatty, J. (2013). Software requirements. Microsoft Press.
32. Gross, D.; Yu, E. (2001). From non-functional requirements to design through patterns. Requirements Engineering, Vol. 6, Issue 1, pp. 18-36.
33. Chidamber, S. R.; Kemerer, C. F. (1991). Towards a metrics suite for object oriented design. ACM, Vol. 26, No. 11, pp. 197-211.
34. Chidamber, S. R.; Kemerer, C. F. (1994). A metrics suite for object oriented design. IEEE Transactions on software engineering, Vol. 20, Issue 6, pp. 476-493.
35. Otero, C. E. (2012). *Software Engineering Design: Theory and Practice*. CRC Press, Taylor & Francis Group.
36. Dijkstra, E. W. (1970). Notes on structured programming.
37. Hughes, J. (1989). Why functional programming matters. The computer journal, Vol. 32, No 2, pp. 98-107.
38. Singh, B.; Kannojia, S. P. (2012). Languages and Their Importance in Quality Software. In International Conference on Communication Systems and Network Technologies (CSNT), IEEE, pp. 956-959.
39. Vance, S. (2014). *Quality Code Software Testing Principles, Practices, and Patterns*. Pearson Education, Inc.
40. Fairley, R. E. (1978). Tutorial: Static analysis and dynamic testing of computer software. IEEE Computer, Vol. 11, Issue 4, pp. 14-23.
41. Haley, A.; Zweben, S.; (1984). Development and application of a white box approach to integration testing. Journal of Systems and Software, Vol. 4, Issue 4, pp. 309-315.
42. Myers, G. J.; Sandler, C.; Badgett, T. (2011). The art of software testing. John Wiley & Sons.
43. Ivory, M. Y.; Hearst, M. A. (2001). The state of the art in automating usability evaluation of user interfaces. ACM Computing Surveys (CSUR), Vol. 33, Issue 4, pp. 470-516.
44. Bertoa, M. F.; Troya, J. M.; Vallecillo, A. (2006). Measuring the usability of software components. The Journal of Systems and Software, Issue 79, pp. 427-439.
45. Singh, B.; Gautam, S. (2016). Situational Factors Affecting the Software Process: A Systematic Literature Review. International Conference on Advanced Computing and Software Engineering, ICACSE-16.
46. Svensson, R. B.; Gorschek, T.; Regnell, B.; Torkar, R.; Shahrokni, A.; Feldt, R. (2012). Quality requirements in industrial practice-an extended interview study at eleven companies. Software Engineering, IEEE Transactions on Software Engineering, Vol. 38, No. 4, pp. 923-935.

47. Singh, B.; Gautam, S. (2016). The Impact of Software Development Process on Software Quality: A Review. IEEE 8th International Conference on Computational Intelligence and Communication Networks, CICN 2016.
48. Bevan, N.; Macleod, M. (1994). Usability measurement in context. Behaviour and Information Technology, Vol. 13, pp. 132-145.
49. Bevan, N. (2006). International Standards for HCI and Usability. International Journal of Human Computer Studies, Vol. 55, Issue 4, pp. 533-552.
50. ISO, http://www.iso.org/
51. Karlsson, J.; Ryan, K. (1997). A cost-value approach for prioritizing requirements. IEEE software, Vol. 14, Issue 5, pp. 67-74.
52. Liu, X.; Veera, C. S.; Sun, Y.; Noguchi, K.; Kyoya, Y. (2004). Priority assessment of software requirements from multiple perspectives. In Proceedings of the 28th Annual International Computer Software and Applications Conference, COMPSAC'04, IEEE, pp. 410-415.
53. Sutcliffe, A. (1995). Requirements rationales: integrating approaches to requirement analysis. In Proceedings of the 1st conference on Designing interactive systems: processes, practices, methods, & techniques, ACM, pp. 33-42.
54. Folmer, E.; Bosch, J. (2004). Architecting for usability: a survey. The Journal of Systems and Software, Vol. 70, pp. 61-78.
55. Zhang, Z. (2001). Overview of usability evaluation methods.
56. Lewis, J. R. (2001). Introduction: Current issues in usability evaluation. International Journal of Human-Computer Interaction, Vol. 13, No. 4, pp. 343-349.
57. Nielsen, J. (1994). Usability inspection methods. In Conference companion on Human factors in computing systems, ACM, pp. 413-414.
58. Desurvire, H. W. (1994). Faster, cheaper!! Are usability inspection methods as effective as empirical testing?
59. Gediga, G.; Hamborg, K. C.; Düntsch, I. (1999). The IsoMetrics usability inventory: an operationalization of ISO 9241-10 supporting summative and formative evaluation of software systems. Behaviour & Information Technology, Vol. 18, No. 3, pp. 151-164.
60. Holzinger, A. (2005). Usability engineering methods for software developers. Communications of the ACM, Vol. 48, Issue 1, pp. 71-74.
61. Bias, R. G.; Moon, B. M.; Hoffman, R. R. (2015). Concept Mapping Usability Evaluation: An Exploratory Study of a New Usability Inspection Method. International Journal of Human-Computer Interaction, Vol. 31, No. 9, pp. 571-583.
62. Hassenzahl, M. (2004). The interplay of beauty, goodness, and usability in interactive products. Human-computer interaction, Vol. 19, Issue 4, pp. 319-349.
63. Hornbæk, K. (2006). Current practice in measuring usability: Challenges to usability studies and research. Int. J. Human-Computer Studies, Vol. 64, pp. 79-102.
64. Chin, J. P.; Diehl, V. A.; Norman, K. L. (1988). Development of an instrument measuring user satisfaction of the human-computer interface. In Proceedings of the SIGCHI conference on Human factors in computing systems, ACM, pp. 213-218.
65. Frekjmr, E., Hertzum, M., Hornbmk, K., 2000. Measuring Usability: Are Effectiveness, Efficiency, and Satisfaction Really Correlated? ACM CHI'2000, pp. 345-352.
66. Bevan, N. (1995). Measuring usability as quality of use. Software Quality Journal, Vol. 4, pp. 115-130.
67. Kortum, P.; Peres, S. C.; (2014). The relationship between system effectiveness and subjective usability scores using the System Usability Scale. International Journal of Human-Computer Interaction, Vol. 30, No. 7, pp. 575-584.
68. Sauro, J. (2012). 10 Things To Know About The Single Ease Question (SEQ), MeasuringU, www.measuringu.com/blog/seq10.php
69. Etezadi-Amoli, J.; Farhoomand, A. F. (1996). A structural model of end user computing satisfaction and user performance. Information & management, Vol. 30, Issue 2, pp.65-73.
70. Higginbotham, J.; McNaughton, D.; Caves, K. (2002). Performance and Usability. SIG 12, Perspectives on Augmentative and Alternative Communication, Vol. 11, Issue 2, pp. 17-19.

71. Reinecke, K.; Bernstein, A. (2011). Improving Performance, Perceived Usability, and Aesthetics with Culturally Adaptive User Interfaces. ACM Transactions on Computer-Human Interaction, Vol. 18, Issue 2, Article No. 8.
72. Smith, C. U.; Woodside, M. (2000). *Performance Validation at Early Stages of Software Development*. CRC Press, System Performance Evaluation: Methodologies and Applications, pp. 383-392.
73. Buschek, D.; Luca, A. D.; Alt, F. (2015). Improving Accuracy, Applicability and Usability of Keystroke Biometrics on Mobile Touchscreen Devices. In Proceedings of the 33rd Annual ACM Conference on Human Factors in Computing Systems, pp. 1393-1402.
74. Tsakonas, G.; Papatheodorou, C. (2008). Exploring usefulness and usability in the evaluation of open access digital libraries. Information Processing and Management, Vol. 44, Issue 3, pp. 1234-1250.
75. Bouwers, E.; Deursen, A.; Visser, J. (2013). Evaluating Usefulness of Software Metrics- an Industrial Experience Report. In Proceedings of the IEEE 2013 International Conference on Software Engineering, pp. 921-930.
76. Mifsud, J. (2015). Usability Metrics- A Guide To Quantify The Usability Of Any System http://usabilitygeek.com/usability-metrics-a-guide-to-quantify-system-usability/

Chapter - 12

SOFTWARE TESTING AND TOOLS

Software testing is a process and activity to verify whether the real results match the expected results and to certify that the software system is defect free. It involves execution of a system component or software component to evaluate one or more properties of significance. There are various types of software testing. Each and every used as per need and requirement. Software testing tools are used to find error and fix bug as soon possible to ensure software quality. Section I describes the introduction of this chapter. Section II describes the software testing fundamental. Section III describes the test planning. Section IV describes the test process. Section V describes the test tools. Section VI describes the testing and quality improvement. Section VII describes the summary of this chapter.

12.1 INTRODUCTION

Software testing is an examination conducted to give stakeholders with information about the quality of the software product or examination under test. Software testing also offers an objective, independent view of the software to permit the business to understand and recognize the risks of software development and implementation. Test techniques consist of the process of executing a software program or application by the aim of finding software errors, bugs, and other defects, and verifying that the software product fulfil the software requirement and fit for use. Software testing involves the execution of a software component or system component to assess one or more properties of software requirement. In general, these properties specify the degree to which the component or system under test, such as:

- Meets the software requirements that guided its design and development,
- Responds accurately to all kinds of inputs,
- Performs its functions within a satisfactory time,
- Is suitably usable,
- Can be installed and run in its planned environments, and
- Achieves the universal result its stakeholder's need.

As the number of probable tests for even effortless software components is practically infinite, all software testing uses various strategies to pick tests that are feasible for the offered cost, time, and resources [1]. As an outcome, software testing typically (but not absolutely) attempts to implement a program or application with the target of finding software bugs, errors, or other defects. The work of testing is an iterative process as when one bug is fixed, it can clarify other, deeper bugs, or can still develop new ones. Software testing can give objective, independent information about the quality of software products and risk of its failure to users or sponsors. Software testing can be conducted as soon as executable software (even if partly complete) exists. The overall approach to software development regularly determines when and how software testing is conducted. For example: in a phased process, most software testing occurs once system or software requirements have been defined and then implemented in testable programs or software. In compare, under an agile approach, software requirements, coding (programming), and testing are often done simultaneously [2].

Software testing also helps to recognize errors, gaps or missing requirements in different to the actual software requirements. It can be whichever done manually or using automated tools. Some favour saying software testing as a white box and black box testing. Testing is imperative because software bugs could be costly or even dangerous. Software bugs can potentially cause economic and human loss, history is full of such examples, such as:

- In 1985, Canada's Therac-25 radiation therapy machine malfunctioned due to software bug and delivered lethal radiation doses to patients in which 3 people dead and 3 critically injuring.
- China Airlines Airbus A300 crashed due to software bug on April 26, 1994, killing 264 live.
- In May of 1996, a software bug caused the bank accounts of 823 customers of a major US bank to be credited with 920 million US dollars.
- In April of 1999, a software bug caused the crash of a $1.2 billion military satellite launch, the costliest disaster in history.
- Starbucks was forced to shut about 60 percent of stores in the US and Canada due to software failure in its POS system. At one point store served coffee for free as they not capable to process the transaction.
- Some of the Amazon's third party retailers saw their product price is decreased to 1p due to a software malfunction. They were left with serious losses.
- Nissan cars have to recollect over 1 million cars from the market due to software failure in the airbag sensory detectors. There have been reported two accidents due to this software failure.
- Vulnerability in Window 10. This bug enables users to escape from security sandboxes throughout a flaw in the win32k system.
- In April 2015, Bloomberg terminal in London stopped due to software glitch affected more than 300,000 traders on financial markets. It forced the government to delay a 3bn pound debt sale.
- In 2015 fighter plane F-35 fell sufferer to a software bug, making it incapable to detect targets perfectly.

Software can also be unsuccessful by not fulfilling environmental constraints that fall outside the software specification [3]. For example: if the software code takes too much memory, executes too slowly, or if the software product works on one operating system but not a different, so all these are considered failures. Software testing is classified according to the method in which testers execute the first two phases of the software testing process. The objective of the first phase, modelling the software's environment, determines whether the tester is doing unit, integration, and system testing. Unit testing tests individual software components or a collection of software components. Software testers define the input domain for the units in question and ignore the rest of the software system. Unit testing occasionally requires the development of throwaway driver code and stubs and is frequently performed in a debugger. Integration testing, tests several components that have every received prior and separate unit testing. In general, the focus is on the subset of the domain that represents communication between the software components. System testing, tests a collection of components that constitutes a deliverable product. Usually, the whole domain must be measured to satisfy the criteria for a system test [4].

The second phase of software testing is test selection which determines what type of testing is being done. There are two major types: functional testing requires the selection of test scenarios lacking regard to source code structure. Thus, test selection methods and test data competence criteria must be based on software attributes of the software specification or operational environment and not on attributes of the software code or data structures. Functional testing also called specification based testing, behavioral testing, and black-box testing. Structural testing requires that inputs be based exclusively on the structure of the source code or its data structures. Structural testing is also called code based testing and white-box testing [5-6].

12.2 SOFTWARE TESTING FUNDAMENTAL

Software testing is the practice of evaluating a system or its component(s) through the aim to locate whether it satisfies the specified software requirements or not. In other words, testing is executing a system in order to recognize any gaps, errors, or missing software requirements in contrary to the actual requirements. It depends on the process and the associated stakeholders of the project(s). In the Information Technology (IT) industry, large companies have a software team with responsibilities to assess the developed software in context of the specified requirements. Furthermore, software developers also perform testing which is called unit testing.

An early commence to software testing reduces the cost and time to modify and develop error free software product that is delivered to the customer. However in software development life cycle (SDLC), software testing can be underway from the requirements gathering and specification phase and sustained till the deployment of the software product. It also depends on the software development model that is being used. For example: in the waterfall model, formal testing is conducted in the software testing phase but in the incremental model, testing is performed at the finish of every iteration or increment and the entire application is tested at the finish [7].

Software testing is done in diverse forms at every phase of SDLC:

- During the requirement gathering and specification phase, the analysis and verification of requirements are also considered as testing.
- Reviewing the design in the design phase with the purpose to advance the design is also considered as testing.
- Testing performed by a developer on achievement of the software code is also considered as testing.

Software Testing Life Cycle (STLC) consists of series of activities agreed out by testers methodologically to test software product. Though STLC uses the term testing it does not occupy just testers, at several instances, they have to engage developers as well. In STLC test cases are executed [8].

Fundamentally there are manual and automated software testing. Manual testing is performed by an individual sitting in face of a computer carefully executing the test steps. Manual testing is a type of software testing where software testers manually execute test cases without using any automation tools. Manual testing is the mainly primitive of all testing types and helps find bugs in the software system. Any new function must be manually tested before its testing can be automated. Manual testing requires additional effort but is compulsory to check automation feasibility. Manual testing does not necessitate knowledge of any testing tool. One of the software testing fundamental is 100% automation is not achievable. This makes manual testing very important. Figure 12.1 shows the type of manual testing. In fact any type of software testing can be executed both manually as well using an automation tool.

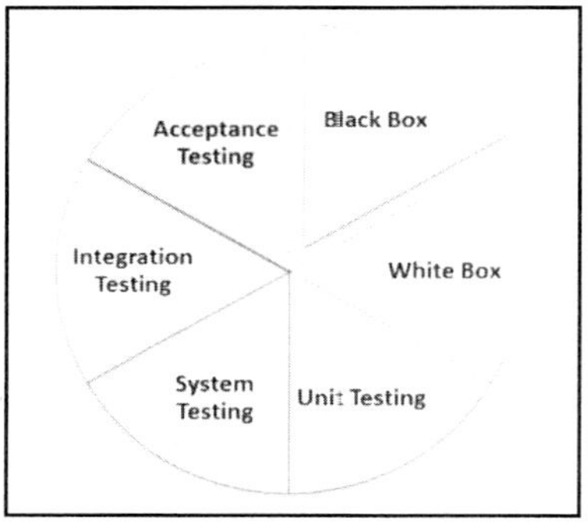

Figure 12.1 Type of Manual Testing

Automation testing uses automation tool to execute test case suite. The automation software can also go through test data into the system under test, evaluate expected and actual results, and produce detailed test reports. Test automation load significant investments of cost and resources. Successive development cycles will necessitate execution of similar test suite continually. Using a test automation tool, it's likely to record this test suite and replay it as mandatory. Once the test suite is automated, no individual involvement is required. This improved return of investment (ROI) of test automation. The goal of automation is to decrease the number of test cases to be run manually and not to reduce manual testing altogether [9].

List of tools to automate manual testing:

- Selenium
- QTP
- Jmeter
- Loadrunner
- TestLink
- Quality Center (ALM)

There are two basics of software testing: black box testing and white box testing. Black box testing is a testing technique that ignores the internal means of the system and focuses on the output generated by input and execution of the system. It is also called functional testing. White box testing is a testing technique that takes into account the internal means of a system. It is also called structural testing and glass box testing. Black box testing is frequently used for validation and white box testing is regularly used for verification.

System testing is the testing of an entire and completely integrated software product. Generally software is simply one element of a larger computer based system. Eventually, software is interfaced through other software or hardware systems. System testing is truly a series of diverse tests whose only purpose is to implement the complete computer based system. System test falls in the black box testing category of software testing. Acceptance testing is a beta testing of the software product completed by the genuine end users. In integration testing, individual software modules are incorporated logically and tested as a set. A typical software project consists of numerous software modules, coded by diverse programmers. Integration testing focuses on assessment data communication between these modules. Hence it is also termed as I & T (Integration and Testing), string testing and occasionally thread testing [10].

There are various types of testing such as:

- Unit Testing
- Integration Testing
- System Testing
- Functional Testing
- Acceptance Testing
- Stress Testing
- Performance Testing
- Usability Testing
- Regression Testing
- Beta Testing

Unit testing is the testing of a person unit or group of related units. It falls under the category of white box testing. It is regularly done by the software programmer to test that the unit implemented and is producing accepted output against specified input.

Integration testing is testing in which a group of components are combined to generate output. Also the communication between software and hardware is tested in integration testing if software and hardware components have some relation. It may fall under both white box testing and black box testing [11].

System testing is the testing to certify that by putting the software in different environments (operating systems) it still works. System testing is finished with complete system implementation and environment. It falls under the category of black box testing.

Functional testing is the testing to ensure that the particular functionality required in the system requirements works. It falls under the category of black box testing.

Acceptance testing is repeatedly done by the user to guarantee that the delivered product meets the software requirements and works as the user expected. It falls under the category of black box testing.

Stress testing is the testing to assess how system behaves under critical conditions. Testing is conducted at beyond limits of the software specifications. It falls under the category of black box testing.

Performance testing is the testing to measure the speed and effectiveness of the system and to make sure it is generating results within a specific time as in performance requirements. It falls under the category of black box testing.

Usability testing is performed to the perception of the user, to calculate how the GUI is user-friendly? How effortlessly can the user learn? After learning how to use, how expertly can the user perform? How enjoyable is it to use its design? This falls under the category of black box testing.

Regression testing is the testing after modification of a component, system, or a group of related units to certify that the modification is working accurately and is not damaging or striking other modules to create unexpected results. It falls under the category of black box testing.

Beta testing is the testing which is completed by end users, a software team outside development, or publicly releasing full pre-version of the software product which is known as beta version. The objective of beta testing is to wrap unexpected bugs and errors. It falls under the category of black box testing [12].

12.3 TEST PLANNING

Specific planning provides more than half the work is done, this is the significance of test planning. Spending a most favourable time on planning the test helps to accomplish goals in predetermined time and cost. Test planning involves creating an outline of the testing process. Software testing experts analyze the testing requirements and generate a specific test plan. The overall test plan document will basically called test plan. The extent of the test plan is the entire testing strategy: what application(s) will be tested and why, which maintain layers will be tested and why, and how much (if not all) of the software development life cycle will necessitate testing [13]. A characteristic test plan should include the following:

- Application(s)/system(s) to be tested
- Testing objectives and their basis (risk and requirements)
- Scope and limitations of the test plan
- Testing strategy
- Sources of business proficiency for test planning and execution
- Source of software development knowledge for test planning and execution
- Test environments and their management
- Sources of test data
- Testing information for each software development phase
- Overall testing schedule

During software development process, test planning is an extremely vital step. It ensures to confirm a specific path on the course of which are going to develop the software. Test planning determines the lists of milestones and tasks which will be used to follow the progress of software process. A test plan is the certified documentation that these scope and activities. It gives a baseline although formally testing a software product or a module [14]. The international software testing qualifications board (ISTQB) definition of test plan is: A document describing the scope, approach, resources, and schedule of planned test activities. It identifies between others test items, the features to be tested, the testing tasks, who will do all task, the test environment, degree of tester independence, the test design techniques, entry and exit criteria to be used, rationale for their selection, and some risks requiring contingency planning. It is a documentation of the test planning process.

Test plans can be defined at different stages of software project. As such there are different types of test plans:

1. Master Test Plan: This is a high level documentation that unifies all additional sub-modules test plans. This is usually a single master test plan for a particular project.
2. Testing Level Specific: These test plans are exact to the level of testing that is undergoing. They may be unit, integration, functional, and system testing or user acceptance testing (UAT) test plans.
3. Testing Type: This may be based on the form of testing such as: load/performance testing, security test plan, and UI/UX testing etc.

The content of a test plan is not defined. It can differ depending upon the processes or as per the standards set up by the software company. Sometimes the test management tools also affect the contents of a test plan [15]. A universal format as described by IEEE standard for software test documentation describes test plan as:

1. Test plan identifier: Unique identifying reference.
2. Introduction: A brief introduction about the software project and to the document.
3. Test items: A test item is a software item that is the purpose under test.
4. Features to be tested: A feature that needs to tested on the application under test (AUT).
5. Features not to be tested: Identify the features and the reasons for not including as part of software testing.
6. Approach: Details about the general approach to software testing.
7. Item pass/fail criteria: Documented whether a software item has passed or failed its test.
8. Test deliverables: The deliverables that are delivered as part of the testing process such as: test plans, test specifications, and test review reports.
9. Testing tasks: All tasks for planning and executing the software testing.
10. Environment needs: Defining the environmental requirements such as: software, hardware, OS, network configurations, and tools required.
11. Responsibilities: Lists the roles and responsibilities of the software team members.

12. Staffing and training needs: Captures the definite staffing requirements and any particular skills and training requirements.
13. Schedule: States the main project delivery dates and key milestones.
14. Risks and Mitigation: High-level project risks and assumptions and a mitigating plan for all identified risk.
15. Approvals: Captures all approvers of the document, their titles, and the sign off date.

Software product typically, in that stage, a senior quality assurance manager will determine effort, time, and cost estimates for the software project and would organize and finalize the test plan. In this phase, test strategy is also determined [16].

Activities:

- Preparation of test plan or strategy document for various types of software testing
- Test tool selection
- Test effort estimation
- Resource planning and determining roles and responsibilities
- Training requirement

Deliverables:

- Test plan or strategy document
- Effort estimation document

The purpose of IEEE Std 829-1998 (Revision of IEEE Std 829-1983) standard is to explain a set of basic software test documents. A standardized test document is able to facilitate communication by providing a general frame of reference (e.g. a user and a developer have the equal definition for a test plan). The content definition of a standardized test document is capable to provide a complete checklist for the connected testing process. A standardized set can also provide a baseline for the assessment of existing test documentation practices. In many software organizations, the purpose of this document significantly increases the manageability of software testing. Increased manageability results from the significantly increased visibility of every phase of the software testing process [17].

The purpose of IEEE Std 829-2008 (Revision of IEEE Std 829-1998) standard is to create a universal framework for test processes, activities, and tasks in sustain of all software life cycle processes including acquisition, development, operation, maintenance processes, and supply. Define the required inputs, test tasks, and required outputs. Identify the recommended minimum test tasks corresponding to integrity levels for a four-level integrity scheme. Define the purpose and contents of the master test plan and the level test plan(s) (e.g. for component, system, integration, and acceptance test). Define the use and contents of related test documentation (such as: test case, test design, test procedure, test log, anomaly report, level test report, interim test report, and master test report). The master test plan is considered by IEEE 829-1998 to have the similar structure as the test plan IEEE 829-2008. But there are some differences which is the need for current scenario and technology [18].

12.4 TEST PROCESS

Software testing is the process of executing software to verify whether it matches its software specification and executes in its planned environment. The information that the system is being executed distinguishes software testing from code reviews, in which un-compiled source code is read and analyzed statically (generally by software developers). Software testing requires running executable software. A software specification is a critical artifact to support software testing. It defines exact behavior so that wrong behavior is easier to classify. Wrong behavior is a software failure. Failures are caused by faults in the source code which are frequently referred to as defects, error, or bugs. Software testers compare actual output with particular output to recognize failures. Generally, the code developer diagnoses the fundamental fault [19-21].

To plan and execute tests the software testers must consider the software, its inputs and how they can be combined, the function it computes, and the environment in which the software will finally operate. This complex, time-consuming process requires technological sophistication and appropriate planning. Testers must not simply have excellent development skills, testing frequently require a great deal of software coding but also be familiar in formal languages, graph theory, and algorithms. Indeed, creative software testers have brought numerous related computing disciplines to allow on software testing problems, regularly with impressive results. Even easy software presents testers with obstacles. To get a clearer view of several of software testing's inherent difficulties, we can approach testing in four phases such as:

- Modelling the software's environment
- Selecting test scenarios
- Running and evaluating test scenarios
- Measuring testing progress

These phases offer software testers a structure in which to group related problems that they must resolve before moving on to the next phase [22-23].

Phase 1: Modelling the Software's Environment

A tester's task is to reproduce interaction among software and its environment. Testers must recognize and create the interfaces that a software system uses and specify the inputs that can cross each interface. This might be the mainly essential issue that testers face and it can be hard, considering the dissimilar file formats, communication protocols, and third party such as application programming interfaces (API) available. Four common interfaces are as follows:

- Human interfaces contain all general methods for people to communicate with software. Most prominent is the graphical user interface (GUI) but older designs like the command line interface (CLI) and the menu driven interface (MDI) are still in exercise. Possible input mechanisms to consider are keyboard events, mouse clicks, and input from other devices. Testers then choose how to organize that data to recognize how to collect it into an effective test.

- Software interfaces called APIs, are how software uses an operating system, database, or runtime library. The services these applications give modelled as test inputs. The issue for testers is to verify not only the expected but also the unexpected services. For example: all developers expect the operating system to keep files for them. The service that they ignore is the operating system's informing them that the storage medium is full. Even fault messages must be tested.
- File system interfaces exist when software reads or writes data to external files. Developers must write lots of error checking code to verify if the file contains suitable data and formatting. Thus, testers must build or generate files with content that is equally legal and illegal, and files that contain a diversity of text and formatting.
- Communication interfaces permit direct access to physical devices (such as controllers, device drivers, and other embedded systems) and need a communication protocol. To test such software, testers must be capable to produce both valid and invalid protocol streams. Testers must assemble and submit to the software under test, numerous different combinations of commands and data, in the appropriate packet format.

Next, testers must recognize the user interaction that falls external the control of the software under test, since the consequences can be serious if the software is not ready [24-27]. Examples of situations testers should tackle are as follows:

- Using the operating system, one user deletes a file that a different user has open. What will occur the next time the software tries to access that file?
- A device gets rebooted in the middle of a stream of communication. Will the software recognize this and respond properly or just hang?
- Two software systems struggle for duplicate services from an API. Will the API properly service both? Each application's exclusive environment can result in an important number of user interactions to test.

Phase 2: Selecting Test Scenarios

Several domain models and variable partitions characterize an infinite number of test scenarios each of which costs time and money. Only a subset can be applied in any practical software development schedule so how does a smart tester choose? How many times a filename is selected by pressing the open button? These questions have many answers, are being dynamically researched. Testers prefer an answer that relates to coverage of source code or its input domain. Software testers strive for coverage: covering code statements (executing each source line at least one time) and covering inputs (applying every externally generated event). These are the lowest criteria that testers use to review the completeness of their work therefore, the test set that numerous testers prefer is the one that meets their coverage goals. But if code and input coverage were satisfactory, released software products would have few bugs. Concerning the software code, it isn't individual code statements that concern testers but execution paths: sequences of code statements representing an execution of the software product. Unfortunately, there are a countless number of paths. Concerning the input domain, it isn't the individual inputs that interest testers but input sequences that taken as a complete, represent scenarios to which the software must react. There are an unlimited number of these too. Testers sort throughout these infinite sets to appear at the greatest possible test data adequacy criteria, which are intended to adequately and economically characterize any of the infinite sets. Many users and quality assurance professionals are concerned in having testers assess typical use scenarios, things that will happen most often in the field. Such testing ensures that the software works as specified and that the most frequently occurring bugs will have been detected [28-29].

- Execution path test criteria: Test data adequacy criteria focus on either execution path coverage or input sequence coverage but infrequently both. The most general execution path selection criteria focus on paths that cover control structures. However, control flow is simply one feature of the source code. What software really does is move data from one location to a different. The dataflow family of test data adequacy criteria describes coverage of this data. Finally, fault seeding which claims more concentration from researchers than practitioners. In this method, errors are purposely inserted (seeded) into the source code. Test scenarios are next designed to discover those errors. Ideally, by discovery seeded errors, the tester will also discover real errors.
- Input domain test criteria: Criteria for input domain coverage range from easy coverage of an interface to additional complex statistical measurement.

- The discrimination criterion requires arbitrary selection of input sequences until they statistically symbolize the entire infinite input domain.

Phase 3: Running and Evaluating Test Scenarios

Having recognized suitable tests, testers exchange them to executable form regularly as code so that the resulting test scenarios recommend typical user action. Because manually applying test scenarios is labor intensive and error prone, testers aim to automate the test scenarios as much as achievable. In many environments, automated application of inputs using software code that simulates users is promising, and tools are available to assist. Complete automation requires recreation of each input source and output destination of the complete operational environment. Testers regularly include data gathering code in the simulated environment as test hooks or asserts. This code provides information regarding internal variables, object properties, and so forth. These hooks are removed when the software is released but during test scenario execution they provide important information that helps testers recognize failures and isolate faults. Scenario evaluation, the next part of this phase, is easily stated but complex to do (much less automate). Evaluation involves the assessment of the software's actual output, resulting from test scenario execution, to its estimated output as documented by a software specification. The specification is assumed correct; deviations are failures. In practice, this evaluation is difficult to achieve. Theoretically comparison (to determine equivalence) of two subjective, Turing computable functions is impossible [30].

Phase 4: Measuring Testing Progress

Suppose manager comes to tester and asks, what's the status of software testing? Testers are regularly asked this question but are not well prepared to answer it. The reason is that the state of observe in test measurement is to count things. We count the number of inputs we've applied, the percentage of software code we've covered, and the number of times we've invoked the application. We count the number of times we've completed the application successfully, the number of failures we found, and so on. Interpreting such counts is difficult. A high bug calculation could mean that testing was systematic and very few bugs stay. Or, it could mean that the software basically has lots of bugs and even though many have been exposed, lots of them stay. Since counting measures defer very little insight about the progress of testing, many testers enhance this data by answering questions designed to establish structural and functional testing completeness.

These questions essentially, test data adequacy criteria are helpful to testers; however, determining when to end software testing, determining when a product is prepared to release, is more difficult. Testers want quantitative measures of the amount of bugs left in the software and of the possibility that any of these bugs will be exposed in the field. If testers can accomplish such a measure, they recognize to end software testing. We can approach the quantitative problem structurally and functionally [31].

12.5 TEST TOOLS

Tools from a software testing perspective can be defined as a product that supports one or more test activities from planning, software requirements, creating a build, test execution, defect logging, and test analysis. Several tools are available to assist the execution of tests procedures in a simple (manual), automated way, whatever the technical environment or circumstance of the software might be [32].

These tools are used in two major cases:

- To automate the execution of repetitive, tedious, or meticulous tasks, such as the comparison of files of numerous thousands of lines, so that testers can focus on tasks of higher importance;
- To execute or monitor tasks that cannot be simply carried out by individuals, such as performance testing or the analysis of response times (usually measured in hundred of seconds).

Tools can be used to maintain most test activities, whether they are design, execution, test monitoring or control, management, or test reporting activities. Normally, for each type of tool there are commercial and open source tools. In general, the utilize of automated tools increases test reliability, enables us to take out execution tasks without any individual presence in the room, and improves the effectiveness of repetitive tasks such as: regression tests [32-33].

Test tools can be classified in various ways. The classification is according to the tasks they support:

- Requirement management tools;
- Tools supporting test management;
- Modelling tools;
- Reviewing and static analysis tools;
- Test design tools;
- Support tools for test environments;

- Tools for test design and test basis analysis;
- Tools for test execution;
- Data comparison tools;
- Performance testing, load and dynamical analysis tools;
- Coverage measurement tools;
- Security tools;
- Other tools.

Now days lots of software testing tools in the market. Selection of tools is completely based on the software project requirements and commercial (proprietary or commercial tools) or free tools (open source tools) as interested. Off course, free testing tools may have some restriction in the features list of the product, so it's thoroughly based on what are you looking for and is that software requirement fulfil in free version or go for paid software testing tools. Test automation is one of the most established software testing segments. Currently, the increase of good open source options and the push toward DevOps and multi-channel models are transforming that segment [34-35]. The test automation tools categorize into two categories:

- Functional testing
- Integration/API testing

1. **Functional software testing tools:** Functional testing tools make automated tests by interacting with the UI layer of an application. These tools typically provide a record and playback interface, making it easier for non-technical testers to offer automation coverage. However, these tests can be slow and weak [36-37].

Katalon Studio: Katalon Studio is a cross platform tool to transform manual testing into automation that requires least program background. A full web and mobile automation framework built on top of Selenium and Appium, Katalon Studio is filled with necessary features that only available in commercial tools while remaining as a free solution.

Features:

- Dual scripting interface for novel and advanced users
- Out of the box integration with qTest and JIRA
- Support agile team with CI workflow [38]

HP Unified Functional Testing (UFT): Formerly known as QuickTest Professional (QTP), HP UFT automates GUI functionality and back-end service testing. It also offers reusable test components, embraces the move left to make testing part of the agile development cycle, helps change manual testing assets into automated ones, and allows for testing across multiple machines and devices. Plans start at $600 per month with three pricing tiers. Although HP remains the market share leader in paid test automation, it has just struggled to focus on newer platforms, ceding position to competitors.

Features:

- Increasing sustain for newer browsers and mobile (though HP does fall behind other market leaders)
- Well-built partner network
- Integrated offering that includes other test tools [39]

IBM Rational Functional Tester: Component of IBM's larger rational testing platform, the rational functional tester uses data-driven testing to automate functional and regression testing. It supports a diversity of applications and allows for equally storyboard testing and test scripting. Plans start at $3,400 per month with 6 pricing tiers. Once a proper market leader, IBM has had to focus on standing its position against rising competition from HP and up and coming players in current time.

Features:

- Supports legacy platforms and frameworks
- Integrated offering that includes other test tools
- Offers established and dependable tools [40]

Tricentis Tosca Testsuite: Provides model based test automation focused on test case design against problem solving. The Tosca Testsuite also offers end-to-end testing and recording capabilities. With the Tosca Testsuite, Tricentis has decided on creating model based automation and a mini-suite that goes beyond pure test automation to offer risk coverage and test case design as well.

Features:

- Includes a more absolute set of tools around test data management and orchestration
- Offers support for agile
- Delivers model based automation support that allows for easier reuse or less maintenance [41]

Worksoft Certify: Automates functional testing of business processes across the whole application landscape. Worksoft Certify offers capabilities about measurement, discovery, validation, compliance, and scale. Although Certify works with a broad variety of applications and practice built code or extensions, its focus is on supporting package software such as: Oracle, SAP, Salesforce, etc.

Features:

- Takes a script less approach
- Easy to use
- Compatible with package software as well as custom built code and extensions
- Well supported [42]

TestPlant eggPlant Functional: An automated functional testing tool that uses an image based approach, which looks at the display of any device the similar way a user would. This image based UI testing differs from the conventional approach because it tests from the perception of a user rather than by looking at code. eggPlant functional works across every device types. eggPlant functional uses of an image based approach to automation can assist to resolve complex automation problems. This approach allows testers with less to no automation skills to accomplish robust automation overage and for teams that have struggled to implement automation.

Features:

- Fast development
- Allows for reuse of functional automation for performance testing
- Offers lab management and CI integration
- Supports all main platforms including web, mobile, and POS systems or anything behind a pane of glass [43]

Ranorex: Allows for setup and execution of test automation scripts and offers reusable code modules, GUI object recognition, and record/play back functionality. Ranorex also offers script free functionality. Plans start at € 1,990 per license with 2 pricing tiers. Ranorex provides comparable functionality to integrated suite vendors with lesser cost licenses and maintenance.

Features:

- Delivers well-built object recognition
- Developer friendly
- Offers a straightforward and affordable pricing model [44]

2. **Integration/API Software Testing Tools:** Integration/API testing tools forward various types of payloads through APIs to make sure they acknowledge valid inputs and discard improper ones with suitable messages. Some of these software testing tools can also offer testing for load. Integration/API testing is mainly significant for cloud, mobile, and PaaS.

CA Technologies Application Test: Uses a declarative workflow model that focuses on building tests over scripting. CA application test embraces the move left by allowing for API testing by UI development. It also offers automated mobile testing, customized load testing, visual tests, and advanced analytics. CA technologies are mainly focused on competing in the mainframe space and testing desktop applications and composite infrastructure using LISA.

Features:

- Allows for integration of mobile testing
- Offers robust mainframe support
- Integrates with Selenium for enhanced visual editing
- Allows for reuse of LISA test cases and infrastructure
- Creates a load on an internal network through few resources [45]

IBM Rational Test Workbench: A inclusive test automation tool with integration testing capabilities as well as regression, mobile, performance, and scalability testing capabilities. In terms of integration testing, IBM rational test workbench allows for service level testing, automatic scheduling, and execution of testing via integration with the IBM rational collaborative lifecycle management tool. Plans start at $3,730 per year with 12 pricing tiers. IBM rational test workbench provides a absolute set of enterprise ready testing tools that cover testing needs all over the software development lifecycle.

Features:

- Allows for continuous integration testing
- Provides lifecycle collaboration
- Integrates with the larger IBM rational testing suite for extended capabilities
- Supports automaton [46]

Parasoft SOAtest: Provides test automation for end-to-end scenarios across many endpoints with support for REST, web services, and over 120 protocols or message types. Parasoft SOAtest creates extensible and reusable tests. In addition to API testing, it also offers SOA testing, service virtualization, web and performance testing, runtime error testing, web UI testing, API security testing, and development testing. Parasoft SOAtest offers automated API testing as well as numerous other testing capabilities. Its main exclusive feature is the extensive list of environments, including protocols, platforms, and messaging formats that it supports.

Features:

- Offers capabilities beyond integration/API testing
- Supports a wide range of messaging/protocols [47]

SmartBear Ready! API: A entire API testing platform with support for API functional testing, API security testing, API load testing, API testing in code, API performance management, service virtualization, and defining, building, and managing APIs. SmartBear Ready! API provides project management, metrics and reporting, discovery, script support, and continuous integration across all of these API testing capabilities. SmartBear Ready! API is a inclusive API testing tool that covers API testing across functional, performance, and post-deployment testing. It also integrates with the larger SmartBear testing environment.

Features:

- Provides a broad variety of API testing capabilities
- Supports continuous integration [48]

Crosscheck Networks SOAPSonar: Provides API service testing, API emulation and virtualization, and API security gateway technologies. Crosscheck networks uses a dynamic mutation technology to send functional automation, performance, and security testing. It also offers API simulated behaviour modelling with support for a variety of protocols and messages. Plans start at $799 per year with 2 pricing tiers. Crosscheck Networks SOAPSonar offers a range of API testing capabilities, although it emphasizes performance, functional, and security testing for service endpoints. It runs these tests by a patented mutation technology.

Features:

- Provides virtualization and security testing as well as API service testing
- Supports a diversity of protocols and messages, including REST, JSON, and SOAP [49]

Performance Testing Tools: Performance testing tools take an easy automation script and run it through hundreds or thousands of machines to simulate how an application will execute under load. These tests are characteristically expensive and are run less commonly than functional automation tests. Performance testing is mainly important for cloud applications and there are two main types to consider:

- Pure Play
- Extension Tools

Pure Play Performance Testing Tools:

Pros:

- Creates less friction throughout purchase and implementation processes
- Typically cheaper
- Offers quicker development with best of breed tools
- Supports more integration and open source use

Cons:

- Requires different scripts/languages to train and maintain

BlazeMeter: Runs open source based performance tests as well as scalable load testing. BlazeMeter is friendly with Apache JMeter, offers real-time reporting and mobile capture, simulates network circumstances and allows developers to generate and edit performance tests. Free plan available and paid plans start at $99 per mo with 3 pricing tiers. BlazeMeter is an emergent player in the cloud performance testing space. Although it has restricted use cases and does not provide monitoring or post-production capabilities, it is popular with software developers.

Features:

- Scalable
- Uses open source and AWS
- DevOps ready
- Offers strong support for APIs and customization [50]

Neotys NeoLoad: NeoLoad uses an intuitive GUI for use case creation, target identification, and test execution. It allows to simulate load locally or using the Neotys Cloud Platform, supports web and mobile to monitors all servers in load testing.

Neotys is a small but fast growing player with cross platform support for web and mobile (via NeoLoad) and post-production monitoring capabilities (via NeoSense).

Features:

- Provides end-to-end testing
- More reasonable pricing than legacy solutions
- Offers cross platform capabilities
- Cloud and on premise enabled [51]

Radview WebLOAD: Simulates a diversity of load conditions for an unlimited number of users and distributes the load across any number of machines. Radview WebLOAD can also examine server performance and offers cross platform support for web, desktop, and mobile (via PerfectoMobile). Plans start at € 1,990 per license with 3 pricing tiers. Radview is a fledgling player in performance testing that was previously free listed but has since been delisted and is not performing very fine currently (it has scaled down from 100 employees to about 25).

Features:

- Some cross platform support
- Established tool
- Some open source support (supports Selenium) [52]

Extension Performance Testing Tools

Pros:

- Uses well-known languages
- Allows for reuse of past automation scripts
- Offers a only vendor relationship across multiple tools

Cons:

- Provides a closed ecosystem that limits access to most excellent of breed tools
- Follows slower, less focused software development
- Typically more costly and geared toward large organizations

Automation Anywhere Testing Anywhere: Uses robotic process automation to automate every type of testing and offers cross platform support for mobile, web, and desktop. Automation anywhere can change scripts into load scripts and allows for machine learning and analytics. Automation anywhere is a lesser end competitor to HP that offers some test case management capabilities as well as extra robust automation capabilities.

Features:

- Less costly but similar to HP

- Offers simple conversion from automation
- Includes a TestLab to manage testing devices [53]

HP LoadRunner, Performance Center & StormRunner: HP offers a mixture of three tools for performance and load testing. LoadRunner provides complete load testing with interactive simulations and root cause analysis capabilities, while Performance Center creates a center of excellence for reusing best practices and resources across testing for several applications. Both LoadRunner and Performance Center support continuous and mobile testing. Finally, StormRunner extends testing capabilities to the SaaS world. LoadRunner and StormRunner Load are priced at $.56 per virtual user per day but they have a open edition as well. With LoadRunner and Performance Center as well as StormRunner to extend to the cloud, HP is a head in that space.

Features:

- Robust and established tools
- Supports legacy and newer technologies
- Offers documentation and playback or scripting
- Cloud enabled
- Allows for reuse of unified functional testing tests [54-56]

IBM Rational Performance Tester: Offers code free testing and real-time reporting as well as root cause analysis tools. IBM Rational Performance Tester also compares test results to SLA targets, integrates with IBM Rational Quality Manager and includes an analysis recorder. IBM Rational Performance Tester is cloud enabled and offers web and desktop support but its mobile support is very weak.

Features:

- Supports legacy and newer technologies
- Offers record and playback or scripting
- Cloud enabled
- Allows for reuse of rational functional tester tests [57-58]

12.6 TESTING AND QUALITY IMPROVEMENT

Software testing is the process of assessment a software item to identify differences between given input and accepted output. It is also used to measure the feature of a software item. Testing assesses the quality of the software product. Software testing is a process that should be done through the software development process. In other words software testing is a verification and validation process. Verification is the process to create sure the product satisfies the circumstances imposed at the start of the development phase. In other words, to make ensure the product behaves the way we want it to. Validation is the process to make sure the product satisfy the specified requirements at the conclusion of the development phase. In other words, to make sure the product is built as per user requirements. Quality improvement (QI) is essential to the success of software development organizations. Automated tools can greatly assist in the implementation of quality improvement [59-62].

Software testing may be considered a part of a software quality assurance (SQA) process. In SQA, software process specialists and auditors are worried with the software development process rather than just the artefacts such as documentation, code, and systems. They observe and change the software process itself to decrease the number of faults that end up in the delivered software: the so called defect rate. Although there are close links with SQA, testing departments regularly exist independently, and there may be no SQA function in some companies. Software testing is an action to investigate software under test in order to give quality related information to stakeholders. By contrast, QA (quality assurance) is the execution of policies and procedures planned to prevent defects from reaching users [63-64].

The significance of software quality assurance and process improvement is broadly recognized. A number of models for quality improvement and assessment have been developed, and some of them are specially targeted for the software industry, such as the Software Capability Maturity Model (SW-CMM) and ISO/IEC 15504. The objective of quality improvement is to get better development processes to better meet user requirements and expectations. Software process improvement (SPI) has been adopted by numerous organizations trying to advance their competitiveness. The benefits of SPI include improving the efficiency and effectiveness of various organizational processes, increasing user satisfaction, improving the quality of software products and services, reducing process costs, errors and cycle time, and improving the predictability of software process. Many tools for quality improvement have been developed. They can improve the efficiency and effectiveness of process improvement implementation by providing numerous functions that decrease the effort involved in data collection, change management, root cause analysis, document management, etc. However, for a process practitioner planning to purchase SPI tools, the availability of so many tools poses a tool selection problem [65-66].

Quality improvement models present a systematic, formal framework for establishing QI processes in practice such as:

- The PDCA model: The PDCA model, also known as the "Plan Do Check Act cycle" was first developed by Walter Shewhart in the 1920s, and was popularized later by Edwards Deming. It can be used as an approach to developing, implementing, and improving the quality management system of an organization. The PDCA model has four phases and 22 steps. The PDCA is a cycle of continuous improvement. Every improvement realized by executing one PDCA cycle will become the baseline for improvement targets of the subsequently PDCA cycle.
- Six-Sigma DMAIC model originally, six-sigma was a statistical term. Sigma(s) is the representation for standard deviation. Nowadays, six-sigma focuses on the decline of defects in a product or process and is used as a measurement of quality. It can also be viewed as a management philosophy for reducing costs while increasing user satisfaction. The Define Measure Analyze Improve Control (DMAIC) model is the most universally used model for achieving six-sigma. It has five phases and 26 steps.

- The IDEAL^{SM} model: The IDEAL model is an organizational improvement model developed by Software Engineering Institute. It provides a disciplined engineering approach to improvement, focuses on managing the improvement program, and establishes the foundation for a long term improvement strategy. The model defines five phases:

1. Initiating
2. Diagnosing
3. Establishing
4. Acting
5. Leveraging

Each phase consists of numerous tasks which may be subdivided into subtasks [67]. Quality improvement tools are individual strategies or processes that can assist to better understand, analyze, and communicate QI efforts such as:

- Cause and effect diagram (also called Ishikawa or fishbone chart): Identifies many promising causes for an effect or problem and sorts ideas into useful categories.
- Check sheet: A structured, prepared form for collecting and analyzing data; a general tool that can be adapted for a broad variety of purposes.
- Control charts: Graphs used to study how a process changes over time.
- Histogram: The most commonly used graph for showing frequency distributions, or how often each different value in a set of data occurs.
- Pareto chart: Shows on a bar graph which factors are more significant.
- Scatter diagram: Graphs pairs of numerical data, one variable on every axis, to look for a relationship.
- Stratification: A technique that separates data gathered from a diversity of sources so that patterns can be seen (some lists replace stratification with flowchart or run-chart).

Deming's 14 Quality Principles: Deming outlined 14 quality principles which must be used concurrently in order to achieve quality. Although these principles were applied to industry, influencing government, hospitals, and schools, many are also applicable to achieving software quality from an information technology viewpoint [68].

1. Create Constancy of Purpose
2. Adopt the New Philosophy
3. Cease Dependence on Mass Inspection
4. End the Practice of Awarding Business on Price Tag Alone
5. Improve Constantly and Forever the System of Production and Service
6. Institute Training and Retraining
7. Institute Leadership
8. Drive out Fear
9. Break Down Barriers between Staff Areas
10. Eliminate Slogans, Exhortations, and Targets for the Workforce
11. Eliminate Numerical Goals

12. Remove Barriers to Pride of Workmanship
13. Institute a Vigorous Program of Education and Retraining
14. Take Action to Accomplish the Transformation

It is significant not to underestimate the people factors such as: culture in selecting a quality improvement approach. Any improvement (change) takes time to implement, gain acceptance, and stabilize as established practice. Improvement must permit pauses between implementing new changes so that the alteration is stabilized and assessed as an actual improvement, before the subsequently improvement is made (hence continual improvement not continuous improvement). Improvements that change the culture acquire longer as they have to overcome greater resistance to change. It is easier and often additional effective to work within the existing cultural boundaries and make tiny improvements (that is Kaizen, it is the Japanese word for improvement) than to build major transformational changes. Use of Kaizen in Japan was a main reason for the creation of Japanese industrial and economic strength. On the other hand transformational alter works best when an enterprise faces a catastrophe and needs to build major changes in order to carry on. In Japan the land of Kaizen, Carlos Ghosn led a transformational change at Nissan motor company which was in a financial and operational crisis. So well organized quality improvement programs obtain all these factors into account when selecting the quality improvement methods [69-70].

12.7 SUMMARY

Software testing is a method or process of verifying a computer system or program to choose whether it meets the specified software requirements and produces the required results. As a result, we identify error or bugs in software program, product, and project. Software testing is essential to provide a quality product without any bug or issue. Specific planning provides more than half the work is done, this is the significance of test planning. Spending a most favourable time on planning the test helps to accomplish goals in predetermined time and cost. So test planning involves creating a blue print of the testing process. There are various software testing tool which is used by as per requirement, testing objective, and application. Usually there is commercial and open source software testing tools. Various software testing tools for quality improvement have been developed. Software testing may be considered a component of a software quality assurance. Therefore software testing is significant to the success of software development.

12.8 REFERENCES

1. Myers, G. J.; Sandler, C.; Badgett, T. (2011). *The art of software testing*. John Wiley & Sons.
2. Kuhn, D. R.; Wallace, D. R.; Gallo, A. M. (2004). Software fault interactions and implications for software testing. IEEE transactions on software engineering, Vol. 30, No. 6, pp. 418-421.
3. Bertolino, A. (2007). Software testing research: Achievements, challenges, dreams. In IEEE Future of Software Engineering, pp. 85-103.
4. Basili, V. R.; Selby, R. W. (1987). Comparing the effectiveness of software testing strategies. IEEE transactions on software engineering, Vol. 12, pp. 1278-1296.
5. Whittaker, J. A. (2000). What is software testing? And why is it so hard?. IEEE Software, Vol. 17, No. 1, pp. 70-79.

6. Melo, S. M.; Carver, J. C.; Souza, P. S.; Souza, S. R. (Sep 2018). Empirical Research on Concurrent Software Testing: A Systematic Mapping Study. Information and Software Technology.
7. McCaffrey, J. D. (2009). *Software Testing: Fundamental Principles and Essential Knowledge*. BookSurge Publishing.
8. Chen, H. Y.; Tse, T. H. (2009). Automatic generation of normal forms for testing object-oriented software. In IEEE 9th International Conference on Quality Software, QSIC'09, pp. 108-116.
9. Hutcheson, M. L. (2003). *Software testing fundamentals: Methods and metrics*. John Wiley & Sons.
10. Yuan, X.; Cohen, M. B.; Memon, A. M. (2011). GUI interaction testing: Incorporating event context. IEEE Transactions on Software Engineering, Vol. 37, No. 4, pp. 559-574.
11. Everett, G. D.; McLeod Jr, R. (2007). *Software Testing. Testing Across the Entire Software Development Life Cycle*. Wiley & Sons.
12. Zhang, B.; Shen, X.; Wang, J.; Chen, Y. (2013). Measurement of effectiveness of software testing. In Fifth International Conference on Machine Vision (ICMV 2012), International Society for Optics and Photonics: Computer Vision, Image Analysis and Processing, Vol. 8783.
13. Afzal, W. (2007). Metrics in software test planning and test design processes.
14. Karlström, D.; Runeson, P.; Nordén, S. (2005). A minimal test practice framework for emerging software organizations. Software Testing, Verification and Reliability, Vol. 15, No. 3, pp. 145-166.
15. Paige, M. (1980). A metric for software test planning. In Conference Proceedings of COMPSAC, Vol. 80, pp. 499-504.
16. Afzal, W.; Torkar, R. (2008). Incorporating metrics in an organizational test strategy. In IEEE International Conference on Software Testing Verification and Validation Workshop, ICSTW'08, pp. 304-315.
17. 829-1998 - IEEE Standard for Software Test Documentation
18. 829-2008 - IEEE Standard for Software and System Test Documentation
19. Burnstein, I.; Suwanassart, T.; Carlson, R. (1996). Developing a testing maturity model for software test process evaluation and improvement. In IEEE International Proceedings onTest Conference, pp. 581-589.
20. Cangussu, J. W.; DeCarlo, R. A.; Mathur, A. P. (2002). A formal model of the software test process. IEEE Transactions on Software Engineering, Vol. 28, No. 8, pp. 782-796.
21. Cangussu, J. W.; DeCarlo, R. A.; Mathur, A. P. (2003). Using sensitivity analysis to validate a state variable model of the software test process. IEEE Transactions on Software Engineering, Vol. 29, No. 5, pp. 430-443.
22. Cangussu, J. W.; Mathur, A. P.; DeCarlo, R. A. (2001). Feedback control of the software test process through measurements of software reliability. In IEEE Proceedings 12th International Symposium on Software Reliability Engineering, ISSRE 2001, pp. 232-241.
23. Li, Q.; Li, M.; Yang, Y.; Wang, Q.; Tan, T.; Boehm, B.; Hu, C. (2009). Bridge the gap between software test process and business value: a case study. In International Conference on Software Process, pp. 212-223, Springer, Berlin, Heidelberg.
24. Cangussu, J. W.; DeCarlo, R. A.; Mathur, A. P. (2001). A state model for the software test process with automated parameter identification. In IEEE International Conference on Systems, Man, and Cybernetics 2001, Vol. 2, pp. 706-711.
25. Abdou, T.; Grogono, P.; Kamthan, P. (2012). A conceptual framework for open source software test process. In IEEE 36th Annual Computer Software and Applications Conference Workshops, COMPSACW 2012, pp. 458-463.
26. Hooda, I.; Chhillar, R. S. (2015). Software test process, testing types and techniques. International Journal of Computer Applications, Vol. 111, No. 13.

27. Abu, G.; Cangussu, J. W.; Turi, J. (2005). A quantitative learning model for software test process. In IEEE Proceedings of the 38th Annual Hawaii International Conference on System Sciences, HICSS'05, pp. 78b-78b.
28. Duran, D. E. S.; Builes, J. J.; Gamboa, A. X. R. (2017). A Knowledge Management Model for Improving the Software Test Process. In European Conference on Knowledge Management, Academic Conferences International Limited, pp. 922-929.
29. Afzal, W.; Alone, S.; Glocksien, K.; Torkar, R. (2016). Software test process improvement approaches: A systematic literature review and an industrial case study. Journal of Systems and Software, Vol. 111, pp. 1-33.
30. Garousi, V.; Mäntylä, M. V. (2016). When and what to automate in software testing? A multi-vocal literature review. Information and Software Technology, Vol. 76, pp. 92-117.
31. Whittaker, J. A. (2000). What is software testing? And why is it so hard?. IEEE Software, Vol. 17, No. 1, pp. 70-79.
32. Chang-Ai, S.; Chao, L.; Mao-Zhong, J.; Mei, Z. (2000). Architecture framework for software test tool. In IEEE Proceedings 36th International Conference on Technology of Object-Oriented Languages and Systems, TOOLS-Asia 2000, pp. 40-47.
33. Whyte, G.; Mulder, D. L. (2011). Mitigating the impact of software test constraints on software testing effectiveness. Electronic Journal of Information Systems Evaluation, Vol. 14, No. 2, p. 254.
34. Chillarege, R. (1999). Software testing best practices. IBM Thomas J. Watson Research Division.
35. Freeman, H. (2002). Software testing. IEEE instrumentation & measurement magazine, Vol. 5, No. 3, pp. 48-50.
36. Mullin, F. J. (2004). Software test tools. In Proc. TR W Symp. on Reliable, Cost Effective, Secure Software, pp. 6-47.
37. Brown, J. R. (1974). Software Test Tools: Technology. In Proceedings of the TRW Symposium on Reliable, Cost-Effective, Secure Software, TRW-SS-74-14.
38. https://www.katalon.com/?utm_source=QASymphony&utm_campaign=top%20testing%20 tools&utm_medium=article
39. https://software.microfocus.com/en-us/products/unified-functional-automated-testing/overview
40. https://www.ibm.com/in-en/marketplace/rational-functional-tester
41. https://www.tricentis.com/model-based-test-automation/#
42. https://www.worksoft.com/products/worksoft-certify
43. https://eggplant.io/eggplant-functional/
44. https://www.ranorex.com/
45. https://www.ca.com/us/products/ca-application-test.html
46. https://www.ibm.com/in-en/marketplace/rational-test-workbench
47. https://www.parasoft.com/products/soatest
48. https://smartbear.com/product/ready-api/overview/
49. http://www.crosschecknet.com/products/soapsonar.php
50. https://www.blazemeter.com/
51. https://www.neotys.com/neoload/overview
52. https://www.radview.com/
53. https://www.automationanywhere.com/
54. https://software.microfocus.com/en-us/products/loadrunner-load-testing/overview
55. https://software.microfocus.com/en-us/products/performance-center/overview
56. https://software.microfocus.com/en-us/products/stormrunner-load-agile-cloud-testing/overview
57. https://www.ibm.com/in-en/marketplace/ibm-rational-performance-tester
58. https://www.qasymphony.com/blog/100-plus-best-software-testing-tools/
59. Khoshgoftaar, T. M.; Seliya, N.; Gao, K. (2005). Assessment of a new three-group software quality classification technique: An empirical case study. Empirical Software Engineering, Vol. 10, No. 2, pp. 183-218.

60. Khoshgoftaar, T. M.; Seliya, N. (2004). Three-Group software quality classification modeling using an Automated Reasoning approach. In Artificial Intelligence Methods in Software Testing, pp. 133-174.
61. Hermans, J. (2006). A Software Testing Approach Supported by a Tool Environment for the Development of Component Tests. Technische Universiteit Eindhoven.
62. Elamin, R.; Osman, R. (2018). Implementing Traceability Repositories as Graph Databases for Software Quality.
63. Fly, H. (2004). Software quality improvement. Computer, Vol. 37, No. 4, p. 6.
64. Mizuno, Y. (1983). Software quality improvement. Computer, Vol. 3, pp. 66-72.
65. Liu, X. F.; Kane, G.; Bambroo, M. (2006). An intelligent early warning system for software quality improvement and project management. Journal of Systems and Software, Vol. 79, No. 11, pp. 1552-1564.
66. Redzic, C.; Baik, J. (2006). Six sigma approach in software quality improvement. In IEEE Fourth International Conference on Software Engineering Research, Management and Applications, 2006, pp. 396-406.
67. Leung, H. K.; Liao, L.; Qu, Y. (2007). Automated support of software quality improvement. International Journal of Quality & Reliability Management, Vol. 24, No. 3, pp. 230-243.
68. Lewis, W. E. (2005). *Software testing and continuous quality improvement*. Auerbach publications.
69. Gelperin, D.; Hetzel, B. (1988). The growth of software testing. Communications of the ACM, Vol. 31, No. 6, pp. 687-695.
70. Nidagundi, P.; Novickis, L. (2016). Introduction to lean canvas transformation models and metrics in software testing. Applied Computer Systems, Vol. 19, Issue 1, pp. 30-36.

Appendix
Java Code for Library Management System

1. About.java

Purpose: This code is used to provide the information about software (Library Management System, LMS).

```
public class About extends javax.swing.JFrame {
/** Creates new form About */
public About() {
initComponents();
}
// <editor-fold defaultstate="collapsed" desc="Generated Code">
private void initComponents() {
jLabel1 = new javax.swing.JLabel();
CmdClose = new javax.swing.JButton();
setDefaultCloseOperation(javax.swing.WindowConstants.EXIT_ON_CLOSE);
setTitle("P2P Library : About");
addWindowListener(new java.awt.event.WindowAdapter() {
public void windowOpened(java.awt.event.WindowEvent evt) {
formWindowOpened(evt);
}
});
jLabel1.setFont(new java.awt.Font("Tahoma", 1, 18));
jLabel1.setForeground(new java.awt.Color(51, 0, 255));
jLabel1.setText("P2P Library System");
CmdClose.setText("Close");
CmdClose.addActionListener(new java.awt.event.ActionListener() {
public void actionPerformed(java.awt.event.ActionEvent evt) {
CmdCloseActionPerformed(evt);
}
});
javax.swing.GroupLayout layout = new javax.swing.GroupLayout(getContentPane());
getContentPane().setLayout(layout);
layout.setHorizontalGroup(
layout.createParallelGroup(javax.swing.GroupLayout.Alignment.LEADING)
.addGroup(layout.createSequentialGroup()
.addGap(110, 110, 110)
.addComponent(jLabel1)
.addContainerGap(111, Short.MAX_VALUE))
.addGroup(javax.swing.GroupLayout.Alignment.TRAILING, layout.createSequentialGroup()
.addContainerGap(291, Short.MAX_VALUE)
.addComponent(CmdClose, javax.swing.GroupLayout.PREFERRED_SIZE, 76,
javax.swing.GroupLayout.PREFERRED_SIZE).addGap(27, 27, 27))
);
layout.setVerticalGroup(
layout.createParallelGroup(javax.swing.GroupLayout.Alignment.LEADING)
.addGroup(layout.createSequentialGroup()
.addGap(30, 30, 30)
.addComponent(jLabel1)
.addPreferredGap(javax.swing.LayoutStyle.ComponentPlacement.RELATED, 112,
Short.MAX_VALUE)
.addComponent(CmdClose)
.addContainerGap())
```

```
);
pack();
}// </editor-fold>
private void CmdCloseActionPerformed(java.awt.event.ActionEvent evt) {
// Close project
this.setVisible(false);
}
private void formWindowOpened(java.awt.event.WindowEvent evt) {
this.setLocationRelativeTo(null);
}
/*
* @param args the command line arguments
*/
public static void main(String args[]) {
java.awt.EventQueue.invokeLater(new Runnable() {
public void run() {
new About().setVisible(true);
}
});
}
// Variables declaration
private javax.swing.JButton CmdClose;
private javax.swing.JLabel jLabel1;
// End of variables declaration
}
```

2. BookMaster.java

Purpose: Book master used to add and update the detail of books and manage the book information using first, previous, next, last, new, edit, save, and close command. Basically there are book id, book title, author, publisher, year publish, category, and search keywords entry to manage book information.

```
import java.sql.Statement;
import java.sql.Connection;
import java.sql.DriverManager;
import java.sql.PreparedStatement;
import java.sql.ResultSet;
import javax.swing.JOptionPane;
public class BookMaster extends javax.swing.JFrame {
MainClass mc=new MainClass();
ResultSet rsbook;
int rspos=0;
int rscount=0;
boolean chadd=false;
/** Creates new form BookMaster */
public BookMaster() {
initComponents();
}
// <editor-fold defaultstate="collapsed" desc="Generated Code">
private void initComponents() {
jLabel1 = new javax.swing.JLabel();
jLabel2 = new javax.swing.JLabel();
TxtBookID = new javax.swing.JTextField();
jLabel3 = new javax.swing.JLabel();
```

```
TxtBookTitle = new javax.swing.JTextField();
jLabel4 = new javax.swing.JLabel();
TxtAuthor = new javax.swing.JTextField();
jLabel5 = new javax.swing.JLabel();
TxtPublisher = new javax.swing.JTextField();
jLabel6 = new javax.swing.JLabel();
TxtYear = new javax.swing.JTextField();
jLabel7 = new javax.swing.JLabel();
TxtKeywords = new javax.swing.JTextField();
jLabel8 = new javax.swing.JLabel();
CmbCategory = new javax.swing.JComboBox();
CmdFirst = new javax.swing.JButton();
CmdPrevious = new javax.swing.JButton();
CmdNext = new javax.swing.JButton();
CmdLast = new javax.swing.JButton();
CmdNew = new javax.swing.JButton();
CmdEdit = new javax.swing.JButton();
CmdSave = new javax.swing.JButton();
CmdClose = new javax.swing.JButton();
LblPosition = new javax.swing.JLabel();
setDefaultCloseOperation(javax.swing.WindowConstants.DISPOSE_ON_CLOSE);
setTitle("P2P Library : Book Master");
addWindowListener(new java.awt.event.WindowAdapter() {
public void windowOpened(java.awt.event.WindowEvent evt) {
formWindowOpened(evt);
}
});
jLabel1.setFont(new java.awt.Font("Tahoma", 1, 12));
jLabel1.setForeground(new java.awt.Color(51, 0, 255));
jLabel1.setText("Library Management System - Book Master");
jLabel2.setText("Book ID:");
TxtBookID.setEditable(false);
TxtBookID.setText("jTextField1");
jLabel3.setText("Book Title:");
TxtBookTitle.setText("jTextField1");
jLabel4.setText("Author:");
TxtAuthor.setText("jTextField1");
jLabel5.setText("Publisher:");
TxtPublisher.setText("jTextField1");
jLabel6.setText("Year Publish:");
TxtYear.setText("jTextField1");
jLabel7.setText("Search Keyword:");
TxtKeywords.setText("jTextField1");
jLabel8.setText("Category:");
CmbCategory.setModel(new javax.swing.DefaultComboBoxModel(new String[] { "Item 1",
"Item 2",
"Item 3", "Item 4" }));
CmdFirst.setText("First");
CmdFirst.addActionListener(new java.awt.event.ActionListener() {
public void actionPerformed(java.awt.event.ActionEvent evt) {
CmdFirstActionPerformed(evt);
}
});
CmdPrevious.setText("Previous");
CmdPrevious.addActionListener(new java.awt.event.ActionListener() {
```

```
public void actionPerformed(java.awt.event.ActionEvent evt) {
CmdPreviousActionPerformed(evt);
}
});
CmdNext.setText("Next");
CmdNext.addActionListener(new java.awt.event.ActionListener() {
public void actionPerformed(java.awt.event.ActionEvent evt) {
CmdNextActionPerformed(evt);
}
});
CmdLast.setText("Last");
CmdLast.addActionListener(new java.awt.event.ActionListener() {
public void actionPerformed(java.awt.event.ActionEvent evt) {
CmdLastActionPerformed(evt);
}
});
CmdNew.setText("New");
CmdNew.addActionListener(new java.awt.event.ActionListener() {
public void actionPerformed(java.awt.event.ActionEvent evt) {
CmdNewActionPerformed(evt);
}
});
CmdEdit.setText("Edit");
CmdEdit.addActionListener(new java.awt.event.ActionListener() {
public void actionPerformed(java.awt.event.ActionEvent evt) {
CmdEditActionPerformed(evt);
}
});
CmdSave.setText("Save");
CmdSave.addActionListener(new java.awt.event.ActionListener() {
public void actionPerformed(java.awt.event.ActionEvent evt) {
CmdSaveActionPerformed(evt);
}
});
CmdClose.setText("Close");
CmdClose.addActionListener(new java.awt.event.ActionListener() {
public void actionPerformed(java.awt.event.ActionEvent evt) {
CmdCloseActionPerformed(evt);
}
});
LblPosition.setText("Book ID:");
javax.swing.GroupLayout layout = new javax.swing.GroupLayout(getContentPane());
getContentPane().setLayout(layout);
layout.setHorizontalGroup(
layout.createParallelGroup(javax.swing.GroupLayout.Alignment.LEADING)
.addGroup(layout.createSequentialGroup()
.addGroup(layout.createParallelGroup(javax.swing.GroupLayout.Alignment.LEADING)
.addGroup(layout.createSequentialGroup()
.addGroup(layout.createParallelGroup(javax.swing.GroupLayout.Alignment.LEADING)
.addGroup(layout.createSequentialGroup()
.addGap(87, 87, 87)
.addGroup(layout.createParallelGroup(javax.swing.GroupLayout.Alignment.LEADING)
.addComponent(jLabel4)
.addComponent(jLabel5)
.addComponent(jLabel6)
```

```
.addComponent(jLabel8)
.addComponent(jLabel2)
.addComponent(jLabel3)
.addComponent(jLabel7))
.addGap(27, 27, 27)
.addGroup(layout.createParallelGroup(javax.swing.GroupLayout.Alignment.LEADING)
.addGroup(layout.createParallelGroup(javax.swing.GroupLayout.Alignment.LEADING, false)
.addComponent(TxtYear)
.addComponent(TxtPublisher)
.addComponent(TxtAuthor)
.addComponent(TxtBookTitle)
.addComponent(TxtBookID, javax.swing.GroupLayout.DEFAULT_SIZE, 198,
Short.MAX_VALUE)
.addComponent(CmbCategory, 0, javax.swing.GroupLayout.DEFAULT_SIZE,
Short.MAX_VALUE))
.addGroup(layout.createParallelGroup(javax.swing.GroupLayout.Alignment.TRAILING)
.addComponent(CmdNew)
.addComponent(TxtKeywords, javax.swing.GroupLayout.PREFERRED_SIZE, 197,
javax.swing.GroupLayout.PREFERRED_SIZE))))
.addGroup(layout.createSequentialGroup()
.addGap(25, 25, 25)
.addComponent(CmdFirst)
.addPreferredGap(javax.swing.LayoutStyle.ComponentPlacement.RELATED)
.addComponent(CmdPrevious)
.addPreferredGap(javax.swing.LayoutStyle.ComponentPlacement.RELATED)
.addComponent(CmdNext)
.addPreferredGap(javax.swing.LayoutStyle.ComponentPlacement.RELATED)
.addComponent(CmdLast)))
.addPreferredGap(javax.swing.LayoutStyle.ComponentPlacement.RELATED)
.addComponent(CmdEdit)
.addGap(6, 6, 6)
.addComponent(CmdSave)
.addPreferredGap(javax.swing.LayoutStyle.ComponentPlacement.RELATED)
.addComponent(CmdClose))
.addGroup(layout.createSequentialGroup()
.addContainerGap()
.addComponent(jLabel1))
.addGroup(layout.createSequentialGroup()
.addGap(143, 143, 143)
.addComponent(LblPosition)))
.addContainerGap(39, Short.MAX_VALUE))
);
layout.setVerticalGroup(layout.createParallelGroup(javax.swing.GroupLayout.Alignment.LEA
DING)
.addGroup(layout.createSequentialGroup()
.addGroup(layout.createParallelGroup(javax.swing.GroupLayout.Alignment.LEADING)
.addGroup(layout.createSequentialGroup()
.addContainerGap()
.addComponent(jLabel1)
.addGap(26, 26, 26)
.addComponent(jLabel2)
.addPreferredGap(javax.swing.LayoutStyle.ComponentPlacement.UNRELATED)
.addComponent(jLabel3)
.addGap(18, 18, 18)
.addComponent(jLabel4)
```

```
.addPreferredGap(javax.swing.LayoutStyle.ComponentPlacement.UNRELATED)
.addComponent(jLabel5)
.addPreferredGap(javax.swing.LayoutStyle.ComponentPlacement.UNRELATED)
.addComponent(jLabel6)
.addGap(19, 19, 19)
.addComponent(jLabel8))
.addGroup(layout.createSequentialGroup()
.addGap(47, 47, 47)
.addComponent(TxtBookID, javax.swing.GroupLayout.PREFERRED_SIZE, 23,
javax.swing.GroupLayout.PREFERRED_SIZE)
.addPreferredGap(javax.swing.LayoutStyle.ComponentPlacement.RELATED)
.addComponent(TxtBookTitle, javax.swing.GroupLayout.PREFERRED_SIZE, 23,
javax.swing.GroupLayout.PREFERRED_SIZE)
.addPreferredGap(javax.swing.LayoutStyle.ComponentPlacement.RELATED)
.addComponent(TxtAuthor, javax.swing.GroupLayout.PREFERRED_SIZE, 23,
javax.swing.GroupLayout.PREFERRED_SIZE)
.addPreferredGap(javax.swing.LayoutStyle.ComponentPlacement.RELATED)
.addComponent(TxtPublisher, javax.swing.GroupLayout.PREFERRED_SIZE, 23,
javax.swing.GroupLayout.PREFERRED_SIZE)
.addPreferredGap(javax.swing.LayoutStyle.ComponentPlacement.RELATED)
.addComponent(TxtYear, javax.swing.GroupLayout.PREFERRED_SIZE, 23,
javax.swing.GroupLayout.PREFERRED_SIZE)
.addGap(7, 7, 7)
.addComponent(CmbCategory, javax.swing.GroupLayout.PREFERRED_SIZE, 25,
javax.swing.GroupLayout.PREFERRED_SIZE)))
.addGap(19, 19, 19)
.addGroup(layout.createParallelGroup(javax.swing.GroupLayout.Alignment.LEADING)
.addComponent(TxtKeywords, javax.swing.GroupLayout.PREFERRED_SIZE, 45,
javax.swing.GroupLayout.PREFERRED_SIZE)
.addComponent(jLabel7))
.addPreferredGap(javax.swing.LayoutStyle.ComponentPlacement.RELATED, 43,
Short.MAX_VALUE)
.addComponent(LblPosition)
.addPreferredGap(javax.swing.LayoutStyle.ComponentPlacement.RELATED)
.addGroup(layout.createParallelGroup(javax.swing.GroupLayout.Alignment.BASELINE)
.addComponent(CmdFirst)
.addComponent(CmdPrevious)
.addComponent(CmdNext)
.addComponent(CmdLast)
.addComponent(CmdNew)
.addComponent(CmdEdit)
.addComponent(CmdSave)
.addComponent(CmdClose))
.addGap(39, 39, 39))
);
pack();
}// </editor-fold>

private void CmdCloseActionPerformed(java.awt.event.ActionEvent evt) {
// TODO add your handling code here:
this.setVisible(false);
}
private void formWindowOpened(java.awt.event.WindowEvent evt) {
// TODO add your handling code here:
try
```

```
{
this.setLocationRelativeTo(null);
ClearText();
CmbCategory.removeAllItems();
Connection connection;
connection=DriverManager.getConnection(mc.StrUrl,mc.StrUid,mc.StrPwd);
ResultSet rs;
//fill book category
PreparedStatement stmt=connection.prepareStatement("select * from lib_book_category order
by
book_category");
rs = stmt.executeQuery();
while (rs.next()){
CmbCategory.addItem(rs.getString(1));
}
rs.close();
//get book count
stmt=connection.prepareStatement("select count(*) from lib_book_master");
rs = stmt.executeQuery();
rs.first();
rscount=rs.getInt(1);
//get book details
Statement stmt1 = connection.createStatement(ResultSet.TYPE_SCROLL_SENSITIVE,
ResultSet.CONCUR_UPDATABLE);
String sqlQuery = "select * from lib_book_master order by book_id";
rsbook = stmt1.executeQuery(sqlQuery);
//display first record
rsbook.first();
Display();
}
catch (Exception e)
{
System.err.println(e);
//System.exit(1);
}
}
private void CmdNewActionPerformed(java.awt.event.ActionEvent evt) {
chadd=true;
ClearText();
}
private void CmdNextActionPerformed(java.awt.event.ActionEvent evt) {
try
{
if (rsbook.isLast()==true)
{
JOptionPane.showMessageDialog(null,"Already in last record.");
return;
}
rsbook.next();
Display();
}
catch (Exception e)
{
System.err.println(e);
}
```

```
}
private void CmdFirstActionPerformed(java.awt.event.ActionEvent evt) {
try
{
rsbook.first();
Display();
}
catch (Exception e)
{
System.err.println(e);
}
}
private void CmdPreviousActionPerformed(java.awt.event.ActionEvent evt) {
try
{
if (rsbook.isFirst()==true)
{
JOptionPane.showMessageDialog(null,"Already in first record.");
return;
}
rsbook.previous();
Display();
}
catch (Exception e)
{
System.err.println(e);
}
}
private void CmdLastActionPerformed(java.awt.event.ActionEvent evt) {
try
{
rsbook.last();
Display();
}
catch (Exception e)
{
System.err.println(e);
//System.exit(1);
}
}
private void CmdEditActionPerformed(java.awt.event.ActionEvent evt) {
chadd=false;
}
private void CmdSaveActionPerformed(java.awt.event.ActionEvent evt) {
//validate entry and save record
try
{
if (TxtBookTitle.getText().trim().length()==0 )
{
JOptionPane.showMessageDialog(null,"Enter book title.");
return;
}
if (TxtAuthor.getText().trim().length()==0 )
{
JOptionPane.showMessageDialog(null,"Enter author.");
```

```
return;
}
if (TxtPublisher.getText().trim().length()==0)
{
JOptionPane.showMessageDialog(null,"Enter publisher name.");
return;
}
if (CmbCategory.getSelectedItem().equals(""))
{
JOptionPane.showMessageDialog(null,"Select book category.");
return;
}
//String strqr="";
Connection connection;
connection=DriverManager.getConnection(mc.StrUrl,mc.StrUid,mc.StrPwd);
if (TxtBookID.getText().trim().length()==0 )
{
//strqr="insert into lib_book_master (book_title, book_author, book_publisher,
book_publish_year, book_category, book_keyword) values ('" + TxtBookTitle.getText() + "','" +
TxtAuthor.getText() +"','"+ TxtPublisher.getText() +"','" + TxtYear.getText() + "','" +
CmbCategory.getSelectedItem().toString() +"','" + TxtKeywords.getText() +"')";
rsbook.moveToInsertRow();
rsbook.updateString("book_title",TxtBookTitle.getText());
rsbook.updateString("book_author",TxtAuthor.getText());
rsbook.updateString("book_publisher",TxtPublisher.getText());
rsbook.updateString("book_publish_year",TxtYear.getText());
rsbook.updateString("book_category",CmbCategory.getSelectedItem().toString());
rsbook.updateString("book_keyword",TxtKeywords.getText());
rsbook.insertRow();
rsbook.last();
Display();
}
else
{
rsbook.updateString("book_title",TxtBookTitle.getText());
rsbook.updateString("book_author",TxtAuthor.getText());
rsbook.updateString("book_publisher",TxtPublisher.getText());
rsbook.updateString("book_publish_year",TxtYear.getText());
rsbook.updateString("book_category",CmbCategory.getSelectedItem().toString());
rsbook.updateString("book_keyword",TxtKeywords.getText());
rsbook.updateRow();
rsbook.refreshRow();
}
JOptionPane.showMessageDialog(null,"Record updated.");
}
catch (Exception e)
{
System.err.println(e);
JOptionPane.showMessageDialog(null,"Unable to save record.");
return;
}
}
private void Display()
{
try
```

```
{
//clear text of fields for entry/display old data
TxtBookID.setText(rsbook.getString("book_id"));
TxtBookTitle.setText(rsbook.getString("book_title"));
TxtAuthor.setText(rsbook.getString("book_author"));
TxtPublisher.setText(rsbook.getString("book_publisher"));
TxtYear.setText(rsbook.getString("book_publish_year"));
CmbCategory.setSelectedItem(rsbook.getString("book_category"));
TxtKeywords.setText(rsbook.getString("book_keyword"));
LblPosition.setText("Record position : " + rsbook.getRow() + "/" + String.valueOf(rscount));
}
catch (Exception e)
{
System.err.println(e);
}
}
private void ClearText()
{
//clear text of fields for entry/display old data
TxtBookID.setText("");
TxtBookTitle.setText("");
TxtAuthor.setText("");
TxtPublisher.setText("");
TxtYear.setText("");
CmbCategory.setSelectedItem(null);
TxtKeywords.setText("");

LblPosition.setText("");
}
/**
* @param args the command line arguments
*/
public static void main(String args[]) {
java.awt.EventQueue.invokeLater(new Runnable() {
public void run() {
new BookMaster().setVisible(true);
}
});
}
// Variables declaration
private javax.swing.JComboBox CmbCategory;
private javax.swing.JButton CmdClose;
private javax.swing.JButton CmdEdit;
private javax.swing.JButton CmdFirst;
private javax.swing.JButton CmdLast;
private javax.swing.JButton CmdNew;
private javax.swing.JButton CmdNext;
private javax.swing.JButton CmdPrevious;
private javax.swing.JButton CmdSave;
private javax.swing.JLabel LblPosition;
private javax.swing.JTextField TxtAuthor;
private javax.swing.JTextField TxtBookID;
private javax.swing.JTextField TxtBookTitle;
private javax.swing.JTextField TxtKeywords;
private javax.swing.JTextField TxtPublisher;
```

```
private javax.swing.JTextField TxtYear;
private javax.swing.JLabel jLabel1;
private javax.swing.JLabel jLabel2;
private javax.swing.JLabel jLabel3;
private javax.swing.JLabel jLabel4;
private javax.swing.JLabel jLabel5;
private javax.swing.JLabel jLabel6;
private javax.swing.JLabel jLabel7;
private javax.swing.JLabel jLabel8;
// End of variables declaration
}
```

3. BookStatus.java

Purpose: Book status depicts the status of book using refresh command which generate a table that shows the book id, title, category, and status (issue and available) of book.

```
import java.sql.Connection;
import java.sql.DriverManager;
import java.sql.PreparedStatement;
import java.sql.ResultSet;
import javax.swing.table.DefaultTableModel;
import javax.swing.table.AbstractTableModel;
import javax.swing.JOptionPane;
public class BookStatus extends javax.swing.JFrame {
/** Creates new form BookStatus */
public BookStatus() {
initComponents();
}
// <editor-fold defaultstate="collapsed" desc="Generated Code">
private void initComponents() {
jScrollPane1 = new javax.swing.JScrollPane();
jTable1 = new javax.swing.JTable();
jButton1 = new javax.swing.JButton();
CmdClose = new javax.swing.JButton();
jLabel1 = new javax.swing.JLabel();
setDefaultCloseOperation(javax.swing.WindowConstants.DISPOSE_ON_CLOSE);
setTitle("P2P Library : Book Status");
addWindowListener(new java.awt.event.WindowAdapter() {
public void windowOpened(java.awt.event.WindowEvent evt) {
formWindowOpened(evt);
}
});
jTable1.setModel(new javax.swing.table.DefaultTableModel(
new Object [][] {
{null, null, null, null},
{null, null, null, null},
{null, null, null, null},
{null, null, null, null}
},
new String [] {
"Title 1", "Title 2", "Title 3", "Title 4"
}
));
jTable1.setAutoResizeMode(javax.swing.JTable.AUTO_RESIZE_ALL_COLUMNS);
```

```
jScrollPane1.setViewportView(jTable1);
jButton1.setText("Refresh Status");
jButton1.addActionListener(new java.awt.event.ActionListener() {
public void actionPerformed(java.awt.event.ActionEvent evt) {
jButton1ActionPerformed(evt);
}
});
CmdClose.setText("Close");
CmdClose.addActionListener(new java.awt.event.ActionListener() {
public void actionPerformed(java.awt.event.ActionEvent evt) {
CmdCloseActionPerformed(evt);
}
});
jLabel1.setFont(new java.awt.Font("Tahoma", 1, 12));
jLabel1.setForeground(new java.awt.Color(51, 0, 255));
jLabel1.setText("Library Management System  - Book Status");
javax.swing.GroupLayout layout = new javax.swing.GroupLayout(getContentPane());
getContentPane().setLayout(layout);
layout.setHorizontalGroup(
layout.createParallelGroup(javax.swing.GroupLayout.Alignment.LEADING)
.addGroup(layout.createSequentialGroup()
.addContainerGap()
.addGroup(layout.createParallelGroup(javax.swing.GroupLayout.Alignment.LEADING)
.addComponent(jScrollPane1)
.addComponent(jLabel1)
.addGroup(layout.createSequentialGroup()
.addComponent(jButton1, javax.swing.GroupLayout.PREFERRED_SIZE, 138,
javax.swing.GroupLayout.PREFERRED_SIZE)
.addPreferredGap(javax.swing.LayoutStyle.ComponentPlacement.RELATED)
.addComponent(CmdClose, javax.swing.GroupLayout.PREFERRED_SIZE, 75,
javax.swing.GroupLayout.PREFERRED_SIZE)))
.addContainerGap())
);
layout.setVerticalGroup(
layout.createParallelGroup(javax.swing.GroupLayout.Alignment.LEADING)
.addGroup(layout.createSequentialGroup()
.addContainerGap()
.addComponent(jLabel1)
.addGap(15, 15, 15)
.addGroup(layout.createParallelGroup(javax.swing.GroupLayout.Alignment.BASELINE)
.addComponent(CmdClose)
.addComponent(jButton1))
.addPreferredGap(javax.swing.LayoutStyle.ComponentPlacement.UNRELATED)
.addComponent(jScrollPane1, javax.swing.GroupLayout.DEFAULT_SIZE, 280,
Short.MAX_VALUE)
.addContainerGap())
);
pack();
}// </editor-fold>
private void jButton1ActionPerformed(java.awt.event.ActionEvent evt) {
// TODO add your handling code here:
try {
MainClass mc=new MainClass();
Connection connection;
connection=DriverManager.getConnection(mc.StrUrl,mc.StrUid,mc.StrPwd);
```

```
ResultSet rs;
PreparedStatement stmt=connection.prepareStatement("select book_id, book_title,
book_category,
book_status from lib_book_master order by book_id");
rs = stmt.executeQuery();
jTable1.getColumnModel().getColumn(0).setHeaderValue("Book ID");
jTable1.getColumnModel().getColumn(1).setHeaderValue("Title");
jTable1.getColumnModel().getColumn(2).setHeaderValue("Category");
jTable1.getColumnModel().getColumn(3).setHeaderValue("Status");
jTable1.getTableHeader().resizeAndRepaint();
// Removing Previous Data
while (jTable1.getRowCount() > 0) {
((DefaultTableModel) jTable1.getModel()).removeRow(0);
}
//Creating Object []rowData for jTable's Table Model
int columns = rs.getMetaData().getColumnCount();
while (rs.next()) {
Object[] row = new Object[columns];
for (int i = 1; i <= columns; i++) {
row[i - 1] = rs.getObject(i); // 1
}
((DefaultTableModel) jTable1.getModel()).insertRow(rs.getRow() - 1,row);
}
} catch (Exception e) {
System.err.println(e);
//System.exit(1);
}
}
private void CmdCloseActionPerformed(java.awt.event.ActionEvent evt) {
this.setVisible(false);
}
private void formWindowOpened(java.awt.event.WindowEvent evt) {
this.setLocationRelativeTo(null);
jTable1.getColumnModel().getColumn(0).setHeaderValue("Book ID");
jTable1.getColumnModel().getColumn(1).setHeaderValue("Title");
jTable1.getColumnModel().getColumn(2).setHeaderValue("Category");
jTable1.getColumnModel().getColumn(3).setHeaderValue("Status");
jTable1.getTableHeader().resizeAndRepaint();
// Removing Previous Data
while (jTable1.getRowCount() > 0) {
((DefaultTableModel) jTable1.getModel()).removeRow(0);
}
}
/**
* @param args the command line arguments
*/
public static void main(String args[]) {
java.awt.EventQueue.invokeLater(new Runnable() {
public void run() {
new BookStatus().setVisible(true);
}
});
}
// Variables declaration
private javax.swing.JButton CmdClose;
```

```
private javax.swing.JButton jButton1;
private javax.swing.JLabel jLabel1;
private javax.swing.JScrollPane jScrollPane1;
private javax.swing.JTable jTable1;
// End of variables declaration
}
```

4. ChangePassword.java

Purpose: Change password used to modify login password using old password, new password, and confirm password entry.

```
import java.sql.Connection;
import java.sql.DriverManager;
import java.sql.PreparedStatement;
import java.sql.ResultSet;
import javax.swing.JOptionPane;
public class ChangePassword extends javax.swing.JFrame {
/** Creates new form ChangePassword */
public ChangePassword() {
initComponents();
}
// <editor-fold defaultstate="collapsed" desc="Generated Code">
private void initComponents() {
jLabel3 = new javax.swing.JLabel();
TxtOldPassword = new javax.swing.JPasswordField();
jLabel4 = new javax.swing.JLabel();
TxtNewPassword = new javax.swing.JPasswordField();
jLabel5 = new javax.swing.JLabel();
TxtConfirmPassword = new javax.swing.JPasswordField();
CmdOk = new javax.swing.JButton();
CmdClose = new javax.swing.JButton();
jLabel1 = new javax.swing.JLabel();
setDefaultCloseOperation(javax.swing.WindowConstants.DISPOSE_ON_CLOSE);
setTitle("P2P Library : Change Password");
addWindowListener(new java.awt.event.WindowAdapter() {
public void windowOpened(java.awt.event.WindowEvent evt) {
formWindowOpened(evt);
}
});
jLabel3.setText("Old Password:");
jLabel4.setText("New Password:");
jLabel5.setText("Confirm Password:");
CmdOk.setText("Ok");
CmdOk.addActionListener(new java.awt.event.ActionListener() {
public void actionPerformed(java.awt.event.ActionEvent evt) {
CmdOkActionPerformed(evt);
}
});
CmdClose.setText("Close");
CmdClose.addActionListener(new java.awt.event.ActionListener() {
public void actionPerformed(java.awt.event.ActionEvent evt) {
CmdCloseActionPerformed(evt);
}
});
```

```
jLabel1.setFont(new java.awt.Font("Tahoma", 1, 12));
jLabel1.setForeground(new java.awt.Color(51, 0, 255));
jLabel1.setText("Library Management System - Change Password");
javax.swing.GroupLayout layout = new javax.swing.GroupLayout(getContentPane());
getContentPane().setLayout(layout);
layout.setHorizontalGroup(
layout.createParallelGroup(javax.swing.GroupLayout.Alignment.LEADING)
.addGroup(layout.createSequentialGroup()
.addGap(54, 54, 54)
.addGroup(layout.createParallelGroup(javax.swing.GroupLayout.Alignment.LEADING)
.addComponent(jLabel4)
.addComponent(jLabel3)
.addComponent(jLabel5))
.addPreferredGap(javax.swing.LayoutStyle.ComponentPlacement.RELATED)
.addGroup(layout.createParallelGroup(javax.swing.GroupLayout.Alignment.LEADING)
.addComponent(TxtOldPassword, javax.swing.GroupLayout.DEFAULT_SIZE, 136,
Short.MAX_VALUE)
.addComponent(TxtNewPassword, javax.swing.GroupLayout.DEFAULT_SIZE, 136,
Short.MAX_VALUE)
.addComponent(TxtConfirmPassword, javax.swing.GroupLayout.DEFAULT_SIZE, 136,
Short.MAX_VALUE))
.addGap(140, 140, 140))
.addGroup(layout.createSequentialGroup()
.addGroup(layout.createParallelGroup(javax.swing.GroupLayout.Alignment.LEADING)
.addGroup(layout.createSequentialGroup()
.addContainerGap()
.addComponent(jLabel1))
.addGroup(layout.createSequentialGroup()
.addGap(154, 154, 154)
.addComponent(CmdOk)
.addPreferredGap(javax.swing.LayoutStyle.ComponentPlacement.RELATED)
.addComponent(CmdClose, javax.swing.GroupLayout.PREFERRED_SIZE, 76,
javax.swing.GroupLayout.PREFERRED_SIZE)))
.addContainerGap(javax.swing.GroupLayout.DEFAULT_SIZE, Short.MAX_VALUE))
);
layout.setVerticalGroup(
layout.createParallelGroup(javax.swing.GroupLayout.Alignment.LEADING)
.addGroup(layout.createSequentialGroup()
.addGap(12, 12, 12)
.addComponent(jLabel1)
.addGap(18, 18, 18)
.addGroup(layout.createParallelGroup(javax.swing.GroupLayout.Alignment.BASELINE)
.addComponent(TxtOldPassword, javax.swing.GroupLayout.PREFERRED_SIZE,
javax.swing.GroupLayout.DEFAULT_SIZE, javax.swing.GroupLayout.PREFERRED_SIZE)
.addComponent(jLabel3))
.addGap(18, 18, 18)
.addGroup(layout.createParallelGroup(javax.swing.GroupLayout.Alignment.BASELINE)
.addComponent(jLabel4)
.addComponent(TxtNewPassword, javax.swing.GroupLayout.PREFERRED_SIZE,
javax.swing.GroupLayout.DEFAULT_SIZE, javax.swing.GroupLayout.PREFERRED_SIZE))
.addGap(18, 18, 18)
.addGroup(layout.createParallelGroup(javax.swing.GroupLayout.Alignment.BASELINE)
.addComponent(jLabel5)
.addComponent(TxtConfirmPassword, javax.swing.GroupLayout.PREFERRED_SIZE,
javax.swing.GroupLayout.DEFAULT_SIZE, javax.swing.GroupLayout.PREFERRED_SIZE))
```

```
.addGap(26, 26, 26)
.addGroup(layout.createParallelGroup(javax.swing.GroupLayout.Alignment.BASELINE)
.addComponent(CmdOk)
.addComponent(CmdClose))
.addContainerGap(47, Short.MAX_VALUE))
);
pack();
}// </editor-fold>
private void CmdOkActionPerformed(java.awt.event.ActionEvent evt) {
//validation
try {
if (TxtOldPassword.getText().isEmpty()==true) {
JOptionPane.showMessageDialog(null,"Enter Old Password.");
return;
}
if (TxtNewPassword.getText().isEmpty()==true) {
JOptionPane.showMessageDialog(null,"Enter New Password.");
return;
}
if (TxtNewPassword.getText().equals(TxtConfirmPassword.getText() )==false) {
JOptionPane.showMessageDialog(null,"New and Confirm passwords are not matched.");
return;
}
//get database connection details
MainClass mc=new MainClass();
//open connection
Connection connection;
connection=DriverManager.getConnection(mc.StrUrl,mc.StrUid,mc.StrPwd);
String str="";
str="select * from lib_user where user_name =? and user_password =?";
PreparedStatement pst=connection.prepareStatement(str);
pst.setString(1, MainClass.StrUser);
pst.setString(2, TxtOldPassword.getText());
ResultSet rs;
rs=pst.executeQuery();
if (rs.next()) {
//Update password
String sql = "UPDATE lib_user SET user_password='" + TxtNewPassword.getText() + "'
WHERE user_name = '" + MainClass.StrUser +"'";
pst.executeUpdate(sql);
JOptionPane.showMessageDialog(null,"Password updated.");
this.setVisible(false);
} else {
JOptionPane.showMessageDialog(null,"Old password are not correct.");
return;
}
} catch (Exception e) {
System.err.println(e);
System.exit(1);
}
}
private void CmdCloseActionPerformed(java.awt.event.ActionEvent evt) {
// Close project
this.setVisible(false);
}
```

```
private void formWindowOpened(java.awt.event.WindowEvent evt) {
this.setLocationRelativeTo(null);
}
/**
* @param args the command line arguments
*/
public static void main(String args[]) {
java.awt.EventQueue.invokeLater(new Runnable() {
public void run() {
new ChangePassword().setVisible(true);
}
});
}
// Variables declaration
private javax.swing.JButton CmdClose;
private javax.swing.JButton CmdOk;
private javax.swing.JPasswordField TxtConfirmPassword;
private javax.swing.JPasswordField TxtNewPassword;
private javax.swing.JPasswordField TxtOldPassword;
private javax.swing.JLabel jLabel1;
private javax.swing.JLabel jLabel3;
private javax.swing.JLabel jLabel4;
private javax.swing.JLabel jLabel5;
// End of variables declaration
}
```

5. IssueBook.java

Purpose: Issue book used to issue book to members using member id where book id is used to issue book. Basically there are reset, issue book, details, and close command.

```
import java.sql.Connection;
import java.sql.DriverManager;
import java.sql.PreparedStatement;
import java.sql.ResultSet;
import javax.swing.JOptionPane;
public class IssueBook extends javax.swing.JFrame {
/** Creates new form IssueBook */
public IssueBook() {
initComponents();
}
// <editor-fold defaultstate="collapsed" desc="Generated Code">
private void initComponents() {
jLabel3 = new javax.swing.JLabel();
TxtMemberID = new javax.swing.JTextField();
jLabel1 = new javax.swing.JLabel();
CmdMemberDetails = new javax.swing.JButton();
jLabel2 = new javax.swing.JLabel();
TxtMemberName = new javax.swing.JTextField();
TxtMemberStatus = new javax.swing.JTextField();
jLabel4 = new javax.swing.JLabel();
TxtBookID = new javax.swing.JTextField();
CmdBookDetails = new javax.swing.JButton();
jLabel5 = new javax.swing.JLabel();
TxtBookStatus = new javax.swing.JTextField();
```

```
TxtBookTitle = new javax.swing.JTextField();
CmdReset = new javax.swing.JButton();
CmdIssueBook = new javax.swing.JButton();
CmdClose = new javax.swing.JButton();
setDefaultCloseOperation(javax.swing.WindowConstants.DISPOSE_ON_CLOSE);
setTitle("P2P Library : Issue Book");
addWindowListener(new java.awt.event.WindowAdapter() {
public void windowOpened(java.awt.event.WindowEvent evt) {
formWindowOpened(evt);
}
});
jLabel3.setText("Member ID :");
TxtMemberID.setText("jTextField1");
jLabel1.setFont(new java.awt.Font("Tahoma", 1, 12));
jLabel1.setForeground(new java.awt.Color(51, 0, 255));
jLabel1.setText("Library Management System - Issue Book");
CmdMemberDetails.setText("Details");
CmdMemberDetails.addActionListener(new java.awt.event.ActionListener() {
public void actionPerformed(java.awt.event.ActionEvent evt) {
CmdMemberDetailsActionPerformed(evt);
}
});
jLabel2.setText("Name :");
TxtMemberName.setEditable(false);
TxtMemberName.setText("jTextField1");
TxtMemberStatus.setEditable(false);
TxtMemberStatus.setText("jTextField1");
jLabel4.setText("Book ID :");
TxtBookID.setText("jTextField1");
CmdBookDetails.setText("Details");
CmdBookDetails.addActionListener(new java.awt.event.ActionListener() {
public void actionPerformed(java.awt.event.ActionEvent evt) {
CmdBookDetailsActionPerformed(evt);
}
});

jLabel5.setText("Title :");
TxtBookStatus.setEditable(false);
TxtBookStatus.setText("jTextField1");
TxtBookTitle.setEditable(false);
TxtBookTitle.setText("jTextField1");
CmdReset.setText("Reset");
CmdReset.addActionListener(new java.awt.event.ActionListener() {
public void actionPerformed(java.awt.event.ActionEvent evt) {
CmdResetActionPerformed(evt);
}
});
CmdIssueBook.setText("Issue Book");
CmdIssueBook.addActionListener(new java.awt.event.ActionListener() {
public void actionPerformed(java.awt.event.ActionEvent evt) {
CmdIssueBookActionPerformed(evt);
}
});
CmdClose.setText("Close");
CmdClose.addActionListener(new java.awt.event.ActionListener() {
```

```
public void actionPerformed(java.awt.event.ActionEvent evt) {
CmdCloseActionPerformed(evt);
}
});
javax.swing.GroupLayout layout = new javax.swing.GroupLayout(getContentPane());
getContentPane().setLayout(layout);
layout.setHorizontalGroup(
layout.createParallelGroup(javax.swing.GroupLayout.Alignment.LEADING)
.addGroup(javax.swing.GroupLayout.Alignment.TRAILING, layout.createSequentialGroup()
.addGroup(layout.createParallelGroup(javax.swing.GroupLayout.Alignment.TRAILING)
.addGroup(javax.swing.GroupLayout.Alignment.LEADING, layout.createSequentialGroup()
.addGap(18, 18, 18)
.addComponent(jLabel1))
.addGroup(javax.swing.GroupLayout.Alignment.LEADING, layout.createSequentialGroup()
.addGap(40, 40, 40)
.addGroup(layout.createParallelGroup(javax.swing.GroupLayout.Alignment.LEADING)
.addGroup(javax.swing.GroupLayout.Alignment.TRAILING, layout.createSequentialGroup()
.addComponent(jLabel5)
.addPreferredGap(javax.swing.LayoutStyle.ComponentPlacement.RELATED, 38,
Short.MAX_VALUE)
.addGroup(layout.createParallelGroup(javax.swing.GroupLayout.Alignment.LEADING)
.addComponent(TxtBookStatus, javax.swing.GroupLayout.PREFERRED_SIZE, 229,
javax.swing.GroupLayout.PREFERRED_SIZE)
.addComponent(TxtBookTitle, javax.swing.GroupLayout.PREFERRED_SIZE, 229,
javax.swing.GroupLayout.PREFERRED_SIZE)
.addGroup(layout.createSequentialGroup()
.addComponent(CmdReset)
.addPreferredGap(javax.swing.LayoutStyle.ComponentPlacement.UNRELATED)
.addComponent(CmdIssueBook)
.addPreferredGap(javax.swing.LayoutStyle.ComponentPlacement.UNRELATED)
.addComponent(CmdClose))))
.addGroup(layout.createSequentialGroup()
.addComponent(jLabel3)
.addPreferredGap(javax.swing.LayoutStyle.ComponentPlacement.RELATED)
.addComponent(TxtMemberID, javax.swing.GroupLayout.PREFERRED_SIZE,
javax.swing.GroupLayout.DEFAULT_SIZE, javax.swing.GroupLayout.PREFERRED_SIZE)
.addGap(10, 10, 10)
.addComponent(CmdMemberDetails))
.addGroup(javax.swing.GroupLayout.Alignment.TRAILING, layout.createSequentialGroup()
.addComponent(jLabel2)
.addPreferredGap(javax.swing.LayoutStyle.ComponentPlacement.RELATED, 33,
Short.MAX_VALUE)
.addGroup(layout.createParallelGroup(javax.swing.GroupLayout.Alignment.LEADING)
.addComponent(TxtMemberStatus, javax.swing.GroupLayout.PREFERRED_SIZE, 229,
javax.swing.GroupLayout.PREFERRED_SIZE)
.addComponent(TxtMemberName, javax.swing.GroupLayout.PREFERRED_SIZE, 229,
javax.swing.GroupLayout.PREFERRED_SIZE)))
.addGroup(layout.createSequentialGroup()
.addComponent(jLabel4)
.addGap(18, 18, 18)
.addComponent(TxtBookID, javax.swing.GroupLayout.PREFERRED_SIZE,
javax.swing.GroupLayout.DEFAULT_SIZE, javax.swing.GroupLayout.PREFERRED_SIZE)
.addGap(10, 10, 10)
.addComponent(CmdBookDetails)))))
.addContainerGap(97, javax.swing.GroupLayout.PREFERRED_SIZE))
```

```
);
layout.setVerticalGroup(
layout.createParallelGroup(javax.swing.GroupLayout.Alignment.LEADING)
.addGroup(layout.createSequentialGroup()
.addContainerGap()
.addComponent(jLabel1)
.addGap(13, 13, 13)
.addGroup(layout.createParallelGroup(javax.swing.GroupLayout.Alignment.BASELINE)
.addComponent(jLabel3)
.addComponent(TxtMemberID, javax.swing.GroupLayout.PREFERRED_SIZE, 23,
javax.swing.GroupLayout.PREFERRED_SIZE)
.addComponent(CmdMemberDetails))
.addPreferredGap(javax.swing.LayoutStyle.ComponentPlacement.RELATED)
.addGroup(layout.createParallelGroup(javax.swing.GroupLayout.Alignment.LEADING)
.addComponent(jLabel2)
.addComponent(TxtMemberName, javax.swing.GroupLayout.PREFERRED_SIZE, 23,
javax.swing.GroupLayout.PREFERRED_SIZE))
.addPreferredGap(javax.swing.LayoutStyle.ComponentPlacement.RELATED)
.addComponent(TxtMemberStatus, javax.swing.GroupLayout.PREFERRED_SIZE, 23,
javax.swing.GroupLayout.PREFERRED_SIZE)
.addGap(18, 18, 18)
.addGroup(layout.createParallelGroup(javax.swing.GroupLayout.Alignment.BASELINE)
.addComponent(jLabel4)
.addComponent(TxtBookID, javax.swing.GroupLayout.PREFERRED_SIZE, 23,
javax.swing.GroupLayout.PREFERRED_SIZE)
.addComponent(CmdBookDetails))
.addPreferredGap(javax.swing.LayoutStyle.ComponentPlacement.RELATED)
.addGroup(layout.createParallelGroup(javax.swing.GroupLayout.Alignment.BASELINE)
.addComponent(TxtBookTitle, javax.swing.GroupLayout.PREFERRED_SIZE, 23,
javax.swing.GroupLayout.PREFERRED_SIZE)
.addComponent(jLabel5))
.addPreferredGap(javax.swing.LayoutStyle.ComponentPlacement.RELATED)
.addComponent(TxtBookStatus, javax.swing.GroupLayout.PREFERRED_SIZE, 23,
javax.swing.GroupLayout.PREFERRED_SIZE)
.addPreferredGap(javax.swing.LayoutStyle.ComponentPlacement.RELATED, 38,
Short.MAX_VALUE)
.addGroup(layout.createParallelGroup(javax.swing.GroupLayout.Alignment.BASELINE)
.addComponent(CmdReset)
.addComponent(CmdIssueBook)
.addComponent(CmdClose))
.addGap(29, 29, 29))
);
pack();
}// </editor-fold>
private void CmdResetActionPerformed(java.awt.event.ActionEvent evt) {
ClearText();
}
private void CmdIssueBookActionPerformed(java.awt.event.ActionEvent evt) {
//issue book
try
{
if (TxtMemberID.getText().trim().length()==0 ) {
JOptionPane.showMessageDialog(null,"Enter member id.");
return;
}
```

```
if (TxtBookID.getText().trim().length()==0 ) {
JOptionPane.showMessageDialog(null,"Enter member id.");
return;
}
MainClass mc=new MainClass();
Connection connection;
connection=DriverManager.getConnection(mc.StrUrl,mc.StrUid,mc.StrPwd);
String sql = "insert into lib_transaction (trn_mem_id,trn_book_id,trn_issue_dt) values (" +
TxtMemberID.getText() + "," + TxtBookID.getText() + ",sysdate())";
PreparedStatement pst=connection.prepareStatement(sql);
pst.executeUpdate(sql);
sql = "update  lib_book_master set book_status='Issued' where book_id=" +
TxtBookID.getText();
pst=connection.prepareStatement(sql);
pst.executeUpdate(sql);
JOptionPane.showMessageDialog(null,"Book Issued");
ClearText();
}
catch (Exception e) {
System.err.println(e);
System.exit(1);
}
}
private void CmdCloseActionPerformed(java.awt.event.ActionEvent evt) {
this.setVisible(false);
}
private void formWindowOpened(java.awt.event.WindowEvent evt) {
ClearText();
this.setLocationRelativeTo(null);
}
private void CmdMemberDetailsActionPerformed(java.awt.event.ActionEvent evt) {
//get member deatils
try
{
//get database connection details
MainClass mc=new MainClass();
//open connection
Connection connection;
connection=DriverManager.getConnection(mc.StrUrl,mc.StrUid,mc.StrPwd);
String str="";
str="select * from lib_member_master where mem_id =? ";
PreparedStatement pst=connection.prepareStatement(str);
pst.setString(1, TxtMemberID.getText());
ResultSet rs;
rs=pst.executeQuery();
if (rs.next())
{
TxtMemberName.setText(rs.getString("mem_name"));
//get mem status
str="select * from lib_transaction where trn_mem_id =? and (trn_receive_dt is null  or
trn_receive_dt='')";
PreparedStatement pst1=connection.prepareStatement(str);
pst1.setString(1, TxtMemberID.getText());
ResultSet rs1;
rs1=pst1.executeQuery();
```

```
if (rs1.next())
{
TxtMemberStatus.setText("Book already issued.");
CmdIssueBook.setEnabled(false);
CmdBookDetails.setEnabled(false);
}
else
{
TxtMemberStatus.setText("No book issued.");
CmdIssueBook.setEnabled(true);
CmdBookDetails.setEnabled(true);
}
}
else
{
JOptionPane.showMessageDialog(null,"Invalid member id.");
TxtMemberID.setText("");
TxtMemberName.setText("");
TxtMemberStatus.setText("");
CmdIssueBook.setEnabled(false);
return;
}
}
catch (Exception e)
{
System.err.println(e);
System.exit(1);
}
}
private void CmdBookDetailsActionPerformed(java.awt.event.ActionEvent evt) {
//get member deatils
try
{
//get database connection details
MainClass mc=new MainClass();
//open connection
Connection connection;
connection=DriverManager.getConnection(mc.StrUrl,mc.StrUid,mc.StrPwd);
String str="";
str="select * from lib_book_master where book_id =? ";
PreparedStatement pst=connection.prepareStatement(str);
pst.setString(1, TxtBookID.getText());
ResultSet rs;
rs=pst.executeQuery();
if (rs.next())
{
TxtBookTitle.setText(rs.getString("book_title"));
//get mem status
str="select * from lib_transaction where trn_book_id =? and (trn_receive_dt is null  or
trn_receive_dt='')";
PreparedStatement pst1=connection.prepareStatement(str);
pst1.setString(1, TxtBookID.getText());
ResultSet rs1;
rs1=pst1.executeQuery();
if (rs1.next())
```

```
{
TxtBookStatus.setText("Book not available.");
CmdIssueBook.setEnabled(false);
}
else
{
TxtBookStatus.setText("Book available.");
CmdIssueBook.setEnabled(true);
}
}
else
{
JOptionPane.showMessageDialog(null,"Invalid book id.");
TxtBookID.setText("");
TxtBookTitle.setText("");
TxtBookStatus.setText("");
CmdIssueBook.setEnabled(false);
return;
}
}
catch (Exception e)
{
System.err.println(e);
System.exit(1);
}
}
private void ClearText()
{
TxtMemberID.setText("");
TxtMemberName.setText("");
TxtMemberStatus.setText("");
TxtBookID.setText("");
TxtBookTitle.setText("");
TxtBookStatus.setText("");
CmdIssueBook.setEnabled(false);
CmdBookDetails.setEnabled(false);
}
/**
* @param args the command line arguments
*/
public static void main(String args[]) {
java.awt.EventQueue.invokeLater(new Runnable() {
public void run() {
new IssueBook().setVisible(true);
}
});
}
// Variables declaration
private javax.swing.JButton CmdBookDetails;
private javax.swing.JButton CmdClose;
private javax.swing.JButton CmdIssueBook;
private javax.swing.JButton CmdMemberDetails;
private javax.swing.JButton CmdReset;
private javax.swing.JTextField TxtBookID;
private javax.swing.JTextField TxtBookStatus;
```

```
private javax.swing.JTextField TxtBookTitle;
private javax.swing.JTextField TxtMemberID;
private javax.swing.JTextField TxtMemberName;
private javax.swing.JTextField TxtMemberStatus;
private javax.swing.JLabel jLabel1;
private javax.swing.JLabel jLabel2;
private javax.swing.JLabel jLabel3;
private javax.swing.JLabel jLabel4;
private javax.swing.JLabel jLabel5;
// End of variables declaration
}
```

6. Login.java

Purpose: For login user name and password is required to enter LMS.

```
import java.sql.Connection;
import java.sql.DriverManager;
import java.sql.PreparedStatement;
import java.sql.ResultSet;
import javax.swing.JOptionPane;
public class Login extends javax.swing.JFrame {
/** Creates new form Login */
public Login() {
initComponents();
}
// <editor-fold defaultstate="collapsed" desc="Generated Code">
private void initComponents() {

jPanel1 = new javax.swing.JPanel();
jLabel1 = new javax.swing.JLabel();
jLabel2 = new javax.swing.JLabel();
jLabel3 = new javax.swing.JLabel();
TxtUserName = new javax.swing.JTextField();
CmdOk = new javax.swing.JButton();
CmdClose = new javax.swing.JButton();
TxtPassword = new javax.swing.JPasswordField();
setDefaultCloseOperation(javax.swing.WindowConstants.EXIT_ON_CLOSE);
setTitle("P2P - Library System");
setResizable(false);
addWindowListener(new java.awt.event.WindowAdapter() {
public void windowOpened(java.awt.event.WindowEvent evt) {
formWindowOpened(evt);
}
});
jLabel1.setFont(new java.awt.Font("Tahoma", 1, 12));
jLabel1.setHorizontalAlignment(javax.swing.SwingConstants.CENTER);
jLabel1.setText("Library Managemnet System");
jLabel2.setText("User Name:");
jLabel3.setText("Password:");
CmdOk.setText("Ok");
CmdOk.addActionListener(new java.awt.event.ActionListener() {
public void actionPerformed(java.awt.event.ActionEvent evt) {
CmdOkActionPerformed(evt);
}
```

```
});
CmdClose.setText("Close");
CmdClose.addActionListener(new java.awt.event.ActionListener() {
public void actionPerformed(java.awt.event.ActionEvent evt) {
CmdCloseActionPerformed(evt);
}
});
javax.swing.GroupLayout jPanel1Layout = new javax.swing.GroupLayout(jPanel1);
jPanel1.setLayout(jPanel1Layout);
jPanel1Layout.setHorizontalGroup(
jPanel1Layout.createParallelGroup(javax.swing.GroupLayout.Alignment.LEADING)
.addGroup(jPanel1Layout.createSequentialGroup()
.addGroup(jPanel1Layout.createParallelGroup(javax.swing.GroupLayout.Alignment.LEADING
)
.addGroup(jPanel1Layout.createSequentialGroup()
.addContainerGap()
.addComponent(jLabel1, javax.swing.GroupLayout.PREFERRED_SIZE, 348,
javax.swing.GroupLayout.PREFERRED_SIZE))
.addGroup(jPanel1Layout.createSequentialGroup()
.addGap(85, 85, 85)
.addGroup(jPanel1Layout.createParallelGroup(javax.swing.GroupLayout.Alignment.LEADING
)
.addComponent(jLabel3)
.addComponent(jLabel2))
.addGap(26, 26, 26)
.addGroup(jPanel1Layout.createParallelGroup(javax.swing.GroupLayout.Alignment.LEADING
, false)
.addComponent(TxtPassword, javax.swing.GroupLayout.Alignment.TRAILING)
.addComponent(TxtUserName, javax.swing.GroupLayout.Alignment.TRAILING,
javax.swing.GroupLayout.PREFERRED_SIZE, 130,
javax.swing.GroupLayout.PREFERRED_SIZE)
.addGroup(javax.swing.GroupLayout.Alignment.TRAILING,
jPanel1Layout.createSequentialGroup()
.addComponent(CmdOk)
.addPreferredGap(javax.swing.LayoutStyle.ComponentPlacement.RELATED)
.addComponent(CmdClose, javax.swing.GroupLayout.PREFERRED_SIZE, 76,
javax.swing.GroupLayout.PREFERRED_SIZE)))))
.addContainerGap(javax.swing.GroupLayout.DEFAULT_SIZE, Short.MAX_VALUE))
);
jPanel1Layout.setVerticalGroup(
jPanel1Layout.createParallelGroup(javax.swing.GroupLayout.Alignment.LEADING)
.addGroup(jPanel1Layout.createSequentialGroup()
.addComponent(jLabel1)
.addGap(26, 26, 26)
.addGroup(jPanel1Layout.createParallelGroup(javax.swing.GroupLayout.Alignment.TRAILIN
G)
.addComponent(jLabel3)
.addGroup(jPanel1Layout.createSequentialGroup()
.addGroup(jPanel1Layout.createParallelGroup(javax.swing.GroupLayout.Alignment.BASELIN
E)
.addComponent(TxtUserName, javax.swing.GroupLayout.PREFERRED_SIZE,
javax.swing.GroupLayout.DEFAULT_SIZE, javax.swing.GroupLayout.PREFERRED_SIZE)
.addComponent(jLabel2))
.addPreferredGap(javax.swing.LayoutStyle.ComponentPlacement.RELATED)
.addComponent(TxtPassword, javax.swing.GroupLayout.PREFERRED_SIZE,
```

```
javax.swing.GroupLayout.DEFAULT_SIZE, javax.swing.GroupLayout.PREFERRED_SIZE)))
.addGap(18, 18, 18)
.addGroup(jPanel1Layout.createParallelGroup(javax.swing.GroupLayout.Alignment.BASELIN
E)
.addComponent(CmdOk)
.addComponent(CmdClose))
.addContainerGap(31, Short.MAX_VALUE))
);
javax.swing.GroupLayout layout = new javax.swing.GroupLayout(getContentPane());
getContentPane().setLayout(layout);
layout.setHorizontalGroup(
layout.createParallelGroup(javax.swing.GroupLayout.Alignment.LEADING)
.addGroup(layout.createSequentialGroup()
.addGap(18, 18, 18)
.addComponent(jPanel1, javax.swing.GroupLayout.PREFERRED_SIZE,
javax.swing.GroupLayout.DEFAULT_SIZE, javax.swing.GroupLayout.PREFERRED_SIZE)
.addContainerGap(25, Short.MAX_VALUE))
);
layout.setVerticalGroup(
layout.createParallelGroup(javax.swing.GroupLayout.Alignment.LEADING)
.addGroup(layout.createSequentialGroup()
.addContainerGap()
.addComponent(jPanel1, javax.swing.GroupLayout.PREFERRED_SIZE,
javax.swing.GroupLayout.DEFAULT_SIZE, javax.swing.GroupLayout.PREFERRED_SIZE)
.addContainerGap(javax.swing.GroupLayout.DEFAULT_SIZE, Short.MAX_VALUE))
);
pack();
}// </editor-fold>
private void CmdCloseActionPerformed(java.awt.event.ActionEvent evt) {
// Close project
System.exit(1);
}
private void CmdOkActionPerformed(java.awt.event.ActionEvent evt) {
//validation
String stru="";
stru=TxtUserName.getText();
String strp="";
strp=TxtPassword.getText();
if (stru.isEmpty()==true)
{
JOptionPane.showMessageDialog(null,"Enter User Name");
return;
}
if (strp.isEmpty()==true)
{
JOptionPane.showMessageDialog(null,"Enter Password");
return;
}
try
{
//get database connection details
MainClass mc=new MainClass();
//open connection
Connection connection;
connection=DriverManager.getConnection(mc.StrUrl,mc.StrUid,mc.StrPwd);
```

```
String str="";
str="select * from lib_user where user_name =? and user_password =?";
PreparedStatement pst=connection.prepareStatement(str);
pst.setString(1, stru);
pst.setString(2, strp);
ResultSet rs;
rs=pst.executeQuery();
if (rs.next())
{
MainClass.StrUser=TxtUserName.getText();
MainForm m=new MainForm();
m.setVisible(true);
this.setVisible(false);
}
else
{
JOptionPane.showMessageDialog(null,"User name or password are not correct.");
return;
}
}
catch (Exception e)
{
System.err.println(e);
System.exit(1);
}
}
private void formWindowOpened(java.awt.event.WindowEvent evt) {
// TODO add your handling code here:
this.setLocationRelativeTo(null);
}
/**
* @param args the command line arguments
*/
public static void main(String args[]) {
java.awt.EventQueue.invokeLater(new Runnable() {
public void run() {
new Login().setVisible(true);
}
});
}
// Variables declaration
private javax.swing.JButton CmdClose;
private javax.swing.JButton CmdOk;
private javax.swing.JPasswordField TxtPassword;
private javax.swing.JTextField TxtUserName;
private javax.swing.JLabel jLabel1;
private javax.swing.JLabel jLabel2;
private javax.swing.JLabel jLabel3;
private javax.swing.JPanel jPanel1;
// End of variables declaration
}
```

7. MainClass.java

Purpose: This is main Java class used to connect mysql database.

```
public class MainClass
{
//change your mysql database connection here
public String StrUrl="jdbc:mysql://localhost:3306/p2p_library";
public String StrUid="root";
public String StrPwd= "root";

public static String StrUser;
}
```

8. MainForm.java

Purpose: This code creates main form for LMS where file, admin, book, member, transaction, and help menu available. In file there is exit option. In admin, user master and change password option available. In book menu there are book master, search book, and book status available. Member menu provide member master and search member of LMS. Transaction menu used to issue book, receive book, and generate transaction report of LMS. Help menu provide the information about LMS.

```
public class MainForm extends javax.swing.JFrame {
/** Creates new form Main */
public MainForm() {
initComponents();
}
// <editor-fold defaultstate="collapsed" desc="Generated Code">
private void initComponents() {
jPanel2 = new javax.swing.JPanel();
jLabel1 = new javax.swing.JLabel();
jLabel2 = new javax.swing.JLabel();
jMenuBar1 = new javax.swing.JMenuBar();
jMenu1 = new javax.swing.JMenu();
MnuExits = new javax.swing.JMenuItem();
jMenu6 = new javax.swing.JMenu();
jMenuItem7 = new javax.swing.JMenuItem();
jMenuItem8 = new javax.swing.JMenuItem();
jMenu2 = new javax.swing.JMenu();
MnuBookMaster = new javax.swing.JMenuItem();
MnuSearchBook = new javax.swing.JMenuItem();
MnuBookStatus = new javax.swing.JMenuItem();
jMenu3 = new javax.swing.JMenu();
jMenuItem1 = new javax.swing.JMenuItem();
jMenuItem2 = new javax.swing.JMenuItem();
jMenu4 = new javax.swing.JMenu();
jMenuItem3 = new javax.swing.JMenuItem();
jMenuItem4 = new javax.swing.JMenuItem();
jMenuItem6 = new javax.swing.JMenuItem();
jMenu5 = new javax.swing.JMenu();
jMenuItem5 = new javax.swing.JMenuItem();
javax.swing.GroupLayout jPanel2Layout = new javax.swing.GroupLayout(jPanel2);
jPanel2.setLayout(jPanel2Layout);
jPanel2Layout.setHorizontalGroup(
jPanel2Layout.createParallelGroup(javax.swing.GroupLayout.Alignment.LEADING)
.addGap(0, 100, Short.MAX_VALUE)
);
```

```
jPanel2Layout.setVerticalGroup(
jPanel2Layout.createParallelGroup(javax.swing.GroupLayout.Alignment.LEADING)
.addGap(0, 100, Short.MAX_VALUE)
);
setDefaultCloseOperation(javax.swing.WindowConstants.EXIT_ON_CLOSE);
setTitle("P2P - Java Library System");
addWindowListener(new java.awt.event.WindowAdapter() {
public void windowOpened(java.awt.event.WindowEvent evt) {
formWindowOpened(evt);
}
});
jLabel1.setFont(new java.awt.Font("Tahoma", 1, 18));
jLabel1.setForeground(new java.awt.Color(51, 51, 255));
jLabel1.setHorizontalAlignment(javax.swing.SwingConstants.CENTER);
jLabel1.setText("Welcome to Library Management System");
jLabel2.setIcon(new javax.swing.ImageIcon(getClass().getResource("/download (3).jpg")));
jMenu1.setText("File");
MnuExits.setText("Exit");
MnuExits.addActionListener(new java.awt.event.ActionListener() {
public void actionPerformed(java.awt.event.ActionEvent evt) {
MnuExitsActionPerformed(evt);
}
});
jMenu1.add(MnuExits);
jMenuBar1.add(jMenu1);
jMenu6.setText("Admin");
jMenuItem7.setText("User Master");
jMenuItem7.addActionListener(new java.awt.event.ActionListener() {
public void actionPerformed(java.awt.event.ActionEvent evt) {
jMenuItem7ActionPerformed(evt);
}
});
jMenu6.add(jMenuItem7);
jMenuItem8.setText("Change Password");
jMenuItem8.addActionListener(new java.awt.event.ActionListener() {
public void actionPerformed(java.awt.event.ActionEvent evt) {
jMenuItem8ActionPerformed(evt);
}
});
jMenu6.add(jMenuItem8);
jMenuBar1.add(jMenu6);
jMenu2.setText("Book");
MnuBookMaster.setText("Book Master");
MnuBookMaster.addActionListener(new java.awt.event.ActionListener() {
public void actionPerformed(java.awt.event.ActionEvent evt) {
MnuBookMasterActionPerformed(evt);
}
});
jMenu2.add(MnuBookMaster);
MnuSearchBook.setText("Search Book");
MnuSearchBook.addActionListener(new java.awt.event.ActionListener() {
public void actionPerformed(java.awt.event.ActionEvent evt) {
MnuSearchBookActionPerformed(evt);
}
});
```

```
jMenu2.add(MnuSearchBook);
MnuBookStatus.setText("Book Status");
MnuBookStatus.addActionListener(new java.awt.event.ActionListener() {
public void actionPerformed(java.awt.event.ActionEvent evt) {
MnuBookStatusActionPerformed(evt);
}
});
jMenu2.add(MnuBookStatus);
jMenuBar1.add(jMenu2);
jMenu3.setText("Member");
jMenuItem1.setText("Member Master");
jMenuItem1.addActionListener(new java.awt.event.ActionListener() {
public void actionPerformed(java.awt.event.ActionEvent evt) {
jMenuItem1ActionPerformed(evt);
}
});
jMenu3.add(jMenuItem1);
jMenuItem2.setText("Search Member");
jMenuItem2.addActionListener(new java.awt.event.ActionListener() {
public void actionPerformed(java.awt.event.ActionEvent evt) {
jMenuItem2ActionPerformed(evt);
}
});
jMenu3.add(jMenuItem2);
jMenuBar1.add(jMenu3);
jMenu4.setText("Transaction");
jMenu4.addActionListener(new java.awt.event.ActionListener() {
public void actionPerformed(java.awt.event.ActionEvent evt) {
jMenu4ActionPerformed(evt);
}
});
jMenuItem3.setText("Issue Book");
jMenuItem3.addActionListener(new java.awt.event.ActionListener() {
public void actionPerformed(java.awt.event.ActionEvent evt) {
jMenuItem3ActionPerformed(evt);
}
});
jMenu4.add(jMenuItem3);
jMenuItem4.setText("Receive Book");
jMenuItem4.addActionListener(new java.awt.event.ActionListener() {
public void actionPerformed(java.awt.event.ActionEvent evt) {
jMenuItem4ActionPerformed(evt);
}
});
jMenu4.add(jMenuItem4);
jMenuItem6.setText("Transaction Report ");
jMenuItem6.addActionListener(new java.awt.event.ActionListener() {
public void actionPerformed(java.awt.event.ActionEvent evt) {
jMenuItem6ActionPerformed(evt);
}
});
jMenu4.add(jMenuItem6);
jMenuBar1.add(jMenu4);
jMenu5.setText("Help");
jMenuItem5.setText("About");
```

```
jMenuItem5.addActionListener(new java.awt.event.ActionListener() {
public void actionPerformed(java.awt.event.ActionEvent evt) {
jMenuItem5ActionPerformed(evt);
}
});
jMenu5.add(jMenuItem5);
jMenuBar1.add(jMenu5);
setJMenuBar(jMenuBar1);
javax.swing.GroupLayout layout = new javax.swing.GroupLayout(getContentPane());
getContentPane().setLayout(layout);
layout.setHorizontalGroup(
layout.createParallelGroup(javax.swing.GroupLayout.Alignment.LEADING)
.addGroup(layout.createSequentialGroup()
.addGap(194, 194, 194)
.addComponent(jLabel2, javax.swing.GroupLayout.DEFAULT_SIZE, 486,
Short.MAX_VALUE)
.addContainerGap())
.addGroup(layout.createSequentialGroup()
.addGap(151, 151, 151)
.addComponent(jLabel1, javax.swing.GroupLayout.PREFERRED_SIZE, 386,
javax.swing.GroupLayout.PREFERRED_SIZE)
.addContainerGap(javax.swing.GroupLayout.DEFAULT_SIZE, Short.MAX_VALUE))
);
layout.setVerticalGroup(
layout.createParallelGroup(javax.swing.GroupLayout.Alignment.LEADING)
.addGroup(layout.createSequentialGroup()
.addGap(56, 56, 56)
.addComponent(jLabel1, javax.swing.GroupLayout.PREFERRED_SIZE, 68,
javax.swing.GroupLayout.PREFERRED_SIZE)
.addPreferredGap(javax.swing.LayoutStyle.ComponentPlacement.RELATED)
.addComponent(jLabel2, javax.swing.GroupLayout.PREFERRED_SIZE, 227,
javax.swing.GroupLayout.PREFERRED_SIZE)
.addContainerGap(124, Short.MAX_VALUE))
);
pack();
}// </editor-fold>
private void MnuExitsActionPerformed(java.awt.event.ActionEvent evt) {
System.exit(1);
}
private void MnuBookMasterActionPerformed(java.awt.event.ActionEvent evt) {
BookMaster b=new BookMaster();
b.setVisible(true);
}
private void formWindowOpened(java.awt.event.WindowEvent evt) {
this.setLocationRelativeTo(null);
}
private void MnuSearchBookActionPerformed(java.awt.event.ActionEvent evt) {
SearchBook b=new SearchBook();
b.setVisible(true);
}
private void MnuBookStatusActionPerformed(java.awt.event.ActionEvent evt) {
BookStatus b=new BookStatus();
b.setVisible(true);
}
private void jMenuItem1ActionPerformed(java.awt.event.ActionEvent evt) {
```

```
MemberMaster b=new MemberMaster();
b.setVisible(true);
}
private void jMenuItem2ActionPerformed(java.awt.event.ActionEvent evt) {
SearchMember b=new SearchMember();
b.setVisible(true);
}
private void jMenuItem7ActionPerformed(java.awt.event.ActionEvent evt) {
UserMaster b=new UserMaster();
b.setVisible(true);
}
private void jMenuItem8ActionPerformed(java.awt.event.ActionEvent evt) {
ChangePassword b=new ChangePassword();
b.setVisible(true);
}
private void jMenuItem5ActionPerformed(java.awt.event.ActionEvent evt) {
About b=new About();
b.setVisible(true);
}
private void jMenuItem3ActionPerformed(java.awt.event.ActionEvent evt) {
IssueBook b=new IssueBook();
b.setVisible(true);
}
private void jMenuItem4ActionPerformed(java.awt.event.ActionEvent evt) {
ReceiveBook b=new ReceiveBook();
b.setVisible(true);
}
private void jMenu4ActionPerformed(java.awt.event.ActionEvent evt) {
}
private void jMenuItem6ActionPerformed(java.awt.event.ActionEvent evt) {
TransactionReport b=new TransactionReport();
b.setVisible(true);
}
/**
* @param args the command line arguments
*/
public static void main(String args[]) {
java.awt.EventQueue.invokeLater(new Runnable() {
public void run() {
new MainForm().setVisible(true);
}
});
}
// Variables declaration
private javax.swing.JMenuItem MnuBookMaster;
private javax.swing.JMenuItem MnuBookStatus;
private javax.swing.JMenuItem MnuExits;
private javax.swing.JMenuItem MnuSearchBook;
private javax.swing.JLabel jLabel1;
private javax.swing.JLabel jLabel2;
private javax.swing.JMenu jMenu1;
private javax.swing.JMenu jMenu2;
private javax.swing.JMenu jMenu3;
private javax.swing.JMenu jMenu4;
private javax.swing.JMenu jMenu5;
```

```
private javax.swing.JMenu jMenu6;
private javax.swing.JMenuBar jMenuBar1;
private javax.swing.JMenuItem jMenuItem1;
private javax.swing.JMenuItem jMenuItem2;
private javax.swing.JMenuItem jMenuItem3;
private javax.swing.JMenuItem jMenuItem4;
private javax.swing.JMenuItem jMenuItem5;
private javax.swing.JMenuItem jMenuItem6;
private javax.swing.JMenuItem jMenuItem7;
private javax.swing.JMenuItem jMenuItem8;
private javax.swing.JPanel jPanel2;
// End of variables declaration
}
```

9. MemberMaster.java

Purpose: Member master used to add and update the detail of members and manage the member information using first, previous, next, last, new, edit, save, and close command. Basically there are member id, name, address, email id, and mobile no. to manage member information. Check box is used to provide information about active or non-active member.

```
import java.sql.Statement;
import java.sql.Connection;
import java.sql.DriverManager;
import java.sql.PreparedStatement;
import java.sql.ResultSet;
import javax.swing.JOptionPane;
public class MemberMaster extends javax.swing.JFrame {
MainClass mc=new MainClass();
ResultSet rsmem;
int rspos=0;
int rscount=0;
boolean chadd=false;
/** Creates new form MemberMaster */
public MemberMaster() {
initComponents();
}
// <editor-fold defaultstate="collapsed" desc="Generated Code">
private void initComponents() {
jLabel1 = new javax.swing.JLabel();
jLabel2 = new javax.swing.JLabel();
TxtMemID = new javax.swing.JTextField();
jLabel3 = new javax.swing.JLabel();
TxtName = new javax.swing.JTextField();
jLabel4 = new javax.swing.JLabel();
TxtAddress = new javax.swing.JTextField();
jLabel5 = new javax.swing.JLabel();
TxtEmailID = new javax.swing.JTextField();
jLabel6 = new javax.swing.JLabel();
TxtMobileNo = new javax.swing.JTextField();
LblPosition = new javax.swing.JLabel();
CmdFirst = new javax.swing.JButton();
CmdPrevious = new javax.swing.JButton();
CmdNext = new javax.swing.JButton();
CmdLast = new javax.swing.JButton();
```

```
CmdNew = new javax.swing.JButton();
CmdEdit = new javax.swing.JButton();
CmdSave = new javax.swing.JButton();
CmdClose = new javax.swing.JButton();
ChkMemActive = new javax.swing.JCheckBox();
setDefaultCloseOperation(javax.swing.WindowConstants.DISPOSE_ON_CLOSE);
setTitle("P2P Library : Member Master");
addWindowListener(new java.awt.event.WindowAdapter() {
public void windowOpened(java.awt.event.WindowEvent evt) {
formWindowOpened(evt);
}
});
jLabel1.setFont(new java.awt.Font("Tahoma", 1, 12));
jLabel1.setForeground(new java.awt.Color(51, 0, 255));
jLabel1.setText("Library Management System - Member Master");
jLabel2.setText("Member ID:");
TxtMemID.setEditable(false);
TxtMemID.setText("jTextField1");
jLabel3.setText("Name:");
TxtName.setText("jTextField1");
jLabel4.setText("Address:");
TxtAddress.setText("jTextField1");
jLabel5.setText("Email ID:");
TxtEmailID.setText("jTextField1");
jLabel6.setText("Mobile No:");
TxtMobileNo.setText("jTextField1");
LblPosition.setText("Book ID:");
CmdFirst.setText("First");
CmdFirst.addActionListener(new java.awt.event.ActionListener() {
public void actionPerformed(java.awt.event.ActionEvent evt) {
CmdFirstActionPerformed(evt);
}
});
CmdPrevious.setText("Previous");
CmdPrevious.addActionListener(new java.awt.event.ActionListener() {
public void actionPerformed(java.awt.event.ActionEvent evt) {
CmdPreviousActionPerformed(evt);
}
});
CmdNext.setText("Next");
CmdNext.addActionListener(new java.awt.event.ActionListener() {
public void actionPerformed(java.awt.event.ActionEvent evt) {
CmdNextActionPerformed(evt);
}
});
CmdLast.setText("Last");
CmdLast.addActionListener(new java.awt.event.ActionListener() {
public void actionPerformed(java.awt.event.ActionEvent evt) {
CmdLastActionPerformed(evt);
}
});
CmdNew.setText("New");
CmdNew.addActionListener(new java.awt.event.ActionListener() {
public void actionPerformed(java.awt.event.ActionEvent evt) {
CmdNewActionPerformed(evt);
```

```
}
});
CmdEdit.setText("Edit");
CmdEdit.addActionListener(new java.awt.event.ActionListener() {
public void actionPerformed(java.awt.event.ActionEvent evt) {
CmdEditActionPerformed(evt);
}
});
CmdSave.setText("Save");
CmdSave.addActionListener(new java.awt.event.ActionListener() {
public void actionPerformed(java.awt.event.ActionEvent evt) {
CmdSaveActionPerformed(evt);
}
});
CmdClose.setText("Close");
CmdClose.addActionListener(new java.awt.event.ActionListener() {
public void actionPerformed(java.awt.event.ActionEvent evt) {
CmdCloseActionPerformed(evt);
}
});
ChkMemActive.setText("Member Active");
javax.swing.GroupLayout layout = new javax.swing.GroupLayout(getContentPane());
getContentPane().setLayout(layout);
layout.setHorizontalGroup(
layout.createParallelGroup(javax.swing.GroupLayout.Alignment.LEADING)
.addGroup(layout.createSequentialGroup()
.addGap(18, 18, 18)
.addGroup(layout.createParallelGroup(javax.swing.GroupLayout.Alignment.LEADING)
.addGroup(layout.createSequentialGroup()
.addGroup(layout.createParallelGroup(javax.swing.GroupLayout.Alignment.LEADING, false)
.addGroup(layout.createSequentialGroup()
.addGap(77, 77, 77)
.addGroup(layout.createParallelGroup(javax.swing.GroupLayout.Alignment.LEADING)
.addComponent(jLabel2)
.addComponent(jLabel6)
.addComponent(jLabel5)
.addComponent(jLabel3)
.addComponent(jLabel4))
.addGap(27, 27, 27)
.addGroup(layout.createParallelGroup(javax.swing.GroupLayout.Alignment.LEADING)
.addComponent(TxtMobileNo)
.addComponent(TxtEmailID)
.addComponent(TxtAddress)
.addComponent(TxtName)
.addComponent(TxtMemID, javax.swing.GroupLayout.PREFERRED_SIZE, 198,
javax.swing.GroupLayout.PREFERRED_SIZE)
.addComponent(ChkMemActive)))
.addGroup(layout.createSequentialGroup()
.addGap(15, 15, 15)
.addComponent(CmdFirst)
.addPreferredGap(javax.swing.LayoutStyle.ComponentPlacement.RELATED)
.addComponent(CmdPrevious)
.addPreferredGap(javax.swing.LayoutStyle.ComponentPlacement.RELATED)
.addComponent(CmdNext)
.addPreferredGap(javax.swing.LayoutStyle.ComponentPlacement.RELATED)
```

```
.addComponent(CmdLast)
.addPreferredGap(javax.swing.LayoutStyle.ComponentPlacement.RELATED,
javax.swing.GroupLayout.DEFAULT_SIZE, Short.MAX_VALUE)
.addComponent(CmdNew)))
.addPreferredGap(javax.swing.LayoutStyle.ComponentPlacement.RELATED)
.addComponent(CmdEdit)
.addGap(6, 6, 6)
.addComponent(CmdSave)
.addPreferredGap(javax.swing.LayoutStyle.ComponentPlacement.RELATED)
.addComponent(CmdClose))
.addComponent(jLabel1)
.addGroup(layout.createSequentialGroup()
.addGap(133, 133, 133)
.addComponent(LblPosition)))
.addContainerGap(18, Short.MAX_VALUE))
);
layout.setVerticalGroup(
layout.createParallelGroup(javax.swing.GroupLayout.Alignment.LEADING)
.addGroup(layout.createSequentialGroup()
.addContainerGap()
.addGroup(layout.createParallelGroup(javax.swing.GroupLayout.Alignment.LEADING)
.addGroup(layout.createSequentialGroup()
.addGap(36, 36, 36)
.addComponent(TxtMemID, javax.swing.GroupLayout.PREFERRED_SIZE, 23,
javax.swing.GroupLayout.PREFERRED_SIZE)
.addPreferredGap(javax.swing.LayoutStyle.ComponentPlacement.RELATED)
.addComponent(TxtName, javax.swing.GroupLayout.PREFERRED_SIZE, 23,
javax.swing.GroupLayout.PREFERRED_SIZE)
.addPreferredGap(javax.swing.LayoutStyle.ComponentPlacement.RELATED)
.addComponent(TxtAddress, javax.swing.GroupLayout.PREFERRED_SIZE, 23,
javax.swing.GroupLayout.PREFERRED_SIZE))
.addGroup(layout.createSequentialGroup()
.addComponent(jLabel1)
.addGap(26, 26, 26)
.addComponent(jLabel2)
.addGap(18, 18, 18)
.addComponent(jLabel3)
.addPreferredGap(javax.swing.LayoutStyle.ComponentPlacement.UNRELATED)
.addComponent(jLabel4)))
.addPreferredGap(javax.swing.LayoutStyle.ComponentPlacement.RELATED)
.addGroup(layout.createParallelGroup(javax.swing.GroupLayout.Alignment.BASELINE)
.addComponent(TxtEmailID, javax.swing.GroupLayout.PREFERRED_SIZE, 23,
javax.swing.GroupLayout.PREFERRED_SIZE)
.addComponent(jLabel5))
.addPreferredGap(javax.swing.LayoutStyle.ComponentPlacement.RELATED)
.addGroup(layout.createParallelGroup(javax.swing.GroupLayout.Alignment.BASELINE)
.addComponent(TxtMobileNo, javax.swing.GroupLayout.PREFERRED_SIZE, 23,
javax.swing.GroupLayout.PREFERRED_SIZE)
.addComponent(jLabel6))
.addGap(18, 18, 18)
.addComponent(ChkMemActive)
.addGap(98, 98, 98)
.addComponent(LblPosition)
.addPreferredGap(javax.swing.LayoutStyle.ComponentPlacement.RELATED)
.addGroup(layout.createParallelGroup(javax.swing.GroupLayout.Alignment.BASELINE)
```

```
.addComponent(CmdFirst)
.addComponent(CmdPrevious)
.addComponent(CmdNext)
.addComponent(CmdLast)
.addComponent(CmdEdit)
.addComponent(CmdSave)
.addComponent(CmdClose)
.addComponent(CmdNew))
.addContainerGap(javax.swing.GroupLayout.DEFAULT_SIZE, Short.MAX_VALUE))
);
pack();
}// </editor-fold>
private void Display()
{
try
{
//clear text of fields for entry/display old data
TxtMemID.setText(rsmem.getString("mem_id"));
TxtName.setText(rsmem.getString("mem_name"));
TxtAddress.setText(rsmem.getString("mem_address"));
TxtEmailID.setText(rsmem.getString("mem_email_id"));
TxtMobileNo.setText(rsmem.getString("mem_mobile_no"));
if (rsmem.getString("mem_active").equals("Y"))
{
ChkMemActive.setSelected(true);
}
else
{
ChkMemActive.setSelected(false);
}
LblPosition.setText("Record position : " + rsmem.getRow() + "/" + String.valueOf(rscount));
}
catch (Exception e)
{
System.err.println(e);
}
}
private void ClearText()
{
//clear text of fields for entry/display old data
TxtMemID.setText("");
TxtName.setText("");
TxtAddress.setText("");
TxtEmailID.setText("");
TxtMobileNo.setText("");
ChkMemActive.setSelected(false);
LblPosition.setText("");
}
private void CmdFirstActionPerformed(java.awt.event.ActionEvent evt) {
try {
rsmem.first();
Display();
} catch (Exception e) {
System.err.println(e);
}
```

```
}
private void CmdPreviousActionPerformed(java.awt.event.ActionEvent evt) {
try {
if (rsmem.isFirst()==true) {
JOptionPane.showMessageDialog(null,"Already in first record.");
return;
}
rsmem.previous();
Display();
} catch (Exception e) {
System.err.println(e);
}
}
private void CmdNextActionPerformed(java.awt.event.ActionEvent evt) {
try {
if (rsmem.isLast()==true) {
JOptionPane.showMessageDialog(null,"Already in last record.");
return;
}
rsmem.next();
Display();
} catch (Exception e) {
System.err.println(e);
}
}
private void CmdLastActionPerformed(java.awt.event.ActionEvent evt) {
try {
rsmem.last();
Display();
} catch (Exception e) {
System.err.println(e);
//System.exit(1);
}
}
private void CmdNewActionPerformed(java.awt.event.ActionEvent evt) {
chadd=true;
ClearText();
}
private void CmdEditActionPerformed(java.awt.event.ActionEvent evt) {
chadd=false;
}
private void CmdSaveActionPerformed(java.awt.event.ActionEvent evt) {
//validate entry and save record
try {
if (TxtName.getText().trim().length()==0 ) {
JOptionPane.showMessageDialog(null,"Enter member name.");
return;
}
if (TxtAddress.getText().trim().length()==0 ) {
JOptionPane.showMessageDialog(null,"Enter address.");
return;
}
if (TxtEmailID.getText().trim().length()==0) {
JOptionPane.showMessageDialog(null,"Enter email id.");
return;
```

```
}
if (TxtMobileNo.getText().trim().length()==0) {
JOptionPane.showMessageDialog(null,"Enter mobile no.");
return;
}
//String strqr="";
Connection connection;
connection=DriverManager.getConnection(mc.StrUrl,mc.StrUid,mc.StrPwd);
if (TxtMemID.getText().trim().length()==0 ) {
rsmem.moveToInsertRow();
rsmem.updateString("mem_name",TxtName.getText());
rsmem.updateString("mem_address",TxtAddress.getText());
rsmem.updateString("mem_email_id",TxtEmailID.getText());
rsmem.updateString("mem_mobile_no",TxtMobileNo.getText());
if (ChkMemActive.isSelected()==true)
{
rsmem.updateString("mem_active","Y");
}
else
{
rsmem.updateString("mem_active","N");
}
rsmem.insertRow();
rsmem.last();
Display();
} else {
rsmem.updateString("mem_name",TxtName.getText());
rsmem.updateString("mem_address",TxtAddress.getText());
rsmem.updateString("mem_email_id",TxtEmailID.getText());
rsmem.updateString("mem_mobile_no",TxtMobileNo.getText());
if (ChkMemActive.isSelected()==true)
{
rsmem.updateString("mem_active","Y");
}
else
{
rsmem.updateString("mem_active","N");
}
rsmem.updateRow();
rsmem.refreshRow();
}
JOptionPane.showMessageDialog(null,"Record updated.");
} catch (Exception e) {
System.err.println(e);
JOptionPane.showMessageDialog(null,"Unable to save record.");
return;
}
}
private void CmdCloseActionPerformed(java.awt.event.ActionEvent evt) {
// TODO add your handling code here:
this.setVisible(false);
}
private void formWindowOpened(java.awt.event.WindowEvent evt) {
try
{
```

```
this.setLocationRelativeTo(null);
ClearText();
Connection connection;
connection=DriverManager.getConnection(mc.StrUrl,mc.StrUid,mc.StrPwd);
ResultSet rs;
//get mem count
PreparedStatement stmt;
stmt=connection.prepareStatement("select count(*) from lib_member_master");
rs = stmt.executeQuery();
rs.first();
rscount=rs.getInt(1);
//get mem details
Statement stmt1 = connection.createStatement(ResultSet.TYPE_SCROLL_SENSITIVE,
ResultSet.CONCUR_UPDATABLE);
String sqlQuery = "select * from lib_member_master order by mem_id";
rsmem = stmt1.executeQuery(sqlQuery);
//display first record
rsmem.first();
Display();
}
catch (Exception e)
{
System.err.println(e);
//System.exit(1);
}
}
/**
* @param args the command line arguments
*/
public static void main(String args[]) {
java.awt.EventQueue.invokeLater(new Runnable() {
public void run() {
new MemberMaster().setVisible(true);
}
});
}
// Variables declaration
private javax.swing.JCheckBox ChkMemActive;
private javax.swing.JButton CmdClose;
private javax.swing.JButton CmdEdit;
private javax.swing.JButton CmdFirst;
private javax.swing.JButton CmdLast;
private javax.swing.JButton CmdNew;
private javax.swing.JButton CmdNext;
private javax.swing.JButton CmdPrevious;
private javax.swing.JButton CmdSave;
private javax.swing.JLabel LblPosition;
private javax.swing.JTextField TxtAddress;
private javax.swing.JTextField TxtEmailID;
private javax.swing.JTextField TxtMemID;
private javax.swing.JTextField TxtMobileNo;
private javax.swing.JTextField TxtName;
private javax.swing.JLabel jLabel1;
private javax.swing.JLabel jLabel2;
private javax.swing.JLabel jLabel3;
```

```
private javax.swing.JLabel jLabel4;
private javax.swing.JLabel jLabel5;
private javax.swing.JLabel jLabel6;
// End of variables declaration
}
```

10. ReceiveBook.java

Purpose: Receive book used to receive book from member using member id and receive command button.

```
import java.sql.Connection;
import java.sql.DriverManager;
import java.sql.PreparedStatement;
import java.sql.ResultSet;
import javax.swing.JOptionPane;
public class ReceiveBook extends javax.swing.JFrame {
/** Creates new form ReceiveBook */
public ReceiveBook() {
initComponents();
}
// <editor-fold defaultstate="collapsed" desc="Generated Code">
private void initComponents() {
jLabel1 = new javax.swing.JLabel();
jLabel3 = new javax.swing.JLabel();
TxtMemberID = new javax.swing.JTextField();
CmdMemberDetails = new javax.swing.JButton();
jLabel2 = new javax.swing.JLabel();
TxtMemberName = new javax.swing.JTextField();
CmdReset = new javax.swing.JButton();
CmdReceiveBook = new javax.swing.JButton();
CmdClose = new javax.swing.JButton();
TxtBookDetails = new javax.swing.JTextField();
jLabel4 = new javax.swing.JLabel();
LblBookID = new javax.swing.JLabel();
setDefaultCloseOperation(javax.swing.WindowConstants.DISPOSE_ON_CLOSE);
setTitle("P2P Library : Receive Book");
addWindowListener(new java.awt.event.WindowAdapter() {
public void windowOpened(java.awt.event.WindowEvent evt) {
formWindowOpened(evt);
}
});
jLabel1.setFont(new java.awt.Font("Tahoma", 1, 12));
jLabel1.setForeground(new java.awt.Color(51, 0, 255));
jLabel1.setText("Library Management System - Receive Book");
jLabel3.setText("Member ID :");
TxtMemberID.setText("jTextField1");
CmdMemberDetails.setText("Details");
CmdMemberDetails.addActionListener(new java.awt.event.ActionListener() {
public void actionPerformed(java.awt.event.ActionEvent evt) {
CmdMemberDetailsActionPerformed(evt);
}
});
jLabel2.setText("Name :");
TxtMemberName.setEditable(false);
```

```
TxtMemberName.setText("jTextField1");
CmdReset.setText("Reset");
CmdReset.addActionListener(new java.awt.event.ActionListener() {
public void actionPerformed(java.awt.event.ActionEvent evt) {
CmdResetActionPerformed(evt);
}
});
CmdReceiveBook.setText("Receive Book");
CmdReceiveBook.addActionListener(new java.awt.event.ActionListener() {
public void actionPerformed(java.awt.event.ActionEvent evt) {
CmdReceiveBookActionPerformed(evt);
}
});
CmdClose.setText("Close");
CmdClose.addActionListener(new java.awt.event.ActionListener() {
public void actionPerformed(java.awt.event.ActionEvent evt) {
CmdCloseActionPerformed(evt);
}
});
TxtBookDetails.setEditable(false);
TxtBookDetails.setText("jTextField1");
jLabel4.setText("Book :");
LblBookID.setFont(new java.awt.Font("Tahoma", 0, 8));
LblBookID.setForeground(java.awt.Color.lightGray);
javax.swing.GroupLayout layout = new javax.swing.GroupLayout(getContentPane());
getContentPane().setLayout(layout);
layout.setHorizontalGroup(
layout.createParallelGroup(javax.swing.GroupLayout.Alignment.LEADING)
.addGroup(layout.createSequentialGroup()
.addGap(23, 23, 23)
.addGroup(layout.createParallelGroup(javax.swing.GroupLayout.Alignment.TRAILING)
.addComponent(jLabel1, javax.swing.GroupLayout.Alignment.LEADING)
.addGroup(javax.swing.GroupLayout.Alignment.LEADING, layout.createSequentialGroup()
.addGap(22, 22, 22)
.addGroup(layout.createParallelGroup(javax.swing.GroupLayout.Alignment.LEADING)
.addGroup(javax.swing.GroupLayout.Alignment.TRAILING, layout.createSequentialGroup()
.addGap(65, 65, 65)
.addComponent(CmdReset)
.addPreferredGap(javax.swing.LayoutStyle.ComponentPlacement.RELATED)
.addComponent(CmdReceiveBook)
.addPreferredGap(javax.swing.LayoutStyle.ComponentPlacement.RELATED)
.addComponent(CmdClose))
.addGroup(layout.createSequentialGroup()
.addComponent(jLabel3)
.addGap(14, 14, 14)
.addComponent(TxtMemberID, javax.swing.GroupLayout.PREFERRED_SIZE, 53,
javax.swing.GroupLayout.PREFERRED_SIZE))
.addGroup(javax.swing.GroupLayout.Alignment.TRAILING, layout.createSequentialGroup()
.addGroup(layout.createParallelGroup(javax.swing.GroupLayout.Alignment.LEADING)
.addComponent(jLabel4)
.addComponent(jLabel2))
.addPreferredGap(javax.swing.LayoutStyle.ComponentPlacement.RELATED, 37,
Short.MAX_VALUE)
.addGroup(layout.createParallelGroup(javax.swing.GroupLayout.Alignment.LEADING)
```

```
.addComponent(TxtBookDetails, javax.swing.GroupLayout.PREFERRED_SIZE, 229,
javax.swing.GroupLayout.PREFERRED_SIZE)
.addComponent(LblBookID)
.addGroup(layout.createParallelGroup(javax.swing.GroupLayout.Alignment.TRAILING)
.addComponent(CmdMemberDetails)
.addComponent(TxtMemberName, javax.swing.GroupLayout.PREFERRED_SIZE, 229,
javax.swing.GroupLayout.PREFERRED_SIZE))))))
.addGap(60, 60, 60))
);
layout.setVerticalGroup(
layout.createParallelGroup(javax.swing.GroupLayout.Alignment.LEADING)
.addGroup(layout.createSequentialGroup()
.addContainerGap()
.addComponent(jLabel1)
.addGap(13, 13, 13)
.addGroup(layout.createParallelGroup(javax.swing.GroupLayout.Alignment.BASELINE)
.addComponent(jLabel3)
.addComponent(TxtMemberID, javax.swing.GroupLayout.PREFERRED_SIZE, 23,
javax.swing.GroupLayout.PREFERRED_SIZE)
.addComponent(CmdMemberDetails))
.addPreferredGap(javax.swing.LayoutStyle.ComponentPlacement.RELATED)
.addGroup(layout.createParallelGroup(javax.swing.GroupLayout.Alignment.BASELINE)
.addComponent(TxtMemberName, javax.swing.GroupLayout.PREFERRED_SIZE, 23,
javax.swing.GroupLayout.PREFERRED_SIZE)
.addComponent(jLabel2))
.addPreferredGap(javax.swing.LayoutStyle.ComponentPlacement.UNRELATED)
.addGroup(layout.createParallelGroup(javax.swing.GroupLayout.Alignment.BASELINE)
.addComponent(TxtBookDetails, javax.swing.GroupLayout.PREFERRED_SIZE, 23,
javax.swing.GroupLayout.PREFERRED_SIZE)
.addComponent(jLabel4))
.addGap(18, 18, 18)
.addComponent(LblBookID)
.addGap(100, 100, 100)
.addGroup(layout.createParallelGroup(javax.swing.GroupLayout.Alignment.BASELINE)
.addComponent(CmdReset)
.addComponent(CmdReceiveBook)
.addComponent(CmdClose))
.addContainerGap(33, Short.MAX_VALUE))
);
pack();
}// </editor-fold>
private void CmdMemberDetailsActionPerformed(java.awt.event.ActionEvent evt) {
//get member deatils
try {
//get database connection details
MainClass mc=new MainClass();
//open connection
Connection connection;
connection=DriverManager.getConnection(mc.StrUrl,mc.StrUid,mc.StrPwd);
String str="";
str="select * from lib_member_master where mem_id =? ";
PreparedStatement pst=connection.prepareStatement(str);
pst.setString(1, TxtMemberID.getText());
ResultSet rs;
rs=pst.executeQuery();
```

```
if (rs.next()) {
TxtMemberName.setText(rs.getString("mem_name"));
//get mem status
str="select * from lib_transaction, lib_book_master where book_id=trn_book_id and
trn_mem_id =? and (trn_receive_dt is null or trn_receive_dt='')";
PreparedStatement pst1=connection.prepareStatement(str);
pst1.setString(1, TxtMemberID.getText());
ResultSet rs1;
rs1=pst1.executeQuery();
if (rs1.next()) {
TxtBookDetails.setText("Available Book : " + rs1.getString("book_title"));
CmdReceiveBook.setEnabled(true);
LblBookID.setText(rs1.getString("book_id"));
} else {
TxtBookDetails.setText("No book available.");
CmdReceiveBook.setEnabled(false);
LblBookID.setText("");
}
} else {
JOptionPane.showMessageDialog(null,"Invalid member id.");
TxtMemberID.setText("");
TxtMemberName.setText("");
TxtBookDetails.setText("");
CmdReceiveBook.setEnabled(false);
LblBookID.setText("");
return;
}
} catch (Exception e) {
System.err.println(e);
System.exit(1);
}
}
private void CmdResetActionPerformed(java.awt.event.ActionEvent evt) {
ClearText();
}
private void CmdReceiveBookActionPerformed(java.awt.event.ActionEvent evt) {
//issue book
try {
if (TxtMemberID.getText().trim().length()==0 ) {
JOptionPane.showMessageDialog(null,"Enter member id.");
return;
}
MainClass mc=new MainClass();
Connection connection;
connection=DriverManager.getConnection(mc.StrUrl,mc.StrUid,mc.StrPwd);
String sql = "update lib_transaction set trn_receive_dt = sysdate() where trn_mem_id= " +
TxtMemberID.getText() + " and trn_receive_dt is null and trn_book_id=" +
LblBookID.getText();
PreparedStatement pst=connection.prepareStatement(sql);
pst.executeUpdate(sql);
sql = "update lib_book_master set book_status='Available' where book_id=" +
LblBookID.getText();
pst=connection.prepareStatement(sql);
pst.executeUpdate(sql);
JOptionPane.showMessageDialog(null,"Book Received");
```

```
ClearText();
} catch (Exception e) {
System.err.println(e);
System.exit(1);
}
}
private void ClearText()
{
TxtMemberID.setText("");
TxtMemberName.setText("");
TxtBookDetails.setText("");
LblBookID.setText("");
CmdReceiveBook.setEnabled(false);
}
private void CmdCloseActionPerformed(java.awt.event.ActionEvent evt) {
this.setVisible(false);
}
private void formWindowOpened(java.awt.event.WindowEvent evt) {
ClearText();
this.setLocationRelativeTo(null);
}
/**
* @param args the command line arguments
*/
public static void main(String args[]) {
java.awt.EventQueue.invokeLater(new Runnable() {
public void run() {
new ReceiveBook().setVisible(true);
}
});
}
// Variables declaration
private javax.swing.JButton CmdClose;
private javax.swing.JButton CmdMemberDetails;
private javax.swing.JButton CmdReceiveBook;
private javax.swing.JButton CmdReset;
private javax.swing.JLabel LblBookID;
private javax.swing.JTextField TxtBookDetails;
private javax.swing.JTextField TxtMemberID;
private javax.swing.JTextField TxtMemberName;
private javax.swing.JLabel jLabel1;
private javax.swing.JLabel jLabel2;
private javax.swing.JLabel jLabel3;
private javax.swing.JLabel jLabel4;
// End of variables declaration
}
```

11. SearchBook.java

Purpose: Search book used to search book using book id or title or keyword and search command.

```
import java.sql.Statement;
import java.sql.Connection;
import java.sql.DriverManager;
```

```
import java.sql.PreparedStatement;
import java.sql.ResultSet;
import javax.swing.table.DefaultTableModel;
import javax.swing.table.AbstractTableModel;
import javax.swing.JOptionPane;
public class SearchBook extends javax.swing.JFrame {
/** Creates new form SearchBook */
public SearchBook() {
initComponents();
}
// <editor-fold defaultstate="collapsed" desc="Generated Code">
private void initComponents() {
jLabel1 = new javax.swing.JLabel();
jLabel2 = new javax.swing.JLabel();
TxtBookID = new javax.swing.JTextField();
jLabel3 = new javax.swing.JLabel();
TxtTitle = new javax.swing.JTextField();
jLabel4 = new javax.swing.JLabel();
TxtKeyword = new javax.swing.JTextField();
jButton1 = new javax.swing.JButton();
jScrollPane1 = new javax.swing.JScrollPane();
jTable1 = new javax.swing.JTable();
CmdClose = new javax.swing.JButton();
setDefaultCloseOperation(javax.swing.WindowConstants.DISPOSE_ON_CLOSE);
setTitle("P2P Library : Search Book");
addWindowListener(new java.awt.event.WindowAdapter() {
public void windowOpened(java.awt.event.WindowEvent evt) {
formWindowOpened(evt);
}
});
jLabel1.setFont(new java.awt.Font("Tahoma", 1, 12));
jLabel1.setForeground(new java.awt.Color(51, 0, 255));
jLabel1.setText("Library Management System - Searh Book");
jLabel2.setText("Searh by Book ID:");
TxtBookID.setText("jTextField1");
jLabel3.setText("Title:");
TxtTitle.setText("jTextField1");
jLabel4.setText("Keyword :");
TxtKeyword.setText("jTextField1");
jButton1.setText("Search");
jButton1.addActionListener(new java.awt.event.ActionListener() {
public void actionPerformed(java.awt.event.ActionEvent evt) {
jButton1ActionPerformed(evt);
}
});
jTable1.setModel(new javax.swing.table.DefaultTableModel(
new Object [][] {
{null, null, null, null},
{null, null, null, null},
{null, null, null, null},
{null, null, null, null}
},
new String [] {
"Title 1", "Title 2", "Title 3", "Title 4"
}
```

```
));
jTable1.setAutoResizeMode(javax.swing.JTable.AUTO_RESIZE_ALL_COLUMNS);
jScrollPane1.setViewportView(jTable1);
CmdClose.setText("Close");
CmdClose.addActionListener(new java.awt.event.ActionListener() {
public void actionPerformed(java.awt.event.ActionEvent evt) {
CmdCloseActionPerformed(evt);
}
});
javax.swing.GroupLayout layout = new javax.swing.GroupLayout(getContentPane());
getContentPane().setLayout(layout);
layout.setHorizontalGroup(
layout.createParallelGroup(javax.swing.GroupLayout.Alignment.LEADING)
.addGroup(layout.createSequentialGroup()
.addContainerGap()
.addGroup(layout.createParallelGroup(javax.swing.GroupLayout.Alignment.TRAILING)
.addComponent(CmdClose, javax.swing.GroupLayout.PREFERRED_SIZE, 75,
javax.swing.GroupLayout.PREFERRED_SIZE)
.addGroup(javax.swing.GroupLayout.Alignment.LEADING,
layout.createParallelGroup(javax.swing.GroupLayout.Alignment.TRAILING)
.addComponent(jLabel1, javax.swing.GroupLayout.Alignment.LEADING)
.addGroup(javax.swing.GroupLayout.Alignment.LEADING, layout.createSequentialGroup()
.addComponent(jLabel2)
.addPreferredGap(javax.swing.LayoutStyle.ComponentPlacement.UNRELATED)
.addComponent(TxtBookID, javax.swing.GroupLayout.PREFERRED_SIZE,
javax.swing.GroupLayout.DEFAULT_SIZE, javax.swing.GroupLayout.PREFERRED_SIZE)
.addGap(18, 18, 18)
.addComponent(jLabel3)
.addPreferredGap(javax.swing.LayoutStyle.ComponentPlacement.UNRELATED)
.addComponent(TxtTitle, javax.swing.GroupLayout.PREFERRED_SIZE, 112,
javax.swing.GroupLayout.PREFERRED_SIZE)
.addPreferredGap(javax.swing.LayoutStyle.ComponentPlacement.UNRELATED)
.addComponent(jLabel4)
.addPreferredGap(javax.swing.LayoutStyle.ComponentPlacement.RELATED)
.addComponent(TxtKeyword, javax.swing.GroupLayout.PREFERRED_SIZE, 112,
javax.swing.GroupLayout.PREFERRED_SIZE)
.addPreferredGap(javax.swing.LayoutStyle.ComponentPlacement.RELATED)
.addComponent(jButton1, javax.swing.GroupLayout.PREFERRED_SIZE, 75,
javax.swing.GroupLayout.PREFERRED_SIZE))
.addComponent(jScrollPane1, javax.swing.GroupLayout.Alignment.LEADING)))
.addContainerGap(28, Short.MAX_VALUE))
);
layout.setVerticalGroup(
layout.createParallelGroup(javax.swing.GroupLayout.Alignment.LEADING)
.addGroup(layout.createSequentialGroup()
.addContainerGap()
.addComponent(jLabel1)
.addGap(27, 27, 27)
.addGroup(layout.createParallelGroup(javax.swing.GroupLayout.Alignment.BASELINE)
.addComponent(jLabel2)
.addComponent(TxtBookID, javax.swing.GroupLayout.PREFERRED_SIZE, 23,
javax.swing.GroupLayout.PREFERRED_SIZE)
.addComponent(jLabel3)
.addComponent(TxtTitle, javax.swing.GroupLayout.PREFERRED_SIZE, 23,
javax.swing.GroupLayout.PREFERRED_SIZE)
```

```
.addComponent(jLabel4)
.addComponent(TxtKeyword, javax.swing.GroupLayout.PREFERRED_SIZE, 23,
javax.swing.GroupLayout.PREFERRED_SIZE)
.addComponent(jButton1))
.addGap(18, 18, 18)
.addComponent(jScrollPane1, javax.swing.GroupLayout.DEFAULT_SIZE, 231,
Short.MAX_VALUE)
.addGap(7, 7, 7)
.addComponent(CmdClose)
.addContainerGap())
);
pack();
}// </editor-fold>
private void jButton1ActionPerformed(java.awt.event.ActionEvent evt) {
// TODO add your handling code here:
try
{
MainClass mc=new MainClass();
Connection connection;
connection=DriverManager.getConnection(mc.StrUrl,mc.StrUid,mc.StrPwd);
ResultSet rs;
String StrQr="";
if (TxtBookID.getText().trim().length()>0 )
{
StrQr=StrQr + " and book_id = " + TxtBookID.getText().trim() + " ";
}
if (TxtTitle.getText().trim().length()>0 )
{
StrQr=StrQr + " and book_title like '%" + TxtTitle.getText().trim() + "%' ";
}
if (TxtKeyword.getText().trim().length()>0 )
{
StrQr=StrQr + " and book_keyword like '%" + TxtKeyword.getText().trim() + "%' ";
}
if (StrQr.length()==0)
{
JOptionPane.showMessageDialog(null,"Enter search critaria.");
return;
}
PreparedStatement stmt=connection.prepareStatement("select book_id, book_title,
book_category, book_keyword from lib_book_master where 1=1 " + StrQr + " order by
book_id");
rs = stmt.executeQuery();
jTable1.getColumnModel().getColumn(0).setHeaderValue("Book ID");
jTable1.getColumnModel().getColumn(1).setHeaderValue("Title");
jTable1.getColumnModel().getColumn(2).setHeaderValue("Category");
jTable1.getColumnModel().getColumn(3).setHeaderValue("Keyword");
jTable1.getTableHeader().resizeAndRepaint();
// Removing Previous Data
while (jTable1.getRowCount() > 0) {
((DefaultTableModel) jTable1.getModel()).removeRow(0);
}
//Creating Object []rowData for jTable's Table Model
int columns = rs.getMetaData().getColumnCount();
while (rs.next())
```

```
{
Object[] row = new Object[columns];
for (int i = 1; i <= columns; i++)
{
row[i - 1] = rs.getObject(i); // 1
}
((DefaultTableModel) jTable1.getModel()).insertRow(rs.getRow() - 1,row);
}
}
catch (Exception e)
{
System.err.println(e);
//System.exit(1);
}
}
private void CmdCloseActionPerformed(java.awt.event.ActionEvent evt) {
this.setVisible(false);
}
private void formWindowOpened(java.awt.event.WindowEvent evt) {
TxtBookID.setText("");
TxtTitle.setText("");
TxtKeyword.setText("");
this.setLocationRelativeTo(null);
jTable1.getColumnModel().getColumn(0).setHeaderValue("Book ID");
jTable1.getColumnModel().getColumn(1).setHeaderValue("Title");
jTable1.getColumnModel().getColumn(2).setHeaderValue("Category");
jTable1.getColumnModel().getColumn(3).setHeaderValue("Keyword");
jTable1.getTableHeader().resizeAndRepaint();
// Removing Previous Data
while (jTable1.getRowCount() > 0) {
((DefaultTableModel) jTable1.getModel()).removeRow(0);
}
}
/**
* @param args the command line arguments
*/
public static void main(String args[]) {
java.awt.EventQueue.invokeLater(new Runnable() {
public void run() {
new SearchBook().setVisible(true);
}
});
}
// Variables declaration
private javax.swing.JButton CmdClose;
private javax.swing.JTextField TxtBookID;
private javax.swing.JTextField TxtKeyword;
private javax.swing.JTextField TxtTitle;
private javax.swing.JButton jButton1;
private javax.swing.JLabel jLabel1;
private javax.swing.JLabel jLabel2;
private javax.swing.JLabel jLabel3;
private javax.swing.JLabel jLabel4;
private javax.swing.JScrollPane jScrollPane1;
private javax.swing.JTable jTable1;
```

```
// End of variables declaration
}
```

12. SearchMember.java

Purpose: Search member used to search member using member id or name or mobile no. and search command.

```
import java.sql.Statement;
import java.sql.Connection;
import java.sql.DriverManager;
import java.sql.PreparedStatement;
import java.sql.ResultSet;
import javax.swing.table.DefaultTableModel;
import javax.swing.table.AbstractTableModel;
import javax.swing.JOptionPane;
public class SearchMember extends javax.swing.JFrame {
/** Creates new form SearchMember */
public SearchMember() {
initComponents();
}
// <editor-fold defaultstate="collapsed" desc="Generated Code">
private void initComponents() {
jLabel1 = new javax.swing.JLabel();
jLabel2 = new javax.swing.JLabel();
TxtMemID = new javax.swing.JTextField();
jLabel3 = new javax.swing.JLabel();
TxtName = new javax.swing.JTextField();
jLabel4 = new javax.swing.JLabel();
TxtMobile = new javax.swing.JTextField();
jButton1 = new javax.swing.JButton();
jScrollPane1 = new javax.swing.JScrollPane();
jTable1 = new javax.swing.JTable();
CmdClose = new javax.swing.JButton();
setDefaultCloseOperation(javax.swing.WindowConstants.DISPOSE_ON_CLOSE);
setTitle("P2P Library : Search Member");
addWindowListener(new java.awt.event.WindowAdapter() {
public void windowOpened(java.awt.event.WindowEvent evt) {
formWindowOpened(evt);
}
});
jLabel1.setFont(new java.awt.Font("Tahoma", 1, 12));
jLabel1.setForeground(new java.awt.Color(51, 0, 255));
jLabel1.setText("Library Management System - Search Member");
jLabel2.setText("Searh by Member ID:");
TxtMemID.setText("jTextField1");
jLabel3.setText("Name:");
TxtName.setText("jTextField1");
jLabel4.setText("Mobile:");
TxtMobile.setText("jTextField1");
jButton1.setText("Search");
jButton1.addActionListener(new java.awt.event.ActionListener() {
public void actionPerformed(java.awt.event.ActionEvent evt) {
jButton1ActionPerformed(evt);
}
```

```
});
jTable1.setModel(new javax.swing.table.DefaultTableModel(
new Object [][] {
{null, null, null, null},
{null, null, null, null},
{null, null, null, null},
{null, null, null, null}
},
new String [] {
"Title 1", "Title 2", "Title 3", "Title 4"
}
));
jTable1.setAutoResizeMode(javax.swing.JTable.AUTO_RESIZE_ALL_COLUMNS);
jScrollPane1.setViewportView(jTable1);
CmdClose.setText("Close");
CmdClose.addActionListener(new java.awt.event.ActionListener() {
public void actionPerformed(java.awt.event.ActionEvent evt) {
CmdCloseActionPerformed(evt);
}
});
javax.swing.GroupLayout layout = new javax.swing.GroupLayout(getContentPane());
getContentPane().setLayout(layout);
layout.setHorizontalGroup(
layout.createParallelGroup(javax.swing.GroupLayout.Alignment.LEADING)
.addGroup(layout.createSequentialGroup()
.addContainerGap()
.addGroup(layout.createParallelGroup(javax.swing.GroupLayout.Alignment.TRAILING)
.addComponent(CmdClose, javax.swing.GroupLayout.PREFERRED_SIZE, 75,
javax.swing.GroupLayout.PREFERRED_SIZE)
.addGroup(javax.swing.GroupLayout.Alignment.LEADING,
layout.createParallelGroup(javax.swing.GroupLayout.Alignment.TRAILING)
.addComponent(jLabel1, javax.swing.GroupLayout.Alignment.LEADING)
.addGroup(javax.swing.GroupLayout.Alignment.LEADING, layout.createSequentialGroup()
.addComponent(jLabel2)
.addPreferredGap(javax.swing.LayoutStyle.ComponentPlacement.UNRELATED)
.addComponent(TxtMemID, javax.swing.GroupLayout.PREFERRED_SIZE,
javax.swing.GroupLayout.DEFAULT_SIZE, javax.swing.GroupLayout.PREFERRED_SIZE)
.addGap(18, 18, 18)
.addComponent(jLabel3)
.addPreferredGap(javax.swing.LayoutStyle.ComponentPlacement.UNRELATED)
.addComponent(TxtName, javax.swing.GroupLayout.PREFERRED_SIZE, 112,
javax.swing.GroupLayout.PREFERRED_SIZE)
.addPreferredGap(javax.swing.LayoutStyle.ComponentPlacement.UNRELATED)
.addComponent(jLabel4)
.addPreferredGap(javax.swing.LayoutStyle.ComponentPlacement.RELATED)
.addComponent(TxtMobile, javax.swing.GroupLayout.PREFERRED_SIZE, 112,
javax.swing.GroupLayout.PREFERRED_SIZE)
.addPreferredGap(javax.swing.LayoutStyle.ComponentPlacement.RELATED)
.addComponent(jButton1, javax.swing.GroupLayout.PREFERRED_SIZE, 75,
javax.swing.GroupLayout.PREFERRED_SIZE))
.addComponent(jScrollPane1, javax.swing.GroupLayout.Alignment.LEADING)))
.addContainerGap(19, Short.MAX_VALUE))
);
layout.setVerticalGroup(
layout.createParallelGroup(javax.swing.GroupLayout.Alignment.LEADING)
```

```
.addGroup(layout.createSequentialGroup()
.addContainerGap()
.addComponent(jLabel1)
.addGap(27, 27, 27)
.addGroup(layout.createParallelGroup(javax.swing.GroupLayout.Alignment.BASELINE)
.addComponent(jLabel2)
.addComponent(TxtMemID, javax.swing.GroupLayout.PREFERRED_SIZE, 23,
javax.swing.GroupLayout.PREFERRED_SIZE)
.addComponent(jLabel3)
.addComponent(TxtName, javax.swing.GroupLayout.PREFERRED_SIZE, 23,
javax.swing.GroupLayout.PREFERRED_SIZE)
.addComponent(jLabel4)
.addComponent(TxtMobile, javax.swing.GroupLayout.PREFERRED_SIZE, 23,
javax.swing.GroupLayout.PREFERRED_SIZE)
.addComponent(jButton1))
.addGap(18, 18, 18)
.addComponent(jScrollPane1, javax.swing.GroupLayout.DEFAULT_SIZE, 320,
Short.MAX_VALUE)
.addGap(7, 7, 7)
.addComponent(CmdClose)
.addContainerGap())
);
pack();
}// </editor-fold>
private void jButton1ActionPerformed(java.awt.event.ActionEvent evt) {
// TODO add your handling code here:
try {
MainClass mc=new MainClass();
Connection connection;
connection=DriverManager.getConnection(mc.StrUrl,mc.StrUid,mc.StrPwd);
ResultSet rs;
String StrQr="";
if (TxtMemID.getText().trim().length()>0 ) {
StrQr=StrQr + " and mem_id = " + TxtMemID.getText().trim() + " ";
}
if (TxtName.getText().trim().length()>0 ) {
StrQr=StrQr + " and mem_name like '%" + TxtName.getText().trim() + "%' ";
}
if (TxtMobile.getText().trim().length()>0 ) {
StrQr=StrQr + " and mem_mobile_no like '%" + TxtMobile.getText().trim() + "%' ";
}
if (StrQr.length()==0) {
JOptionPane.showMessageDialog(null,"Enter search critaria.");
return;
}
PreparedStatement stmt=connection.prepareStatement("select mem_id, mem_name,
mem_mobile_no, mem_email_id from lib_member_master where 1=1 " + StrQr + " order by
mem_id");
rs = stmt.executeQuery();
jTable1.getColumnModel().getColumn(0).setHeaderValue("Member ID");
jTable1.getColumnModel().getColumn(1).setHeaderValue("Name");
jTable1.getColumnModel().getColumn(2).setHeaderValue("Mobile No");
jTable1.getColumnModel().getColumn(3).setHeaderValue("Email ID");
jTable1.getTableHeader().resizeAndRepaint();
// Removing Previous Data
```

```
while (jTable1.getRowCount() > 0) {
((DefaultTableModel) jTable1.getModel()).removeRow(0);
}
//Creating Object []rowData for jTable's Table Model
int columns = rs.getMetaData().getColumnCount();
while (rs.next()) {
Object[] row = new Object[columns];
for (int i = 1; i <= columns; i++) {
row[i - 1] = rs.getObject(i); // 1
}
((DefaultTableModel) jTable1.getModel()).insertRow(rs.getRow() - 1,row);
}
} catch (Exception e) {
System.err.println(e);
//System.exit(1);
}
}
private void CmdCloseActionPerformed(java.awt.event.ActionEvent evt) {
this.setVisible(false);
}
private void formWindowOpened(java.awt.event.WindowEvent evt) {
TxtMemID.setText("");
TxtName.setText("");
TxtMobile.setText("");
this.setLocationRelativeTo(null);
jTable1.getColumnModel().getColumn(0).setHeaderValue("Member ID");
jTable1.getColumnModel().getColumn(1).setHeaderValue("Name");
jTable1.getColumnModel().getColumn(2).setHeaderValue("Mobile No");
jTable1.getColumnModel().getColumn(3).setHeaderValue("Email ID");
jTable1.getTableHeader().resizeAndRepaint();
// Removing Previous Data
while (jTable1.getRowCount() > 0) {
((DefaultTableModel) jTable1.getModel()).removeRow(0);
}
}
/**
* @param args the command line arguments
*/
public static void main(String args[]) {
java.awt.EventQueue.invokeLater(new Runnable() {
public void run() {
new SearchMember().setVisible(true);
}
});
}
// Variables declaration
private javax.swing.JButton CmdClose;
private javax.swing.JTextField TxtMemID;
private javax.swing.JTextField TxtMobile;
private javax.swing.JTextField TxtName;
private javax.swing.JButton jButton1;
private javax.swing.JLabel jLabel1;
private javax.swing.JLabel jLabel2;
private javax.swing.JLabel jLabel3;
private javax.swing.JLabel jLabel4;
```

```
private javax.swing.JScrollPane jScrollPane1;
private javax.swing.JTable jTable1;
// End of variables declaration
}
```

13. TransactionReport.java

Purpose: Transaction report summarizes the whole transaction on LMS with the detail of member, book, issue date, and receive date using refresh report button.

```
import java.sql.Connection;
import java.sql.DriverManager;
import java.sql.PreparedStatement;
import java.sql.ResultSet;
import javax.swing.table.DefaultTableModel;
public class TransactionReport extends javax.swing.JFrame {
/** Creates new form TransactionReport */
public TransactionReport() {
initComponents();
}
// <editor-fold defaultstate="collapsed" desc="Generated Code">
private void initComponents() {
jLabel1 = new javax.swing.JLabel();
jButton1 = new javax.swing.JButton();
CmdClose = new javax.swing.JButton();
jScrollPane1 = new javax.swing.JScrollPane();
jTable1 = new javax.swing.JTable();
setDefaultCloseOperation(javax.swing.WindowConstants.DISPOSE_ON_CLOSE);
setTitle("P2P Library : Transaction Report");
addWindowListener(new java.awt.event.WindowAdapter() {
public void windowOpened(java.awt.event.WindowEvent evt) {
formWindowOpened(evt);
}
});
jLabel1.setFont(new java.awt.Font("Tahoma", 1, 12));
jLabel1.setForeground(new java.awt.Color(51, 0, 255));
jLabel1.setText("Library Management System - Transcation Report");
jButton1.setText("Refresh Report");
jButton1.addActionListener(new java.awt.event.ActionListener() {
public void actionPerformed(java.awt.event.ActionEvent evt) {
jButton1ActionPerformed(evt);
}
});
CmdClose.setText("Close");
CmdClose.addActionListener(new java.awt.event.ActionListener() {
public void actionPerformed(java.awt.event.ActionEvent evt) {
CmdCloseActionPerformed(evt);
}
});
jTable1.setModel(new javax.swing.table.DefaultTableModel(
new Object [][] {
{null, null, null, null},
{null, null, null, null},
{null, null, null, null},
{null, null, null, null}
```

```
},
new String [] {
"Title 1", "Title 2", "Title 3", "Title 4"
}
));
jTable1.setAutoResizeMode(javax.swing.JTable.AUTO_RESIZE_ALL_COLUMNS);
jScrollPane1.setViewportView(jTable1);
javax.swing.GroupLayout layout = new javax.swing.GroupLayout(getContentPane());
getContentPane().setLayout(layout);
layout.setHorizontalGroup(
layout.createParallelGroup(javax.swing.GroupLayout.Alignment.LEADING)
.addGroup(layout.createSequentialGroup()
.addContainerGap()
.addGroup(layout.createParallelGroup(javax.swing.GroupLayout.Alignment.LEADING)
.addComponent(jScrollPane1, javax.swing.GroupLayout.DEFAULT_SIZE, 567,
Short.MAX_VALUE)
.addComponent(jLabel1)
.addGroup(layout.createSequentialGroup()
.addComponent(jButton1, javax.swing.GroupLayout.PREFERRED_SIZE, 138,
javax.swing.GroupLayout.PREFERRED_SIZE)
.addPreferredGap(javax.swing.LayoutStyle.ComponentPlacement.RELATED)
.addComponent(CmdClose, javax.swing.GroupLayout.PREFERRED_SIZE, 75,
javax.swing.GroupLayout.PREFERRED_SIZE)))
.addContainerGap())
);
layout.setVerticalGroup(
layout.createParallelGroup(javax.swing.GroupLayout.Alignment.LEADING)
.addGroup(layout.createSequentialGroup()
.addContainerGap()
.addComponent(jLabel1)
.addGap(15, 15, 15)
.addGroup(layout.createParallelGroup(javax.swing.GroupLayout.Alignment.BASELINE)
.addComponent(CmdClose)
.addComponent(jButton1))
.addPreferredGap(javax.swing.LayoutStyle.ComponentPlacement.UNRELATED)
.addComponent(jScrollPane1, javax.swing.GroupLayout.PREFERRED_SIZE, 338,
javax.swing.GroupLayout.PREFERRED_SIZE)
.addContainerGap(javax.swing.GroupLayout.DEFAULT_SIZE, Short.MAX_VALUE))
);
pack();
}// </editor-fold>
private void jButton1ActionPerformed(java.awt.event.ActionEvent evt) {
// TODO add your handling code here:
try {
MainClass mc=new MainClass();
Connection connection;
connection=DriverManager.getConnection(mc.StrUrl,mc.StrUid,mc.StrPwd);
ResultSet rs;
PreparedStatement stmt=connection.prepareStatement("SELECT m.mem_name,b.book_title,
l.trn_issue_dt, l.trn_receive_dt FROM lib_transaction l, lib_book_master b, lib_member_master
m where l.trn_book_id=b.book_id and l.trn_mem_id=m.mem_id order by l.trn_id");
rs = stmt.executeQuery();
jTable1.getColumnModel().getColumn(0).setHeaderValue("Member");
jTable1.getColumnModel().getColumn(1).setHeaderValue("Book");
jTable1.getColumnModel().getColumn(2).setHeaderValue("Issue Date");
```

```
jTable1.getColumnModel().getColumn(3).setHeaderValue("Receive Date");
jTable1.getTableHeader().resizeAndRepaint();
// Removing Previous Data
while (jTable1.getRowCount() > 0) {
((DefaultTableModel) jTable1.getModel()).removeRow(0);
}
//Creating Object []rowData for jTable's Table Model
int columns = rs.getMetaData().getColumnCount();
while (rs.next()) {
Object[] row = new Object[columns];
for (int i = 1; i <= columns; i++) {
row[i - 1] = rs.getObject(i); // 1
}
((DefaultTableModel) jTable1.getModel()).insertRow(rs.getRow() - 1,row);
}
} catch (Exception e) {
System.err.println(e);
//System.exit(1);
}
}
private void CmdCloseActionPerformed(java.awt.event.ActionEvent evt) {
this.setVisible(false);
}
private void formWindowOpened(java.awt.event.WindowEvent evt) {
this.setLocationRelativeTo(null);
jTable1.getColumnModel().getColumn(0).setHeaderValue("Member");
jTable1.getColumnModel().getColumn(1).setHeaderValue("Book");
jTable1.getColumnModel().getColumn(2).setHeaderValue("Issue Date");
jTable1.getColumnModel().getColumn(3).setHeaderValue("Receive Date");
jTable1.getTableHeader().resizeAndRepaint();
// Removing Previous Data
while (jTable1.getRowCount() > 0) {
((DefaultTableModel) jTable1.getModel()).removeRow(0);
}
}
/**
* @param args the command line arguments
*/
public static void main(String args[]) {
java.awt.EventQueue.invokeLater(new Runnable() {
public void run() {
new TransactionReport().setVisible(true);
}
});
}
// Variables declaration
private javax.swing.JButton CmdClose;
private javax.swing.JButton jButton1;
private javax.swing.JLabel jLabel1;
private javax.swing.JScrollPane jScrollPane1;
private javax.swing.JTable jTable1;
// End of variables declaration
}
```

14. UserMaster.java

Purpose: Using user master we add and update the detail of users and manage the user information using first, previous, next, last, new, edit, save, and close command. Basically there are user name, password, and user type (user and admin). Check box is used to provide information about active or non-active user.

```
import java.sql.Statement;
import java.sql.Connection;
import java.sql.DriverManager;
import java.sql.PreparedStatement;
import java.sql.ResultSet;
import javax.swing.JOptionPane;
public class UserMaster extends javax.swing.JFrame {
MainClass mc=new MainClass();
ResultSet rsuser;
int rspos=0;
int rscount=0;
boolean chadd=false;
/** Creates new form UserMaster */
public UserMaster() {
initComponents();
}
// <editor-fold defaultstate="collapsed" desc="Generated Code">
private void initComponents() {
jLabel3 = new javax.swing.JLabel();
TxtUserName = new javax.swing.JTextField();
jLabel4 = new javax.swing.JLabel();
TxtPassword = new javax.swing.JTextField();
jLabel1 = new javax.swing.JLabel();
jLabel8 = new javax.swing.JLabel();
CmbUserType = new javax.swing.JComboBox();
ChkUserActive = new javax.swing.JCheckBox();
CmdFirst = new javax.swing.JButton();
CmdPrevious = new javax.swing.JButton();
CmdNext = new javax.swing.JButton();
CmdLast = new javax.swing.JButton();
CmdNew = new javax.swing.JButton();
CmdEdit = new javax.swing.JButton();
CmdSave = new javax.swing.JButton();
CmdClose = new javax.swing.JButton();
LblPosition = new javax.swing.JLabel();
setDefaultCloseOperation(javax.swing.WindowConstants.DISPOSE_ON_CLOSE);
setTitle("P2P Library : User Master");
addWindowListener(new java.awt.event.WindowAdapter() {
public void windowOpened(java.awt.event.WindowEvent evt) {
formWindowOpened(evt);
}
});
jLabel3.setText("User Name :");
TxtUserName.setText("jTextField1");
jLabel4.setText("Password :");
TxtPassword.setText("jTextField1");
jLabel1.setFont(new java.awt.Font("Tahoma", 1, 12));
jLabel1.setForeground(new java.awt.Color(51, 0, 255));
jLabel1.setText("Library Management System - User Master");
```

```
jLabel8.setText("User Type :");
CmbUserType.setModel(new javax.swing.DefaultComboBoxModel(new String[] { "Item 1",
"Item 2", "Item 3", "Item 4" }));
ChkUserActive.setText("User Active");
CmdFirst.setText("First");
CmdFirst.addActionListener(new java.awt.event.ActionListener() {
public void actionPerformed(java.awt.event.ActionEvent evt) {
CmdFirstActionPerformed(evt);
}
});
CmdPrevious.setText("Previous");
CmdPrevious.addActionListener(new java.awt.event.ActionListener() {
public void actionPerformed(java.awt.event.ActionEvent evt) {
CmdPreviousActionPerformed(evt);
}
});
CmdNext.setText("Next");
CmdNext.addActionListener(new java.awt.event.ActionListener() {
public void actionPerformed(java.awt.event.ActionEvent evt) {
CmdNextActionPerformed(evt);
}
});
CmdLast.setText("Last");
CmdLast.addActionListener(new java.awt.event.ActionListener() {
public void actionPerformed(java.awt.event.ActionEvent evt) {
CmdLastActionPerformed(evt);
}
});
CmdNew.setText("New");
CmdNew.addActionListener(new java.awt.event.ActionListener() {
public void actionPerformed(java.awt.event.ActionEvent evt) {
CmdNewActionPerformed(evt);
}
});
CmdEdit.setText("Edit");
CmdEdit.addActionListener(new java.awt.event.ActionListener() {
public void actionPerformed(java.awt.event.ActionEvent evt) {
CmdEditActionPerformed(evt);
}
});
CmdSave.setText("Save");
CmdSave.addActionListener(new java.awt.event.ActionListener() {
public void actionPerformed(java.awt.event.ActionEvent evt) {
CmdSaveActionPerformed(evt);
}
});
CmdClose.setText("Close");
CmdClose.addActionListener(new java.awt.event.ActionListener() {
public void actionPerformed(java.awt.event.ActionEvent evt) {
CmdCloseActionPerformed(evt);
}
});
LblPosition.setText("position");
javax.swing.GroupLayout layout = new javax.swing.GroupLayout(getContentPane());
getContentPane().setLayout(layout);
```

```
layout.setHorizontalGroup(
layout.createParallelGroup(javax.swing.GroupLayout.Alignment.LEADING)
.addGroup(layout.createSequentialGroup()
.addGap(71, 71, 71)
.addGroup(layout.createParallelGroup(javax.swing.GroupLayout.Alignment.LEADING)
.addComponent(jLabel3)
.addComponent(jLabel8)
.addComponent(jLabel4))
.addGap(20, 20, 20)
.addGroup(layout.createParallelGroup(javax.swing.GroupLayout.Alignment.LEADING)
.addGroup(layout.createSequentialGroup()
.addComponent(ChkUserActive)
.addContainerGap())
.addGroup(layout.createSequentialGroup()
.addGroup(layout.createParallelGroup(javax.swing.GroupLayout.Alignment.LEADING)
.addComponent(CmbUserType, 0, 187, Short.MAX_VALUE)
.addComponent(TxtPassword, javax.swing.GroupLayout.DEFAULT_SIZE, 187,
Short.MAX_VALUE)
.addComponent(TxtUserName, javax.swing.GroupLayout.DEFAULT_SIZE, 187,
Short.MAX_VALUE))
.addGap(340, 340, 340))))
.addGroup(javax.swing.GroupLayout.Alignment.TRAILING, layout.createSequentialGroup()
.addContainerGap()
.addComponent(CmdFirst)
.addPreferredGap(javax.swing.LayoutStyle.ComponentPlacement.RELATED)
.addComponent(CmdPrevious)
.addPreferredGap(javax.swing.LayoutStyle.ComponentPlacement.RELATED)
.addComponent(CmdNext)
.addPreferredGap(javax.swing.LayoutStyle.ComponentPlacement.RELATED)
.addComponent(CmdLast)
.addPreferredGap(javax.swing.LayoutStyle.ComponentPlacement.RELATED,
javax.swing.GroupLayout.DEFAULT_SIZE, Short.MAX_VALUE)
.addComponent(CmdNew)
.addPreferredGap(javax.swing.LayoutStyle.ComponentPlacement.RELATED)
.addComponent(CmdEdit)
.addGap(6, 6, 6)
.addComponent(CmdSave)
.addPreferredGap(javax.swing.LayoutStyle.ComponentPlacement.RELATED)
.addComponent(CmdClose)
.addGap(155, 155, 155))
.addGroup(layout.createSequentialGroup()
.addGroup(layout.createParallelGroup(javax.swing.GroupLayout.Alignment.LEADING)
.addGroup(layout.createSequentialGroup()
.addGap(27, 27, 27)
.addComponent(jLabel1))
.addGroup(layout.createSequentialGroup()
.addGap(113, 113, 113)
.addComponent(LblPosition)))
.addContainerGap(javax.swing.GroupLayout.DEFAULT_SIZE, Short.MAX_VALUE))
);
layout.setVerticalGroup(
layout.createParallelGroup(javax.swing.GroupLayout.Alignment.LEADING)
.addGroup(layout.createSequentialGroup()
.addGap(25, 25, 25)
.addComponent(jLabel1)
```

```
.addGap(40, 40, 40)
.addGroup(layout.createParallelGroup(javax.swing.GroupLayout.Alignment.LEADING)
.addGroup(layout.createSequentialGroup()
.addComponent(TxtUserName, javax.swing.GroupLayout.PREFERRED_SIZE, 23,
javax.swing.GroupLayout.PREFERRED_SIZE)
.addPreferredGap(javax.swing.LayoutStyle.ComponentPlacement.RELATED)
.addGroup(layout.createParallelGroup(javax.swing.GroupLayout.Alignment.BASELINE)
.addComponent(TxtPassword, javax.swing.GroupLayout.PREFERRED_SIZE, 23,
javax.swing.GroupLayout.PREFERRED_SIZE)
.addComponent(jLabel4)))
.addGroup(layout.createSequentialGroup()
.addGap(3, 3, 3)
.addComponent(jLabel3)))
.addPreferredGap(javax.swing.LayoutStyle.ComponentPlacement.RELATED)
.addGroup(layout.createParallelGroup(javax.swing.GroupLayout.Alignment.LEADING)
.addComponent(jLabel8)
.addGroup(layout.createSequentialGroup()
.addGap(1, 1, 1)
.addComponent(CmbUserType, javax.swing.GroupLayout.PREFERRED_SIZE, 25,
javax.swing.GroupLayout.PREFERRED_SIZE)))
.addGap(29, 29, 29)
.addComponent(ChkUserActive)
.addGap(61, 61, 61)
.addComponent(LblPosition)
.addPreferredGap(javax.swing.LayoutStyle.ComponentPlacement.RELATED)
.addGroup(layout.createParallelGroup(javax.swing.GroupLayout.Alignment.BASELINE)
.addComponent(CmdFirst)
.addComponent(CmdPrevious)
.addComponent(CmdNext)
.addComponent(CmdLast)
.addComponent(CmdEdit)
.addComponent(CmdSave)
.addComponent(CmdNew)
.addComponent(CmdClose))
.addContainerGap(25, Short.MAX_VALUE))
);
pack();
}// </editor-fold>
private void CmdFirstActionPerformed(java.awt.event.ActionEvent evt) {
try {
rsuser.first();
Display();
} catch (Exception e) {
System.err.println(e);
}
}
private void CmdPreviousActionPerformed(java.awt.event.ActionEvent evt) {
try {
if (rsuser.isFirst()==true) {
JOptionPane.showMessageDialog(null,"Already in first record.");
return;
}
rsuser.previous();
Display();
} catch (Exception e) {
```

```
System.err.println(e);
}
}
private void CmdNextActionPerformed(java.awt.event.ActionEvent evt) {
try {
if (rsuser.isLast()==true) {
JOptionPane.showMessageDialog(null,"Already in last record.");
return;
}
rsuser.next();
Display();
} catch (Exception e) {
System.err.println(e);
}
}
private void CmdLastActionPerformed(java.awt.event.ActionEvent evt) {
try {
rsuser.last();
Display();
} catch (Exception e) {
System.err.println(e);
//System.exit(1);
}
}
private void CmdNewActionPerformed(java.awt.event.ActionEvent evt) {
chadd=true;
ClearText();
}
private void CmdEditActionPerformed(java.awt.event.ActionEvent evt) {
chadd=false;
}
private void CmdSaveActionPerformed(java.awt.event.ActionEvent evt) {
//validate entry and save record
try {
if (TxtUserName.getText().trim().length()==0 ) {
JOptionPane.showMessageDialog(null,"Enter user name.");
return;
}
if (TxtPassword.getText().trim().length()==0 ) {
JOptionPane.showMessageDialog(null,"Enter password.");
return;
}
//String strqr="";
Connection connection;
connection=DriverManager.getConnection(mc.StrUrl,mc.StrUid,mc.StrPwd);
if (chadd==true ) {
rsuser.moveToInsertRow();
rsuser.updateString("user_name",TxtUserName.getText());
rsuser.updateString("user_password",TxtPassword.getText());
rsuser.updateString("user_type",CmbUserType.getSelectedItem().toString());
if (ChkUserActive.isSelected()==true) {
rsuser.updateString("user_active","YES");
} else {
rsuser.updateString("user_active","NO");
}
```

```
rsuser.insertRow();
rsuser.last();
Display();
} else {
rsuser.updateString("user_password",TxtPassword.getText());
rsuser.updateString("user_type",CmbUserType.getSelectedItem().toString());
if (ChkUserActive.isSelected()==true) {
rsuser.updateString("user_active","YES");
} else {
rsuser.updateString("user_active","NO");
}
rsuser.updateRow();
rsuser.refreshRow();
}
chadd=false;
JOptionPane.showMessageDialog(null,"Record updated.");
} catch (Exception e) {
System.err.println(e);
JOptionPane.showMessageDialog(null,"Unable to save record.");
return;
}
}
private void CmdCloseActionPerformed(java.awt.event.ActionEvent evt) {
// TODO add your handling code here:
this.setVisible(false);
}
private void formWindowOpened(java.awt.event.WindowEvent evt) {
try
{
this.setLocationRelativeTo(null);
ClearText();
CmbUserType.removeAllItems();
CmbUserType.addItem("Admin");
CmbUserType.addItem("User");
Connection connection;
connection=DriverManager.getConnection(mc.StrUrl,mc.StrUid,mc.StrPwd);
ResultSet rs;
PreparedStatement stmt;
//get user count
stmt=connection.prepareStatement("select count(*) from lib_user");
rs = stmt.executeQuery();
rs.first();
rscount=rs.getInt(1);
//get user details
Statement stmt1 = connection.createStatement(ResultSet.TYPE_SCROLL_SENSITIVE,
ResultSet.CONCUR_UPDATABLE);
String sqlQuery = "select * from lib_user ";
rsuser = stmt1.executeQuery(sqlQuery);
//display first record
rsuser.first();
Display();
}
catch (Exception e)
{
System.err.println(e);
```

```
//System.exit(1);
}
}
private void Display()
{
try
{
//clear text of fields for entry/display old data
TxtUserName.setText(rsuser.getString("user_name"));
TxtPassword.setText(rsuser.getString("user_password"));
CmbUserType.setSelectedItem(rsuser.getString("user_type"));
if (rsuser.getString("user_active").equals("YES"))
{
ChkUserActive.setSelected(true);
}
else
{
ChkUserActive.setSelected(false);
}
LblPosition.setText("Record position : " + rsuser.getRow() + "/" + String.valueOf(rscount));
}
catch (Exception e)
{
System.err.println(e);
}
}
private void ClearText()
{
//clear text of fields for entry/display old data
TxtUserName.setText("");
TxtPassword.setText("");
ChkUserActive.setSelected(false);
LblPosition.setText("");
}
/**
* @param args the command line arguments
*/
public static void main(String args[]) {
java.awt.EventQueue.invokeLater(new Runnable() {
public void run() {
new UserMaster().setVisible(true);
}
});
}
// Variables declaration
private javax.swing.JCheckBox ChkUserActive;
private javax.swing.JComboBox CmbUserType;
private javax.swing.JButton CmdClose;
private javax.swing.JButton CmdEdit;
private javax.swing.JButton CmdFirst;
private javax.swing.JButton CmdLast;
private javax.swing.JButton CmdNew;
private javax.swing.JButton CmdNext;
private javax.swing.JButton CmdPrevious;
private javax.swing.JButton CmdSave;
```

```
private javax.swing.JLabel LblPosition;
private javax.swing.JTextField TxtPassword;
private javax.swing.JTextField TxtUserName;
private javax.swing.JLabel jLabel1;
private javax.swing.JLabel jLabel3;
private javax.swing.JLabel jLabel4;
private javax.swing.JLabel jLabel8;
// End of variables declaration
}
```

Appendix
Software Requirement Questionnaire for Usability

Usability of any software can be explained in single word "user satisfaction". Usability can be defined as, the ability of user to perform operation/function easily and effectively by specified range of input, environment, and task. Usability refers to a set of multiple concepts such as: effectiveness, execution time, performance, efficiency, user satisfaction, accuracy, and ease of learning. Usability of a software product has recently become a key factor for software quality.

This survey is based on small size software product such as: Library Management System (LMS). The users have UG/PG degree in computer science. The main expectation of user to software (LMS) performs operation effectively. LMS perform according the input given by user, easy to use, navigate, and learn.

The survey should take about 5-10 minutes of your time. Please answer why usability is an important factor for project success.

Part 1: Objective- To find out important aspect of usability. Please tick any one option.

Q. No.	Usability Aspects	Options
1.	Usability defines as: a) Learnability b) Accuracy	 [ ] [ ]
2.	What is the development process do you usually follow for usability? a) Spiral Development b) Prototyping	 [ ] [ ]
3.	Usability satisfaction metric measured by: a) Task Level Satisfaction b) Test Level Satisfaction	 [ ] [ ]
4.	Learnability can be increase by: a) User documentation and help b) System should be easy to learn	 [ ] [ ]
5.	Which kind of interface is more useful? a) Consistent b) Attractive	 [ ] [ ]
6.	Which is more important factor for usability? a) User satisfaction b) Ease of use	 [ ] [ ]
7.	Which method is best to check usability of software? a) Paper based methods such as: Prototyping, Survey, and Walk-Through b) Usability Testing	 [ ] [ ]
8.	To increase operability which is more important? a) Error messages b) User guide	 [ ] [ ]
9.	How error message is helpful to use software easily? a) Provides information about what should be done. b) Suggestion to input a correct data.	 [ ] [ ]
10.	Understandability means: a) Interface elements (menu) should be easy to understand. b) The operation of the system should be easily	 [ ] [ ]

	understandable.	

Part 2: Objective- To find out important attribute of usability. Please rate the following attribute from 1 (High) to 5 (Low).

Q. No.	Usability Attribute	1 (High)	2 (Less High)	3 (Medium)	4 (Less Low)	5 (Low)
11.	Ease of Use					
12.	User Satisfaction					
13.	Usefulness					
14.	Ease of Learning					
15.	Complexity					
16.	Efficiency					
17.	Understandability					
18.	Accuracy					
19.	Learnability					
20.	Consistency					
21.	Effectiveness					
22.	Execution Time					
23.	Reliability					
24.	Response Time					
25.	Performance					
26.	User Interface					
27.	Attractiveness					
28.	Flexibility					
29.	Memorability					
30.	Compatibility					

Part 3: Objective- To find out basic requirement for usability. Please tick yes/no.

Q. No.	Usability Requirement	Yes	No
31.	Help and documentation is useful to use software.		
32.	Simple and easy dialogue saves user time.		
33.	Feedback can improve software performance.		
34.	Minimum memory load encourage user to use software.		
35.	Error messages help user to give correct and required input.		
36.	Consistency of screen and display play a major factor for usability.		
37.	Easy navigation in menu is important for usable software.		
38.	System/software speed affects user satisfaction.		
39.	Using learnability we perform common tasks or operations.		
40.	Understandability increases the satisfaction rate of user.		
41.	Efficiency of software can be measured in terms of task time.		
42.	Frequent execution time is important parameter for consideration.		
43.	Response Time can be measured by in term of turnaround time.		
44.	Traceability defines the relationship between the requirements and design.		
45.	Accuracy of software can be measured by in term of correctness.		

46.	For attractive user interface screen layout and color should be appealing.		
47.	For operability interface actions and elements should be consistent.		
48.	Flexibility of data entry is useful for direct manipulation.		
49.	Task completion rate is important factor for user to perform or work on.		
50.	Help messages useful for using and operating software.		

Please fill the following details:

Name	
Designation	
Institute/Organization	
Contact Number	
Email Address	

(Signature of Participant)

Index